Studies in Systems, Decision and Control

Volume 174

Series editor

Janusz Kacprzyk, Polish Academy of Sciences, Warsaw, Poland
e-mail: kacprzyk@ibspan.waw.pl

The series "Studies in Systems, Decision and Control" (SSDC) covers both new developments and advances, as well as the state of the art, in the various areas of broadly perceived systems, decision making and control–quickly, up to date and with a high quality. The intent is to cover the theory, applications, and perspectives on the state of the art and future developments relevant to systems, decision making, control, complex processes and related areas, as embedded in the fields of engineering, computer science, physics, economics, social and life sciences, as well as the paradigms and methodologies behind them. The series contains monographs, textbooks, lecture notes and edited volumes in systems, decision making and control spanning the areas of Cyber-Physical Systems, Autonomous Systems, Sensor Networks, Control Systems, Energy Systems, Automotive Systems, Biological Systems, Vehicular Networking and Connected Vehicles, Aerospace Systems, Automation, Manufacturing, Smart Grids, Nonlinear Systems, Power Systems, Robotics, Social Systems, Economic Systems and other. Of particular value to both the contributors and the readership are the short publication timeframe and the world-wide distribution and exposure which enable both a wide and rapid dissemination of research output.

More information about this series at http://www.springer.com/series/13304

Andrey E. Gorodetskiy · Irina L. Tarasova
Editors

Smart Electromechanical Systems

Group Interaction

With 140 Figures and 22 Tables

 Springer

Editors
Andrey E. Gorodetskiy
Institute of Problems of Mechanical
  Engineering
Russian Academy of Sciences
Saint Petersburg, Russia

Irina L. Tarasova
Institute of Problems of Mechanical
  Engineering
Russian Academy of Sciences
Saint Petersburg, Russia

ISSN 2198-4182 ISSN 2198-4190 (electronic)
Studies in Systems, Decision and Control
ISBN 978-3-030-07631-3 ISBN 978-3-319-99759-9 (eBook)
https://doi.org/10.1007/978-3-319-99759-9

This Springer imprint is published by the registered company Springer Nature Switzerland AG
The registered company address is: Gewerbestrasse 11, 6330 Cham, Switzerland

Preface

Smart electromechanical systems (SEMS) are used in cyber-physical systems (CPhS). Cyber physical systems the ability to integrate computing, communication and storage of information, monitoring and control of the physical world objects. The main tasks in the field of theory and practice of CPhS are to ensure the efficiency, reliability, and safety of functioning in real time. It is important to bear in mind that recently the task of ensuring functioning of robots as a part of collective performing joint work is set.

The purposes of the publication are to introduce the latest achievements of scientists of the Russian Academy of Sciences in the theory and practice of SEMS Group interaction and development of methods for design and simulation of SEMS Group interaction based on the principles of safety, flexibility, and adaptability in behavior and intelligence and parallelism in information processing, computation, and control.

Topics of primary interest include, but are not limited to the following:

- The Planning Behavior of the SEMS Group;
- Methods and Principles of Designing of Automatic Control Systems;
- Mathematical and Computer Modeling Group Interaction;
- Safety, Flexibility, and Adaptability of SEMS Group;
- Information-Measuring and Control Soft and Hardware;

This book is intended for students, scientists, and engineers specializing in the field of smart electromechanical systems and robotics, and includes many scientific domains such as receipt, transfer and pre-treatment measurement information, decision-making theory, control theory, working bodies of robots that imitate the complexity and adaptability of biological systems.

We are grateful to many people for the support received during the writing of this book. A list of their names cannot be represented here, but all of them we are deeply grateful.

Saint Petersburg, Russia
September 2018

Andrey E. Gorodetskiy
Irina L. Tarasova

Contents

Part I
The Panning Behavior of the SEMS Group

Emerging Issues of Robots to Be Used in Groups

Ivan L. Ermolov

Abstract *Objective* Problem statement: robots and robotic systems are expanding in numbers. This creates an opportunity to use robots in groups. Those groups may count few units and may count thousands of units functioning simultaneously or in sophisticated sequences. Application of robots in groups gives humans opportunity to solve new types of practical tasks, find new application areas for robots, implement new scenarios for them. Robots while in group can work differently in terms of quality and functionality. It creates new challenges for autonomy of robots and their groups. However for robots to be used in groups some technical and scientific tasks need to be solved. An option of robot to be able to be used in group will also affect structure of robot itself. *Purpose of research*: to review and systemize our knowledge about application of robots in groups, to flash on milestones on our way to use robots in groups. *Results* a review of advantages which creates usage of robots in groups. Examples of successful usage of robots in groups are given. List of scientific and technical topics to be addressed by robotists in their pursuit to use robots in groups. Also a brief review of scientific research performed in Russia in this area is presented. *Practical significance* the proposed thoughts could be used to systemize knowledge regarding application of robots in groups. Also it discusses scientific tasks to be solved in order for robots to used in groups.

Keywords Group of robots · Scientific problems · Robots' interaction
Application of robots · SEMS application

I. L. Ermolov (✉)
Ishlinsky Institute for Problems in Mechanics of the Russian
Academy of Sciences, Moscow, Russia
e-mail: ermolov@ipmnet.ru

© Springer Nature Switzerland AG 2019
A. E. Gorodetskiy and I. L. Tarasova (eds.), *Smart Electromechanical Systems*,
Studies in Systems, Decision and Control 174,
https://doi.org/10.1007/978-3-319-99759-9_1

1   Introduction

Human society is being saturated with more and more robots. If some decades ago robots were used in single mode, now we may expect that hundreds and thousands or robots will be introduced to be used simultaneously.

This aspect brings the demand to systemize robotization of society. In order to robots' usage to be effective one must consider thoroughly group usage of robots for various applications.

In fact, group application of robots is a kind of fusion of functionality of agents which form a group. Therefore it resembles data fusion. In [1] it was shown that generally data fusion realizes following scenarios: Time-based data fusion, Reliability-based data fusion, Space-based data fusion, Sensor-type data fusion and finally Data-type data fusion.

Let's list advantages which are given through group application of robots:

once robots are used in group we may form the group from robots with different functionality. In fact, it is accepted that universality in machines deteriorates their performance factor. The same trend can be seen in robotics. Basing of 6 basic functions of robot [2]: transportation, technological, informational, intellectual, energy and communication one may outline 6 basic types of robots: transportation robots, technological robots, inspection robots, robots-coordinators, robots-refuellers, robots-retranslators. It's obvious that quality fulfillment of complex tasks can be ensures only by group containing robots with different key-functionality;

usage of robots in group allow to increase their allowable application environment range. Thus one can use robots of somehow simple structure and low performance (in stand-alone mode) for effective solution of complex tasks. This gives advantages in usage costs and system robustness. E.g. this is especially effective for tasks of inspection, signal retransmission, delivery;

group application of robots permits to use widely effect of overlapping (back-up of functionality). This will increase probability of achievement of goal even in agile and hostile environments. Then even in case of lost of one or even several agents their functionality will be implemented by other members of group;

while being in group robots can in on-line mode redistribute resources among members of the group. This will be especially effective for complex tasks which implementation require a large amount of energy and computational resources;

group application of robots creates difficulties for unauthorized counteraction to them. E.g, in some cases criminal elements may be ready to escape from one or two machines, but not from dozens of them;

another important advantage is that a human is not present on board of such systems. Thus during mission planning once there is a group of robots, some of them may by intentionally "sacrificed" for the benefit of the whole group (e.g. inspection of highly radioactive environment, trials on impassable terrain etc.), however the rest of them may complete mission goal.

time-effective mission performance. Some tasks (e.g. environment exploration) may be performed by single robot, however while in group this task will be done faster almost at multiple decrease.

All these advantages open new opportunities for robots which are not possible for single robot usage.

2 Scientific Problems

Nevertheless despite all these advantages robots are still rarely used in large group simultaneously. This is caused by a fact that most of modern robots are still not ready for group interaction.

This creates new plural challenges to be solved by roboticists in following several years.

Among such key problems we see following:

low level of robots' autonomy. Most of modern robots are still controlled by human-operators. This creates severe restrictions for their usage. One of them is that remotely-controlled robots can not be used in large groups. Despite some experimentation to control group of robots by one operator, or by group of operators, still this has not proved to be practically useful in real life. In first turn intellectual autonomy should be increased. The question of robots' autonomy was discussed e.g. in [2];

new scenarios for robots' applications in group have to be developed. In fact some group applications e.g. group signal retransission or group inspection have been studied well enough [3, 4]. However the rest of application (e.g. group transportation, group technological processing etc.) are only at initial phase of their investigation. This will affect mission planning of robots in group;

these scenarios should generate new specifications for robots to be in group and to group of robots at whole. Especially this will influence on-board interfaces, communication, programming;

discussions on strategies of group application of robots. Researchers study various approaches to group control: centralized, decentralized, combined [5]. Usually these authors outline their own type of robots' group control. However at our opinion all these strategies have their owns advantages. Therefore each of them will be more effective at some specific applications;

effectiveness estimation of robots in group. This topic is important as it allows to balance between necessary quantity of robots and effectiveness of their performance. There were done some attempts to estimate effectiveness of single robot. However author has not met yet any method to effectively estimate performance of robots in group;

trajectory and motion planning for robots also gets some restrictions in group [6] (e.g. collision avoidance, communication interaction). There were received some results for robots-retranslators scenarios [4]. However we believe that once scenarios will become sophisticated and environment more complex more research will need to be done in this area;

multi-agent information processing seems to be relatively understood task, as it is similar to multi-sensor fusion. E.g. it was discussed intensively in [1].

an interesting scientific problem is effective redistribution of resources among members of the group. Right now robots' users have picked a problem of redistribution of communicational resources. Also there exist some effective solutions for redistribution of computational resources. However redistribution of energy and transport resources need more studies.

standardization and unification are of special importance. At first stage information exchange and programming should be standardize. Further on some standard solutions for interfaces and hull construction should be developed.

3 Research Results

In Russian science we may outline following research centers who have achieved interesting results in group robots' applications:

Southern Federal University in cooperation with Southern Scientific Center of Russian Academy of Sciences have been well-known due to their results in developing various control strategies for group of robots [5]. They have achieved challenging results also in redistribution of computational resources among computational agents.

Moscow State Technical Bauman University has interesting results regarding practical simultaneous usage of UGV and UAV [7].

Institute of Problems of Mechanical Engineering of Russian Academy of Sciences has high expertise in SEMS group application. Results of their research is presented in this book.

Moscow Technological University (MIREA) are developing fundamentals of group intelligent systems and new group control algorithms [8];

Russian State Scientific Center for Robotics and Technical Cybernetics studies application of cloud control technologies for group of robots;

St. Petersburg Institute for Informatics and Automation of the Russian Academy of Sciences have received important results [10] in developing tools to reconfigurate robots into various special shapes.

Ishlinsky Institute for Problems in Mechanics of the Russian Academy of Sciences study effects in mechanics for group robots' applications [9].

4 Conclusions

Group application of robots opens new opportunities for robotization. This makes possible to implement cardinally new scenarios for tasks implementation in various areas of human activities.

Still we must confess that group application of robots also creates new challenges for functional capacities of robots and gives new ideas for building new robots.

Presentation of this material was done with support from FASO Russia No. AAAA-A17-117021310384-9.

References

1. Ermolov, I.L.: Hierarchical data fusion architecture for unmanned vehicles. In: Smart Electromechanical Systems: The Central Nervous System. Springer, Berlin (2017)
2. Ermolov, I.L.: Mobile robots' autonomy, its measures and ways towards its increase. In: Ермолов И.Л. Mechatronics, Automation, Control. 6 (2008)
3. Ermolov, I.L., Sobolnikov, S.A.: Distribution of UGVs for relaying mobile reconfigurable communication networks. Vestnik MSTU STANKIN, 4 (2012)
4. Ermolov, I., Gradetsky, V., Knyazkov, M., Sobolnikov, S.: Cooperative motion planning of autonomous UGVs for mobile reconfigurable communcation networks. In: Proc. of IEEE-RAS-IARP Joint Workshop on Technical Challenges for Dependable Robots in Human Environment, IROS2013 WS, 3 Nov 2013, Tokyo Big Sight, Japan
5. Kaliaev, I.A., Gayduk, A.R., Kapustyan, S.G.: Models and algorithms for control in group of robots. PhysMatLit, Moscow (2009)
6. Ivanov, D.Y.: Methods to implement special formation of UAVs. Ph.D. thesis, Taganrog (2016)
7. Anikin, V.A., et al.: Perspective view on UAVs' machine-vision system used on UGVs. Vesnik YuFU 3 (2014)
8. Manko, S., Lokhin, V., Diane, S., Panin, A.: Multi-robot system learning based on evolutionary classification. In: Proceedings of 3rd International Conference on Control, Mechatronics and Automation (ICCMA 2015), MATEC Web of Conferences, vol. 42 (2016)
9. Gradetsky, V., Ermolov, I., Knyazkov, M., Lapin, B., Semenov, E., Sobolnikov, S., Sukhanov, A.: Highly passable propulsive device for ugvs on rugged terrain. 13th International Scientific-Technical Conference on Electromechanics and Robotics "Zavalishin's Readings". 161(03013), 1–5 (2018)
10. Gaponov, V., Dashevsky, V., Ronzhin, A.: Upgrading the hardware and software of RC servos for use in educational robotics. In: International Conference on Mechanical, System and Control Engineering (ICMSC'17), IEEE, pp 235–239 (2017)

Situational Control a Group of Robots Based on SEMS

Andrey E. Gorodetskiy and Irina L. Tarasova

Abstract *Problem statement*: robots and robotic systems based on SEMS are a kind of cyberphysical systems, a distinctive feature of which is parallelism in obtaining and processing information, in calculation and formation of control actions and in performance of various movements on commands from system of automatic control, which is built by analogy with the human Central Nervous System. Since each robot based on the SEMS has the appropriate behavior, group control of such robots becomes not effective centralized control and require a more flexible decentralized strategies. However, the task of implementing control in teams of such intelligent devices is not sufficiently investigated. In particular, the need to select the best scenarios for the development of situations in conditions of incomplete certainty, as well as limitations on the practical feasibility of algorithms for processing data on objects lead to the fact that the scope of state characterization in one way or another requires discretization and fuzzification of data based on linear, nonlinear, or ordinal scales and the transition to a situation of group control on the basis of time slices of trajectories of change of characteristics of control objects in the environment of choice, described as a multi-dimensional space called the configuration space of the robots. *Purpose of research*: development of algorithms for constructing the current dynamic model of the environment and the team of robots, as well as a dynamic space of configurations of the robots for providing the best trajectories for robots (scenarios) in terms of not a complete certainty. *Results:* a review of methods of intellectualization of robots like SEMS. Methods of construction of time slices of trajectories of change of characteristics of control objects in the environment of choice are analyzed. The algorithms for constructing

A. E. Gorodetskiy (✉)
Institute of Problems of Mechanical Engineering Russian Academy
of Sciences, V.O., Bolshoj Pr., 61, Saint-Petersburg 199178, Russia
e-mail: g27764@yandex.ru

I. L. Tarasova
Peter the Great St. Petersburg Polytechnic University, Polytechnicheskaya,
29, Saint-Petersburg 195251, Russia
e-mail: g17265@yandex.ru

© Springer Nature Switzerland AG 2019
A. E. Gorodetskiy and I. L. Tarasova (eds.), *Smart Electromechanical Systems*,
Studies in Systems, Decision and Control 174,
https://doi.org/10.1007/978-3-319-99759-9_2

the current dynamic model of the environment and a team of robots on the basis of an algebraic approach to logical and logical-probabilistic data analysis. The algorithm of construction of dynamic space of configurations of robots on a concrete example of extraction from a room of various subjects by collective of robots is resulted. *Practical significance:* the proposed algorithms for constructing the current dynamic model of the environment and the team of robots, as well as the dynamic space of robot configurations can be used in the planning control systems of a team of robots, providing the choice of the best scenarios of movements in the situational control of a group of robots.

Keywords Cyberphysical systems · Robot · SEMS · Situational control
Central nervous system · Dynamic models · Dynamic configuration space
Behavior planning systems · Logical data analysis · Logical-probabilistic data
analysis · Algebraic approach

1   Introduction

The robots is based on SEMS [1] are a kind of cyber physical systems [2–4]. A distinctive feature of such complexes is parallelism in obtaining and processing information, in calculating and forming control actions and in performing various movements similar to the Central nervous system of human. From the point of view of information control systems, a group of such robots can be considered as a group of "cognitive information control systems built into their environment" [5]. Consequently, such devices should have two-way communication with the outside world and other robots involved in group interaction [6], which is inherent to SEMS. The main feature of robots on based SEMS in solving complex tasks of group control is the ability of intelligent processing of information available to them for making decisions about preferred actions in a changing situation.

In well-known publications on methods of robot intellectualization, for example, using the evolutionary approach [7], self-study [8] or the recently popular algebraic approach [9–11], insufficiently covers the problem of implementing decisions in groups intellectualized devices, although many authors have noted the importance of the task. For example, in [5] it is noted that in order to achieve synergetic effect from interactions of SEMS-type intelligent systems in group control tasks, a unified methodological basis for providing the desired group behavior of robot teams is necessary. Since each robot has an appropriate behavior [12], group control of such robots becomes ineffective in centralized control and more flexible decentralized strategies are required. For example, a team of robots can be regarded as a system of systems (SoS) [13], where the tasks of control planning come to the fore. Planning systems by control (SPC) can be built within the framework of the situational approach [14]. An essential part of the problem is the formalization of the term "situation".

Determination of the situation depends on the subject area and the formalization apparatus used [15, 16]. Usually, the full situation is described in the following main aspects: knowledge about the environment of choice, the current structure of the object, the current state of the control system and the technology (strategies) of control [14]. In addition, the solution of the problem of the matching motion in the group situational control of robots requires taking into account both the physical limitations of the robots themselves and the environment of operation, and the dynamic limitations caused by the movements of robots involved in situational group control [17]. Therefore, the situation can be considered as a time slice of the trajectories of changing the characteristics of control objects in the selection environment described as some multidimensional space. The selection criteria of this space have changed little in comparison with the requirements for the selection of elements of the state vector in the classical theory of the state space [18], developed to control dynamic objects that admit description in the form of differential and difference equations. Namely, in the space of states should include all the time-changing characteristics of the object, significantly affecting the achievement of control objectives. The procedure for constructing a state space, that is, separating the actual control objects from the environment, is usually taken out of the control process, since there are no constructive methods for selecting elements of the state vector for real objects. The need to select the best trajectories (scenarios) in conditions of not complete certainty, as well as limitations on the part of the practical feasibility of algorithms for processing data on objects lead to the fact that the area of determining the state characteristics in one way or another is discretized on the basis of linear, nonlinear or ordinal scales [19]. This complex decision-making problem can be based on the formation of the permitted dynamic space of robot configurations (PDSCR) [20]. In this paper, we consider one of the approaches to the construction of PDSCR.

2 The Construction of the Current Dynamic Model of the Environment

The current PDSCR contains two parts: the current dynamic model of the environment and the current dynamic models of the robots' own state.

When constructing the current dynamic model of the environment, it is necessary to provide robots with the ability to understand the language of sensations formed as a result of solving systems of logical equations. Such systems are similar to the human Central nervous system and therefore can be called Central nervous systems of the robots (CNSR) by analogy [21].

In general, the algorithm for constructing the current dynamic model of the environment can be as follows:

- recording of current measuring information from sensors and external sources into the CNSR database;

- fuzzification of sensor data is formed by fuzzification of logical variables, as attributes which can be of the probability that the measured variable in a given quantum;
- association fuzzification data into groups, forming sense organs of the robot as a human, for example, the vision in the form of Z; the hearing in the form of multiple C; the smell of many O; the taste of many U; touch in the form of a set A;
- the allocation in each of the sets forming a subset, characterizing the properties of the observed or studied object: $Z_i \subset Z, C_i \subset C, O_i \subset O, B_i \subset B, A_i \subset A$;
- allocation of qi images in the environment of the robot space, in the simplest case, this operation is reduced to combining into one set of qi those points of space that have the same set of logical variables with the same attributes and provided that the distance to the nearest neighboring point with the same parameters does not exceed some predetermined value. To do this, as shown in [22], you can use the logical expression algebraization procedure;
- classification of qi images formed by CNSR. This is a recognition operation of qi images generated by the CNSR, which can be described as logical matrices M_i;
- the assignment of these images qi to the different classes of the images of q_1, q_2, q_3 (types of objects) described by the reference Boolean matrices M_{E1}, M_{E2}, M_{E3} [23], for which the previously made optimal solutions are known, for example, for transportation by one of the robots R_1, R_2 or R_3 can rely on the procedure of searching for binary relations $M_i \, g \, M_{Ej}$, where g is a two—place predicate on the analyzed sets, j is 1,2,3. In the simplest case, $g = 1$ if $M_i = M_{Ej}$, otherwise $g = 0$;
- definition of additional qualitative parameters in the selected samples. Additional parameters are obtained in the form of new Boolean variables F_{ij}. They can, for example, be obtained by analyzing the geometric parameters of images (s_i areas, d_i dimensions, gaps between v_{ij} images, the coordinates of the objects x_i, y_i, etc.);
- formation of binary estimates of selected images. Binary estimates of images can be obtained by logical analysis of parameters characterizing images. For this purpose it is necessary to make rules of assignment to this image of this or that binary assessment, for example, for subjects (their images) which should be removed from the room: if the image very close to a door of the room, has small dimensions (the area) and big gaps between neighboring subjects, this image (object) the best for withdrawal from the room.

As a result of applying the described algorithm can be obtained, for example, the following set:

$$\text{Set of objects of type} \quad q_1 Q_1 = \cup q_{1n} \tag{1}$$

$$\text{Set of objects of type} \quad q_2 Q_2 = \cup q_{2m} \tag{2}$$

$$\text{Set of objects of type} \quad q_3 Q_3 = \cup q_{3k} \tag{3}$$

$$\text{A set that covers all the gaps} \quad V = \cup v_{ij} \tag{4}$$

$$\text{The set of coordinates of objects} \quad q_1 X_1 = \cup x_{q1} \tag{5}$$

$$Y_1 = \cup y_{q1} \tag{6}$$

$$\text{The set of coordinates of objects} \quad q_2 X_2 = \cup x_{q2} \tag{7}$$

$$Y_1 = \cup y_{q2} \tag{8}$$

$$\text{The set of coordinates of objects} \quad q_3 X_3 = \cup x_{q3} \tag{9}$$

$$Y_3 = \cup y_{q3} \tag{10}$$

$$\text{A set that characterizes space:} \quad S = \cup l_k h_k \tag{11}$$

where l_k and h_k are the length and width of the k-band.

3 The Construction of the Current Dynamic Models of Robots

When constructing current dynamic models of robots' own States taking into account the physical limitations of the robots themselves, it is necessary to provide the CNSR with "sense organs of their own state" and the possibility of exchanging this information with other participants of the movement. Consequently, it is desirable that robots have the ability to understand the language of self-perception like a human [24]. Then in the solution of systems of Boolean equations formed on the basis of the language of the feelings of the robot, there is a possibility of forming an independent image of your own current condition.

The algorithm for constructing the current dynamic model of self-state in the presence of appropriate sensors of self-perception in General is similar to the previous one.

As a result of the application of this algorithm, for example (see Fig. 1), can be obtained:

Set Sk, characterizing the space of the warehouse, where objects are moved.
Set Sq corresponding to the area of the object.
A set SR_1 corresponding to the area occupied by the first robot.
A set of SR_2 corresponding to the area occupied by the second robot.
A set of SR_3 corresponding to the area occupied by the third robot.
Set of coordinates of the centers of the areas occupied by XSR, YSR robots.
The coordinates of the center of the doorway xd, yd and its width hd.

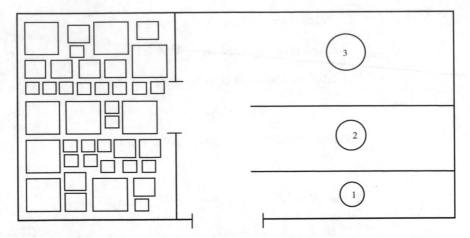

Fig. 1 Illustration for example

4 An Example of the Construction PDSCR for the Task of Extracting from the Premises Items Robots-Loaders

4.1 Statement of the Problem

Let's consider that dynamic models of robots and environment of their functioning are constructed and therefore robots have the following information.

In the room of a rectangular shape length L and width H are U items, of which N items of type q_1 length lq1 and width hq_1, M items of type q_2 length lq_2 and width hq_2 and K items of type q_3 length lq_3 and width hq_3. Figure 1 shows the three types of q_1, q_2, and q_3 objects placed in a room, and the three robots 1, 2 and 3 that are required to vacate the room.

$$U = N + M + K \tag{12}$$

In this case the following conditions are true:

- objects are located so tightly to each other that any gap between them has a size v_{ij}:

$$v_{ij} < \min\{lq_1, lq_2, lq_3, hq_1, hq_2, hq_3\} \tag{13}$$

- objects from the room are unloaded by three robots R_1, R_2, R_3, having a length of lR_1, lR_2, lR_3 and a width of hR_1, hR_2, hR_3, through the doorway width d:

$$d > \max \{lR_1, lR_2, lR_3, hR_1, hR_2, hR_3, lq_1, lq_2, lq_3, hq_1, hq_2, hq_3\} \qquad (14)$$

– the area of the room Sk shall be:

$$S_k > S_q + S_r \qquad (15)$$

where:

$$S_r = \sum (lR_1 * hR_1) + \sum (lR_2 * hR_2) + \sum (lR_3 * hR_3) \qquad (16)$$

$$S_r = \sum (lR_1 * hR_1) + \sum (lR_2 * hR_2) + \sum (lR_3 * hR_3) \qquad (17)$$

It is required to free the room from all objects at the minimum time:

$$\sum_{i=1}^{U} T(i) \to \min (n * t) \qquad (18)$$

where: $T(i) = n * t$ is the time required for removal of the i-th subject from the room, depending on the number of steps n and the duration t of the step of the robot associated with the design of the robots and ambient conditions.

Let's say t = const.

To select or synthesize the algorithm for solving the problem, first of all, it is necessary to obtain an algorithm for constructing the permitted dynamic configuration space.

4.2 Problem Solution

Let's consider the solution of this problem by constructing resolved dynamic spaces of robot configurations. The procedure for extracting objects from the room in this statement consists in step-by-step building allows for dynamic configurations of space, providing sequential extraction of items from the room. This analyzes the possibility of extraction in the minimum time at each step of one object, taking into account the necessary shifts of neighboring objects.

The algorithm allows for dynamic configurations of space that moves from the room to the warehouse of the first object and does not take into account the possibility of shear of adjacent items to increase the clearance in general will be next:

1. Let's number all objects in the room i = 1,..., U, for example, linearly, starting from the lower left corner. Let the number of implementation on removing primary will initially be: p = 0.

2. We calculate for each subject i the ratio of the form:

$$J_i = \min_{i}\{k_1 r_i + k_2/\{\max_{j}\{\min\{v_{ij}\}\}\}\},$$

(19)

where: k_1, k_2—preference coefficients; r_i—the distance from the center of the doorway to the I-th object, v_{ij}—the gap between the I-th object and the adjacent j—th object.

3. Define a bidder's q^P to removal from the room, which will have the minimum value of J_i for the i-th subject.
4. Определим координаты xq^P, yq^P претендента q^P.
5. Find the maximum value of dimensions (width or length) hq or lq of applicant q^P.
6. Select one R_m of the three robots R_1, R_2, R_3, the respective bidder q^P.
7. Find the maximum value of the dimensions (width or length) hR_m or lR_m of the selected robot R_m.
8. Calculate the width Hq (t) of the allowed dynamic space of the current robot configuration: $HR_m = xq^P + yq^P + hR_m \, lR_m$.
9. If the applicant q^P gap $v_{ij} > \max\{hR_m, lR_m\}$, then determine the trajectory of the corresponding R_m robot to remove from the room of this q^P object and remove it.
10. Let's increase the number of implementations to remove the object p = p + 1 and compare the resulting number with the total number of objects U.
11. If p = < U, then go to step 3, removing from the calculation by the formula (19) all previously deleted items.
12. End.

5 Conclusion

The considered algorithm allows to optimize the work of robot loaders to release the room from the objects in it when consistently using robots. However, analysis of the algorithm shows that if it turns out that the objects are q_1, q_2,...q_n—are of different types, they can be shifted by different types of robots working almost in parallel (simultaneously). In this case, to exclude emergency situations and breakdowns of intersections of the trajectories of movements in the robot control must existing current dynamic model of health robots to complement the set of valid instructions of the control.

For this, for example, it is possible, based on the purpose of a particular robot, to make a list of possible instructions. Further, by mathematical and computer simulation of PDSCR to reveal a set of acceptable instructions of group behavior. In solving this problem, it is necessary to take into account the dynamic characteristics of robots, which can be optimized by adjusting the parameters of automatic control systems (ACS) of robots, which is a complex optimization problem with nonlinear constraints [25]. The solution of this problem requires separate consideration.

In addition, in some cases, when implementing systems of group control of robots, some instructions issued by the upper-level ACS may not be clear to the ACS of the robot, although the simulation results they have been assigned to valid. This, for example, may be due to the incomplete adequacy of the models used in the PDSCR. Partially remove such instructions from the permissible possible due to the semantic analysis of the instructions for correctness and are not contradictory, and through the organization of a dialogue between interactive ACS robots.

Acknowledgements This work was financially supported by Russian Foundation for Basic Research, Grant 16-29-04424 and Grant 18-01-00076.

References

1. Gorodetskiy, A.E. (ed.): Smart electromechanical systems. In: Studies in Systems, Decision and Control, vol. 49, 277 p. Springer, Switzerland (2016). http://dx.doi/org/, https://doi.org/10.1007/978-3-319-27547-5_4
2. Lee, E.: Cyber physical systems: design challenges. University of California, Berkeley Technical Report No. UCB/EECS-2008-8. Retrieved 7 June 2008
3. NSF Cyber-Physical Systems Summit. Retrieved 1 Aug 2008
4. NSF Workshop on Cyber-Physical Systems. Retrieved 9 June 2008
5. Fridman, A.Ya.: SEMS-Based control in locally organized hierarchical structures of robots collectives. In: Gorodetskiy, A.E., Kurbanov, V.G. (eds.) Studies in Systems, Decision and Control, vol. 95. Smart Electromechanical Systems: The Central Nervous System 270 p, pp. 31–47. Springer, Switzerland (2017). https://doi.org/10.1007/978-3-319-53327-8_3
6. Gorodetskiy, A.E., Tarasova, I.L., Kurbanov, V.G.: Safe Control of SEMS at Group Interaction of Robots. In: Materialy 10-j Vserossijskoj mul'tikonferencii po problemam upravlenija (Proceedings of the 10th all-Russian multi-conference on governance). Divnomorskoye, Gelendzhik, vol. 2, pp. 259–262 (2017)
7. Shkodyrev, V.P.: Technical systems control: from mechatronics to cyber-physical systems. In: Gorodetskiy, A.E. (ed.) Studies in Systems, Decision and Control, vol. 49, Smart Electromechanical Systems, 277 p, pp. 3–6. Springer, Switzerland (2016)
8. Gorodetskiy A.E.: Smart electromechanical systems modules. In: Gorodetskiy A.E. (ed) Studies in Systems, Decision and Control, vol. 49, Smart Electromechanical Systems, 277 p, pp. 7–15. Springer, Switzerland (2016)
9. Kulik, B.A., Fridman, A.Ya.: Logical analysis of data and knowledge with uncertainties in SEMS. In: Gorodetskiy, A.E. (ed.) Studies in Systems, Decision and Control, vol. 49, Smart Electromechanical Systems, 277 p, pp. 45–59. Springer, Switzerland (2016)
10. Gorodetskiy, A.E., Tarasova, I.L., Kurbanov, V.G.: Logical-mathematical model of decision making in central nervous system SEMS. In: Gorodetskiy A.E., Kurbanov, V.G. (eds.) Smart Electromechanical Systems: The Central Nervous System, 270 p. Springer, AG (2017). https://doi.org/10.1007/978-3-319-53327-8
11. Gorodetskiy, A.E., Tarasova, I.L., Kurbanov, V.G.: Behavioral decisions of a robot based on solving of systems of logical equations. In: Gorodetskiy A.E., Kurbanov, V.G. (eds.) Smart Electromechanical Systems: The Central Nervous System, 270 p. Springer, AG (2017). https://doi.org/10.1007/978-3-319-53327-8

12. Akkof, R., Jemeri, F.: O celeustremlennyh sistemah (About purposeful systems). Moscow, Sov. Radio Publ., 269 p (1974) (in Russian)
13. Gorod, A., Fridman, A., Saucer B.A.: Quantitative approach to analysis of a system of systems operational boundaries. In: Proceedings of International Congress on Ultra Modern Telecommunications and Control Systems (ICUMT-2010), pp. 655–661. 18–20 Oct 2010, Moscow, Russia
14. Pospelov, D.A.: Situacionnoe upravlenie: teorija i praktika (Situational management: theory and practice). Moscow, Nauka Publ., 288 p. (1986) (in Russian)
15. Mishin, S.P.: Optimal'nye ierarhii upravlenija v jekonomicheskih sistemah (Optimal hierarchies of control in economic systems). Moscow, PMSOFT Publ., 190 p. (2004) (in Russian)
16. Melikhov, A.N., Berstein, L.S., Korovin, S.I.: Situacionnye sovetujushhie sistemy s nechetkoj logikoj (*Situational advising systems with fuzzy logic*). Moscow, Science Publ., 272 p. (1990) (in Russian)
17. Sokolov, B., Ivanov, D., Fridman, A.: Situational modelling for structural dynamics control of industry-business processes and supply chains. In: Sgurev, V., Hadjiski, M., Kacprzyk, J. (eds.) Intelligent Systems: From Theory to Practice, pp. 279–308. Springer, Berlin, Heidelberg, London (2010)
18. DeRusso, P.M., Roy, R.J., Close, C.M.: State Variables For Engineers, p. 608. Wiley, Hoboken (1965)
19. Gorodetskiy, A.E., Kurbanov, V.G., Tarasova, I.L.: Methods of synthesis of optimal intelligent control systems SEMS. In: Gorodetskiy, A.E. (ed.) Smart Electromechanical Systems, 277 p. Springer (2016). https://doi.org/10.1007/978-3-319-27547-5
20. Russell, S.J., Norvig, P.: Artificial Intelligence: A Modern Approach, 946 p (2001)
21. Gorodetskiy, A.E., Tarasova, I.L., Kurbanov, V.G.: Challenges related to development of central nervous system of a robot on the bases of SEMS modules. In: Gorodetskiy, A.E., Kurbanov, V.G. (eds) Studies in Systems, Decision and Control, vol. 95: Smart Electromechanical Systems: The Central Nervous System, 270 p, pp. 3–16. Springer, Switzerland (2017)
22. Gorodetsky, A.: Fundamentals of the theory of intelligent control systems (Osnovy teorii intellektual'nyh sistem upravlenija). LAP LAMBERT Academic Publishing GmbH@Co. KG Publ., 313 p (2011)
23. Gorodetsky, A.E., Tarasova, I.L.: Fuzzy mathematical modeling of poorly formalized processes and systems (Nechetkoe matematicheskoe modelirovanie ploho formalizuemyh processov i sistem). SPb. Publishing house of Polytechnical Institute, 336 p. (2010) (in Russian)
24. Gorodetsky, A.E., Tarasova, I.L., Kurbanov, V.G.: Logical and probabilistic methods of formation of a dynamic space configuration of the robot group. In: Materialy 10-j Vserossijskoj mul'tikonferencii po problemam upravlenija (Proceedings of the 10th all-Russian multi-conference on governance) Divnomorskoye, Gelendzhik, vol. 2, pp. 262–265 (2017)
25. Kuchmin, A.Y.: Analysis of polynomial constraints by the solution tree method. Informacionno-upravljajushhie sistemy, 6(91), 6–9 (2017) (in Russian) https://doi.org/10.15217/issn1684-8853.2017.6.9

Evolutionary Algorithms of Stable-Effective Compromises Search in Multi-object Control Problems

Vladimir A. Serov and Evgeny M. Voronov

Abstract In the article the main principles of adaptive evolutionary computing technology of smart electromechanical systems (SEMS) group control optimization under conflict and uncertainty, based on stable-effective compromises (STEC), are discussed. The necessary conditions of STEC existence are formulated in the form of variational principles generalizing the known Ekeland variational principle for a class of multicriteria conflict optimization under uncertainty problems. On the basis of the formulated variational principles the mechanism of adaptive fitness functions formation is developed for using in adaptive genetic algorithms, which allows to search STEC for different types of SEMS conflict group interaction.

Keywords Conflict group interaction · Smart electromechanical system (SEMS) Uncertainty · Multi-object multi-criteria system (MMS) · Stable-effective compromise (STEC) · Genetic algorithm (GA) · Adaptive fitness function (AFF)

1 Introduction

One of the features of the STEC-based SEMS group control problem under uncertainty is that the STEC is based on a combination of different game-theoretic principles of optimality [1, 2]. In addition, in applied problems of group control SEMS implementation of control algorithms in real time requires the representation of control actions in general in the form of parameterized program-corrected control laws. Such cases are characterized by a high dimension of a criterion space and a

V. A. Serov (✉)
Russian Technological University (MIREA), Vernadskogo Av., 78,
Moscow 119454, Russia
e-mail: ser_off@inbox.ru

E. M. Voronov
Bauman Moscow State Technical University, 2nd Baumanskaya St., 5,
Moscow 105005, Russia
e-mail: emvoronov@mail.ru

A. E. Gorodetskiy and I. L. Tarasova (eds.), *Smart Electromechanical Systems*,
Studies in Systems, Decision and Control 174,
https://doi.org/10.1007/978-3-319-99759-9_3

space of control parameters, nonlinearity, nonconvexity, the presence of points of discontinuity of the subsystems vector efficiency index components.

In practical tasks often to the conflict optimal solutions additional requirements are presents, such as:

- reasonably (or necessary, in a case of a lack of optimal solutions) be limited to a search of suboptimal (ε-effective, ε-equilibrium etc.) solutions;
- it's necessary on the set of ε-optimal solutions to find solutions, provided required sensitivity to the change of parameters values (local robust quality);
- it's necessary to find optimal solutions, provide the required global robust quality on the set of an uncertain factor valid values;
- you want to exclude from the set of ε-optimal solutions obviously unacceptable values of vector efficiency index components, in particular, end portions of a Pareto set.

These peculiarities of the problems of SEMS group control, in combination with the problem of global optimization, hinder or make impossible the implementation of known optimization techniques and algorithms for STEC search.

The aim of the article is to develop an adaptive evolutionary computational technology to optimize control in MMS under conditions of uncertainty on the basis of STEC. To do this, formulate necessary conditions for the existence of the SEC in the form of variational principles, generalizing the known variational principle of Ekeland's [3, 4] on a class of problems of multicriteria optimization under conflict and uncertainty. On the basis of the formulated variational principles, the mechanism of adaptive fitness functions formation in adaptive genetic algorithms is developed, which allows to find optimal solutions with specified properties for various types of SEMS conflict group interaction.

2   Problem Statement

Let's consider the model of SEMS group control in the form of a noncooperative game,

$$\Gamma = \left\langle \mathbf{N}, \{\mathbf{Q}_i\}_{i \in \mathbf{N}}, \{\mathbf{J}_i(\mathbf{q})\}_{i \in \mathbf{N}}, \{\Omega_i\}_{i \in \mathbf{N}} \right\rangle \tag{1}$$

In (1) $\mathbf{N} = \{\overline{1,n}\}$—a set of subsystems-players; $\mathbf{q} = \{\mathbf{q}_i | i \in \mathbf{N}\} \in \mathbf{Q} \subset \mathbf{E}^r$—a vector of optimized control parameters of the system, the values of the components of which form the situation in the game Γ; $\mathbf{q}_i \in \mathbf{Q}_i \subset \mathbf{E}^{r_i}$—a vector of optimized control parameters of i-subsystem-player that takes values from a set $\mathbf{Q}_i$, $i \in \mathbf{N}$, $r = \sum_{i \in \mathbf{N}} r_i$, A set $\mathbf{Q}$ is defined by the system of nonlinear constraints-inequalities.

$$\mathbf{Q} = \left\{ \mathbf{q} \in \mathbf{E}^r \middle| \mathbf{G}(\mathbf{q}) \underline{\leq} \mathbf{0}_s \right\} \tag{2}$$

where each $\mathbf{Q}_i$ corresponds to its own group of constraints-inequalities for which the ratio is performed $\mathbf{Q} = \prod_{i \in \mathbf{N}} \mathbf{Q}_i$; $\mathbf{J}(\mathbf{q}) = \{\mathbf{J}_i(\mathbf{q}) | i \in \mathbf{N}\}$—vector efficiency index of the system, where $\mathbf{J}_i(\mathbf{q}) \in \mathbf{E}^{m_i}$-the vector efficiency index of the i-subsystem-player, defined on $\mathbf{Q} = \prod_{i \in \mathbf{N}} \mathbf{Q}_i \subset \mathbf{E}^r$; $\Omega_i \subset \mathbf{E}^{m_i}$—a closed convex polyhedral domination cone.

$$\Omega_i = \left\{ \mathbf{z} \in \mathbf{E}^{m_i} \middle| \mathbf{B}_i \mathbf{z} \underline{\leq} \mathbf{0}_{p_i} \right\}, \tag{3}$$

formalizing a binary preference relation on the set of achievable vector estimations $\mathbf{J}_i(\mathbf{Q}) = \bigcup_{\mathbf{q} \in \mathbf{Q}} \mathbf{J}_i(\mathbf{q}) \subset \mathbf{E}^{m_i}$; $\mathbf{B}_i$—the matrix of a polyhedral domination cone of $[p_i \times m_i]$-dimension. It is necessary to determine the optimal solution that makes sense of a stable-effective compromise (STEC) [1], providing a stable-effective value to the vector efficiency index $\mathbf{J}^* = \mathbf{J}(\mathbf{q}^*)$.

For STEC searching in the game problem (1), in particular, the arbitrage-equilibrium scheme [1] can be used. The construction of the arbitration-equilibrium solution is carried out a combined computational procedure using, which includes the following main stages:

- construction of a set of Nash equilibrium solutions to the problem (1);
- creating a set of Pareto-optimal solutions of problem (1);
- "projecting" Nash equilibria on the Pareto set on the basis of equilibrium and arbitration scheme.

The peculiarity of the problem (1) is that the efficiency indicators of subsystems are functionals that, as a rule, have the properties of nonlinearity, nonconvexity, nondifferentiability, which is due to the structural complexity and nonlinear dynamic nature of the group control model. As a result, the set of achievable vector estimates can contain locally effective and locally-equilibrium solutions. There may be cases of absence of globally equilibrium solutions. These circumstances significantly complicate the STEC search and are likely to lead to unreliable results.

In cases where the equilibrium solutions when the extrema in determining the equilibrium solutions are not attained on the feasible set and also when searching for approximate solutions it is necessary to use concepts of ε-efficiency, ε-equilibrium, and ε-optimality.

Definition 1 Let vector $(-\varepsilon) \in \Omega$, $\Omega = \prod_{i \in \mathbf{N}} \Omega_i$. A valid solution $\mathbf{q}^\varepsilon \in \mathbf{Q}$ will be called $\varepsilon\Omega$-optimal ($\varepsilon\Omega$-efficient) solution of problem (1), if for any valid $\mathbf{q}^\varepsilon \neq \mathbf{q}$ takes place:

$$\mathbf{J}(\mathbf{q}) - (\mathbf{J}(\mathbf{q}^\varepsilon) - \varepsilon) \notin \Omega \tag{4}$$

Definition 2 A valid solution $\mathbf{q}^{e\varepsilon} \in \mathbf{Q}$ is called ε-equilibrium relative to the domination cones $\{\mathbf{\Omega}_i\}$ (ε Ω-equilibrium) of the problem (1), where $\varepsilon = \{\varepsilon_i \in \mathbf{E}^{m_i} | i \in \mathbf{N}\} \in \mathbf{E}^m, \varepsilon_i \in \mathbf{E}^{m_i}, i \in \mathbf{N}$, if for any valid $\mathbf{q}_i \in \mathbf{Q}_i$ takes place:

$$J_i(\mathbf{q}^{e\varepsilon} \| \mathbf{q}_i) - (\mathbf{J}_i(\mathbf{q}^{e\varepsilon}) - \varepsilon_i) \notin \mathbf{\Omega}_i, i \in \mathbf{N}. \tag{5}$$

As shown by numerous studies [3, 5–26], one of the most promising tools for finding globally optimal solutions to multicriteria and conflict problems is the evolutionary computing technology that combines the advantages of local and global search algorithms. At the same time, the efficiency of computational algorithms is largely determined by the degree of study of the relevant optimality conditions. To solve the problem (1), a set of genetic algorithms using the necessary conditions ε-cone optimalities and ε-equilibrium was developed [3, 4, 13, 14, 27]. These conditions generalize the known Ekeland ε-variational principle [15, 28] to the class of multicriteria conflict optimization problems.

3   The Necessary Conditions for ε Ω—Efficiency in Multicriteria Group Control Model

Let's consider the problem of finding ε Ω—effective solutions in the noncooperative game model of group control (1). The following statement has a great importance in the study of effective solutions properties.

Statement 1 Let $\mathbf{Q} \subset \mathbf{E}^r$ be a closed subset of a complete metric space; $\mathbf{J}(\mathbf{q}) \in \mathbf{E}^m$ is a vector of positive bottom semicontinuous function; $\mathbf{B} = [p \times m]$—matrix of a polyhedral domination cone Ω with the elements $b_{ij} \geq 0$, $d(\mathbf{q}, \mathbf{v})$—the distance between the points $\mathbf{q}, \mathbf{v} \in \mathbf{E}^r$. Consider $\mathbf{q}_0 \in \mathbf{Q}$ and $\mathbf{c} > \mathbf{0}_m \in \mathbf{E}^m$.

We define on the set $\mathbf{Q}$ a binary preference relation $\Re$ of the form:

$$\mathbf{q}^2 \Re \mathbf{q}^1 \Leftrightarrow \mathbf{B}\big((\mathbf{J}(\mathbf{q}^2) + \mathbf{c}d(\mathbf{q}^1, \mathbf{q}^2)) - \mathbf{J}(\mathbf{q}^1)\big) \leq \mathbf{0}_p. \tag{6}$$

Then:

(1) there is a point $\bar{\mathbf{q}} \in \mathbf{Q}$ such that the preference relation $\bar{\mathbf{q}} \Re \mathbf{q}^0$ of the type (6) is satisfied;
(2) $\forall \mathbf{q} \in \mathbf{Q}, \mathbf{q} \neq \bar{\mathbf{q}}$ the relation $\mathbf{q} \Re \bar{\mathbf{q}}$ of the type (6) is not fulfilled, i.e. the system of inequalities is not satisfied:

$$\mathbf{B}\big((\mathbf{J}(\mathbf{q}) + \mathbf{c}d(\bar{\mathbf{q}}, \mathbf{q})) - \mathbf{J}(\bar{\mathbf{q}})\big) \leq \mathbf{0}_p. \tag{7}$$

The consequence of Statement 1 is the ε Ω—variational principle.

Statement 2 ($\varepsilon\,\Omega$-variational principle). Let be:

(1) conditions of Statement 1 have been fulfilled;
(2) vector $\varepsilon > \mathbf{0}_m$ is fixed, and $(-\varepsilon) \in \Omega$;
(3) $\mathbf{q}^\varepsilon \in \mathbf{Q}$ feeble $\varepsilon\,\Omega$-optimal solution of the problem (1), i.e. for any $\mathbf{v} \in \mathbf{Q}$ system of inequalities is not fulfilled.

$$\mathbf{B}(\mathbf{J}(\mathbf{v}) - (\mathbf{J}(\mathbf{q}^\varepsilon) - \varepsilon)) < \mathbf{0}_p.$$

We define on the set $\mathbf{Q}$ a binary preference relation $\mathfrak{R}$ of the following form:

$$\mathbf{q}^2\mathfrak{R}\mathbf{q}^1 \Leftrightarrow \mathbf{B}\big((\mathbf{J}(\mathbf{q}^2) + \mathbf{K}\varepsilon d(\mathbf{q}^1, \mathbf{q}^2)) - \mathbf{J}(\mathbf{q}^1)\big) \leqq \mathbf{0}_p, \tag{8}$$

where $\mathbf{K} = \{k_{ij} > 0, i, j = \overline{1, m}\}$—square matrix.
   Then there is $\mathbf{v}^\varepsilon \in \mathbf{Q}$ such that,

(1) $$\mathbf{B}(\mathbf{J}(\mathbf{v}^\varepsilon) - \mathbf{J}(\mathbf{q}^\varepsilon)) \leqq \mathbf{0}_p; \tag{9}$$

(2) $$d(\mathbf{q}^\varepsilon, \mathbf{v}^\varepsilon) \leqq \delta, \tag{10}$$

   Where,

$$\delta = max\left\{ \frac{\mathbf{b}_j^T\varepsilon}{\mathbf{b}_j^T\mathbf{K}\varepsilon}, \quad j = \overline{1, p} \right\}; \tag{11}$$

(3) for any $\mathbf{v} \neq \mathbf{v}^\varepsilon, \mathbf{v} \in \mathbf{Q}$, preference $\mathbf{v}\,\mathfrak{R}\,\mathbf{v}^\varepsilon$ ratio the type (8) is not fulfilled, i.e. the system of inequalities is not fulfilled:

$$\mathbf{B}((\mathbf{J}(\mathbf{v}) + \mathbf{K}\varepsilon d(\mathbf{v}^\varepsilon, \mathbf{v})) - \mathbf{J}(\mathbf{v}^\varepsilon)) \leqq \mathbf{0}_p. \tag{12}$$

The formulated $\varepsilon\,\Omega$-variational principle is a generalization of the known Ekeland ε-variational principle [29] to the class of multicriteria optimization problems.

4   Necessary Conditions for $\varepsilon\,\Omega$-Equilibrium in a Noncooperative Game Model of Multiobject Control

As follows from Definitions 1 and 2, structure of $\varepsilon\,\Omega$-equilibrium in the noncooperative game model of group control (1) is such that for each participant $i \in \mathbf{N}$ the solution $\mathbf{q}^{e\varepsilon} \in \mathbf{Q}$ is $\varepsilon_i\Omega_i$—optimal on the set $\{\mathbf{q}^{e\varepsilon}\|\mathbf{Q}_i\} = \bigcup_{\mathbf{q}_i \in \mathbf{Q}_i}\{\mathbf{q}^{e\varepsilon}\|\mathbf{q}_i\}$. Therefore, the necessary conditions of the $\varepsilon\,\Omega$—equilibrium of the problem (1) can be formulated in the form of the $\varepsilon\Omega$—variational principle, which is a generalization of the $\varepsilon\Omega$—variational principle to the class of noncooperative game control models.

Statement 3 ($\varepsilon\Omega$—variational principle). Let the following conditions be satisfied in the statement of problem (1):

(1) $\mathbf{Q}_i \subset \mathbf{E}^{r_i}$—a closed subset of a complete metric space, $i \in \mathbf{N}$;
(2) $\mathbf{J}_i(\mathbf{q}) \in \mathbf{E}^{m_i}$—vector bottom semicontinuous function, $i \in \mathbf{N}$;
(3) $\mathbf{B}_i = [p_i \times m_i]$—matrix of a closed convex polyhedral domination cone $\Omega_i \in \mathbf{E}^{m_i}$, $i \in \mathbf{N}$ f, with nonnegative elements $b^i_{kj} \geqq 0$;
(4) the vector $\varepsilon > \mathbf{0}_m$ is fixed and its components satisfy the inclusion $(-\varepsilon_i) \in \Omega_i, i \in \mathbf{N}$;
(5) $\mathbf{q}^{e\varepsilon} \in \mathbf{Q}$—$\varepsilon\Omega$-equilibrium in task (1).

Then, for any $i \in \mathbf{N}$ and any matrix $\mathbf{K}_i = \left\{ k^i_{lj} > 0, l, j = \overline{1, m_i} \right\}$, there is a valid solution $\mathbf{v}^\varepsilon = \{\mathbf{v}^{\varepsilon_i}_i, i \in \mathbf{N}\} \in \mathbf{Q}$ that has the following properties:

(1)
$$\mathbf{B}_i\big(\mathbf{J}_i(\mathbf{q}^{e\varepsilon}\|\mathbf{v}^{\varepsilon_i}_i) - \mathbf{J}_i(\mathbf{q}^{e\varepsilon})\big) \leqq \mathbf{0}_{m_i};$$
(13)

(2)
$$d(\mathbf{v}^\varepsilon, \mathbf{q}^{e\varepsilon}) \leqq \delta;$$
(14)

Where,
$$\delta = \sum_{i \in \mathbf{N}} max\left\{ \frac{\mathbf{b}^{iT}_j \varepsilon_i}{\mathbf{b}^{iT}_j \mathbf{K}_i \varepsilon_i} \right\},$$
(15)

$\mathbf{b}^i_j$—vector formed from j—row of the $\mathbf{B}_i$ matrix;
(3)
for any $i \in \mathbf{N}$ and any $\mathbf{v} \neq \mathbf{v}^\varepsilon$, $\mathbf{v} \in \mathbf{Q}$, the system of inequalities is not satisfied.

$$\mathbf{B}_i\big(\big(\mathbf{J}_i(\mathbf{q}^{e\varepsilon}\|\mathbf{v}_i) + \mathbf{K}_i\varepsilon_i d(\mathbf{v}_i, \mathbf{v}^\varepsilon_i)\big) - \mathbf{J}_i(\mathbf{q}^{e\varepsilon}\|\mathbf{v}^{\varepsilon_i}_i)\big) \leqq \mathbf{0}_{p_i}.$$
(16)

5   Adaptive Fitness Functions in Genetic Algorithms of Multicriteria Conflict Optimization

The properties of optimal solutions in the multicriteria conflict optimization (MCO) problem are determined by the choice of the principle of optimality. At the same time, the design of the fitness function used in the genetic algorithm must reflect the logic of the optimality principle, which provides a high degree of adaptability of individuals with specified properties and, as a consequence, a high probability of their selection into the population of parents. The variety of principles of optimality in MCO problems generates a variety of fitness functions structures.

The article deals with the mechanism of formation of adaptive fitness functions (AFF) on the basis of the necessary conditions of $\varepsilon\Omega$-efficiency and $\varepsilon\Omega$-equilibrium in the noncooperative game model of group control optimization.

The construction of the AFF contains explicitly the following parameters, which can be changed during the search:

- ε—the degree of optimality of solutions (ε-efficiency, ε-equilibrium depending on the type of optimization problem);
- matrix $\mathbf{K}$—the required level of robustness of the components of the vector efficiency index on the set of admissible values of the vector of variable parameters of the system;
- matrix $\mathbf{B}$ of polyhedral domination cone Ω, characterizing the preferences of the designer on the set of achievable vector estimates;
- ψ parameter that determines the type of selection scheme and a balance between the random and deterministic components of GA, which after all determines the rate of convergence of GA and the degree of reliability of the globally optimal solution.

5.1 $\varepsilon\Omega$-Variational Principle-Based AFF

Consider the current population of individuals $\tilde{\mathbf{Q}} = \left\{ \mathbf{q}^i \in \mathbf{Q}, i = \overline{1, |\tilde{\mathbf{Q}}|} \right\}$, where $|\tilde{\mathbf{Q}}|$—the number of individuals in the population $\tilde{\mathbf{Q}}$. For each point $\mathbf{q}^j \in \tilde{\mathbf{Q}}$ we form the fitness function $\Phi(\mathbf{q}^j)$ according to the following rule.

1. Fix $\mathbf{q}^j$. For each $\mathbf{q}^i$, $i = \overline{1, p}$, $i \neq j$, check the execution of a binary preference relation $\mathbf{q}^i \Re \mathbf{q}^j$ form (8), i.e. the satisfaction of the inequalities system:

$$\mathbf{B}\big((\mathbf{J}(\mathbf{q}^i) + \mathbf{K}\varepsilon d(\mathbf{q}^j, \mathbf{q}^i)) - \mathbf{J}(\mathbf{q}^j) \big) \leqq 0. \tag{17}$$

Let's define b_j—the number of points $\mathbf{q}^i(t)$ satisfy (17).

2. Calculate the fitness function in the form of:

$$\Phi(\mathbf{q}^j) = \frac{1}{\left(1 + \frac{b_j}{|\tilde{\mathbf{Q}}|-1} \right)^\psi}, \tag{18}$$

where ψ—parameter defined above.

The fitness function (18) has the following properties.

A. The maximum value of the function $\Phi(\mathbf{q}^j) = 1$ is reached when $b_j = 0$. This means that the solution $\mathbf{q}^j$ has the maximum degree of adaptability to preference $\tilde{\mathfrak{R}}$ within the population $\tilde{\mathbf{Q}}$.

B. The minimum value of the function $\Phi(\mathbf{q}^j) = \frac{1}{2^\psi}$ achieved when $b_j(t) = |\tilde{\mathbf{Q}}| - 1$. In this case, the solution $\mathbf{q}^j$ has a minimum degree of adaptability to preference $\tilde{\mathfrak{R}}$ within the population $\tilde{\mathbf{Q}}$.

The content meaning of the function of suitability of the species (17), (18) is that it provides identification of individuals who, within the current population, possess the properties of $\varepsilon\Omega$—optimality and the robustness level of the vector efficiency index components $\mathbf{J}(\mathbf{q})$ specified by the matrix $\mathbf{K}$.

Comment. The construction of the fitness function (17), (18) generalizes the relation of domination with respect to the cone Ω corresponding to the property Ω-optimality in the problem of multicriteria optimization. If put $\mathbf{K}\varepsilon = \mathbf{0}$, the condition check (17) turns into a check of the domination condition with respect to the polyhedral cone Ω with the matrix $\mathbf{B}$.

5.2   *εG-Variation Principle-Based AFF*

Consider populations $\tilde{\mathbf{Q}}_i = \left\{ \mathbf{q}_i^j, j = \overline{1, |\tilde{\mathbf{Q}}_i|} \right\}, i = \overline{1, n}$, corresponding to components of the vector of the varied parameters $\{ \mathbf{q}_i, i = \overline{1, n} \}$. For each point—individual $\tilde{\mathbf{q}}^T = \left[\tilde{\mathbf{q}}_i^T, i = \overline{1, n} \right] \in \tilde{\mathbf{Q}} = \prod_{i=1}^{n} \tilde{\mathbf{Q}}_i$, the fitness function is formed according to the following rule.

1. For each player $i \in \mathbf{N}$, a set of vector assessments is calculated.

$$\mathbf{J}_i(\tilde{\mathbf{q}} \| \tilde{\mathbf{Q}}_i) = \left\{ \mathbf{J}_i(\tilde{\mathbf{q}} \| \mathbf{q}_i^j), j = \overline{1, |\tilde{\mathbf{Q}}_i|}, \mathbf{q}_i^j \neq \tilde{\mathbf{q}}_i \right\}. \tag{19}$$

Calculate function,

$$f_i(\tilde{\mathbf{q}}) = \cfrac{1}{\left(1 + \cfrac{b_i}{|\tilde{\mathbf{Q}}_i| - 1} \right)^\psi}, \tag{20}$$

where b_i—the number of points in the set (19) satisfy the system of inequalities (17):

$$\mathbf{B}_i\Big(\big(\mathbf{J}_i(\tilde{\mathbf{q}} \| \mathbf{q}_i^j) + \mathbf{K}_i \varepsilon_i d(\mathbf{q}_i^j, \tilde{\mathbf{q}}_i) \big) - \mathbf{J}_i(\tilde{\mathbf{q}} \| \tilde{\mathbf{q}}_i) \Big) \leq \mathbf{0}_{p_i}. \tag{21}$$

2. Let's match the point $\tilde{\mathbf{q}}^T = \left[\tilde{\mathbf{q}}_i^T, i = \overline{1, n} \right]$ value of the vector fitness function $\mathbf{f}(\tilde{\mathbf{q}}) = \left[f_i(\tilde{\mathbf{q}}), i = \overline{1, n} \right]$ with the components of the form (20). The function $\mathbf{f}(\tilde{\mathbf{q}})$ has the following properties.

A. If $\tilde{\mathbf{q}}$ is $\varepsilon\Omega$-equilibrium, all components of the fitness function $\mathbf{f}(\tilde{\mathbf{q}})$ are both equal to 1.

B. If $\tilde{\mathbf{q}}$ isn't $\varepsilon\Omega$-equilibrium, that at least one component $\mathbf{f}(\tilde{\mathbf{q}})$ satisfies inequality $\left(\frac{1}{2}\right)^{\psi} \leq f_i(\tilde{\mathbf{q}}) < 1$.

3. Calculate the vector fitness function $\mathbf{f}(\tilde{\mathbf{q}})$ for all points $\tilde{\mathbf{q}} \in \tilde{\mathbf{Q}}$. As a result, we formed set of achievable vector estimates $\mathbf{f}(\tilde{\mathbf{Q}})$.

4. The fitness of each point-individual $\mathbf{f}^i \in \mathbf{f}(\tilde{\mathbf{Q}})$, $i = \overline{1, |\tilde{\mathbf{Q}}|}$ is evaluated using the fitness function:

$$\Phi\left(\mathbf{f}^i\right) = \frac{1}{\left(1 + \frac{c_i}{|\tilde{\mathbf{Q}}|-1}\right)^{\psi}}, \tag{22}$$

where c_i—the number of points $\mathbf{f}^j$, $j = \overline{1, |\tilde{\mathbf{Q}}|}, j \neq i$, satisfy the system of inequalities $\mathbf{f}^i \leq \mathbf{f}^j$, of which at least one is strict.

6 Conclusion

The $\varepsilon\Omega$-variational principle and $\varepsilon\Omega$-variational principle are considered, where the necessary conditions of $\varepsilon\Omega$-optimality for the problem of multicriteria group control optimization and the necessary conditions of $\varepsilon\Omega$-equilibrium for nonco operative game model of group control optimization are formulated.

On the basis of the considered variational principles the mechanism of formation of AFF in GA providing adaptation of GA to accuracy of the decision, sensitivity, domination cone in multicriteria game group control problems is developed.

A complex of adaptive genetic algorithms for multicriteria conflict group control optimization is developed.

The use of AFF in evolutionary computing technologies allows us to search for optimal solutions of game optimization problems with specified properties and significantly improve the efficiency of calculations.

References

1. Voronov, E.M.: Methods of optimization of management of multi-object multi-criteria systems on the basis of stable-effective gaming solutions. E.Egupova. M.: Publishing house of BMSTU. - 576c (2001)
2. Voronov, E.M., Serov, V.A.: A coordinated stable-effective compromises based methodology of design and control in multi-object systems, in this collection

3. Serov, V.A., Klishin, M.A., Borisov, A.B., Kozlov, D.A.: Program complex for implementation of evolutionary algorithms of multicriteria optimization under conflict and uncertainty. Certificate of state registration of computer program No. 2018614102 of 29.03.2018—Federal service for intellectual property (ROSPATENT)—The register of computer programs
4. Serov, V.A.: The conditions of ε-cone optimality in the multicriteria optimization problem. Vestnik RUDN. Ser. Cybernetics, no: 1, pp. 49–54 (1998)
5. Karpenko, A.P.: Modern algorithms of search optimization. Algorithms inspired by nature, 446p. M.: Publishing house of BMSTU (2014)
6. Greiner, D., Periaux, J., Emperador, J., Galván, B., Winter, G.: Game theory based evolutionary algorithms: a review with Nash applications in structural engineering optimization problems. Arch. Comput. Methods Eng. 24(4), 703–750 (2017)
7. Rutkovskaya, D.: Neural networks, genetic algorithms and fuzzy systems, 452p. In: Rutkovskaya, D., Pilinsky, M., Rutkovsky, L.M., Hotline-Telecom (2006)
8. Kureychik, V.V., Kureychik, V.M., Rodzin, S.I.: Theory of evolutionary computation, 260s. – M.: FIZMATLIT (2012)
9. Ashlock, D.: Evolutionary Computation for Modeling and Optimization, p. 571. Springer, Berlin, Germany (2006)
10. Kita, E. (ed.): Evolutionary Algorithms. InTech, 596p (2011)
11. Dos Santos, W.P. (ed.): Evolutionary Computation. InTech, 582p (2009)
12. Zitzler, E., Deb, K., Thiele, L.: Comparison of multiobjective evolutionary algorithms: empirical results. Evol. Comput. 8(2), 173–195 (2000)
13. Serov, V.A.: Adaptive fitness functions in evolutionary game control optimization models in structure complicated systems. Vestnik BMSTU. Ser. Instrument Making 2(113), 111–122 (2017)
14. Serov, V.A.: Genetic algorithms of conflict equilibriums-based multicriteria systems control optimization under uncertainty. Vestnik BMSTU. Ser. Instrument Making. 4(69), 70–80 (2007)
15. Lung, R.I., Dumitrescu, D.: Computing Nash equilibria by means of evolutionary computation. Int. J. Comput. Commun. 3, 364–368 (2008)
16. El Majd, B., Desideri, J., Habbal, A.: Aerodynamic and structural optimization of a business-jet wingshape by a Nash game and an adapted split of variables. Mec. Ind. 1(3–4), 209–214 (2010)
17. Gonzalez, L., Srinivas, K., Seop, D., Lee, C., Periaux, J.: Coupling hybrid-game strategies with evolutionary algorithms for multi-objective design problems in aerospace. In: Evolutionary and deterministic methods for design, optimization and control with applications to industrial and societal problems, CIMNE, pp. 221–248 (2011)
18. D'Amato, E., Daniele, E., Mallozzi, L., Petrone, G.: Equilibrium strategies via GA to Stackelberg games under multiple follower's best reply. Int. J. Intell. Syst. 27, 74–85 (2012)
19. Arias-Montano, A., Coello, C.C., Mezura-Montes, E.: Multiobjective evolutionary algorithms in aeronautical and aerospace engineering. IEEE T Evolut. Comput. 16(5), 662–694 (2012)
20. Coelho, R.: Co-evolutionary optimization for multi-objective design under uncertainty. J. Mech. Des. T. ASME 135(2), 1–8 (2013)
21. Periaux, J., Gonzalez, F., Lee, D.: Multi-objective EAs and game theory. In: Evolutionary Optimization and Game Strategies for Advanced Multi-Disciplinary Design. Intelligent Systems, Control and Automation: Science and Engineering, vol 75, pp. 21–38. Springer, Dordrecht (2015)
22. Greiner, D., Periaux, J., Emperador, J.M., Galvan, B., Winter, G.: A study of Nash-evolutionary algorithms for reconstruction inverse problems in structural engineering. In: Greiner, D. et al. (eds.) Advances in Evolutionary and Deterministic Methods for Design, Optimization and Control in Engineering and Sciences. Computational Methods in Applied Sciences, vol 36. Springer, New York, pp. 321–333 (2015)
23. Lee, D.S., Gonzalez, F., Periaux, J., Srinivas, K.: Efficient hybrid-game strategies coupled to evolutionary algorithms for robust multidisciplinary design optimization in aerospace engineering. IEEE T. Evolut. Comput. 15(2), 133–150 (2011)

24. Leskinen, J., Périaux, J.: Distributed evolutionary optimization using Nash games and GPUs-applications to CFD design problems. Comput. Fluids, 80, 190–201 (2013)
25. Sinha, A., Malo, P., Frantsev, A., Deb, K.: Finding optimal strategies in a multi-period multi-leader-follower Stackelberg game using an evolutionary algorithm. Comput. Oper. Res. **41**, 374–385 (2014)
26. Tang, Z., Desideri, J.A., Periaux, J.: Multi-criteria aerodynamic shape design optimization and inverse problems using control theory and Nash games. J. Optimiz. Theory App. **135**(1), 599–622 (2007)
27. Serov, V.A.: On the variational principle in of multicriteria optimization and decision-making problems. Actual problems of the theory and applications of engineering research: SB. scientific papers. – M.: Mechanical Engineering, pp. 18–22 (1999)
28. Auben, J.-P., Ekland, I.: Applied nonlinear analysis. M. World, 512p (1988)
29. Ekeland, I.: On the variational principle. J. Math. Anal. Appl. **47**(2), 324 (1974)

Logical Control a Group
of Mobile Robots

Stanislav L. Zenkevich, Anaid V. Nazarova and Hua Zhu

Abstract *Problem statement*: group interaction of the smart electromechanical systems (SEMS) play an important role in solving complex problems using robots. It is becoming more relevant that design and research the mechanism for implementing effective, reliable and safe cooperative operation. This article presents method for solving logical control tasks for a group of mobile robots moving in the convoy type formation. This problem arises when it is needed to change the topology of group, for example, when it is needed to merge/split two convoys, insert a new robot into the convoy or detach the convoy member, make other transformations of the SEMS associated with the implementation of the technological task. It is noted that the necessity of design the logical level of control system of a group mobile robots. *Purpose of research*: the article presents method for solving logical control tasks for a group of mobile robots using the theory of finite-state machine, A mechanism for planning and coordinating the behavior of robots in the group has been developed. Finite-state machine is used to describe the logical level of control system for each robot in the group. *Results*: The structure of the logical level of the group robots control system is developed, which includes a network of interacting finite-state machines and provides a simple scaling of the system when the number of robots changes. In the ros_stage environment the simulation result about the group behavior with various changes of its topology is presented. *Practical significance*: The proposed solution provides high functionality of the group robot, taking into account the SEMS ideology, which allows perform tasks related to change the dislocation of mobile robots during the transportation of cargos.

S. L. Zenkevich (✉) · A. V. Nazarova · H. Zhu
Bauman Moscow State Technical University, Moscow, Russia
e-mail: zenkev@bmstu.ru

A. V. Nazarova
e-mail: avn@bmstu.ru

H. Zhu
e-mail: zhuhua1302@gmail.com

© Springer Nature Switzerland AG 2019
A. E. Gorodetskiy and I. L. Tarasova (eds.), *Smart Electromechanical Systems*,
Studies in Systems, Decision and Control 174,
https://doi.org/10.1007/978-3-319-99759-9_4

Keywords Group of mobile robots · SEMS · Logical control · Finite-state machine · Convoy coordinator · Simulation

1 Introduction

Swarm robotic systems based on SEMS [1] (smart electromechanical systems) are a kind of cyber physical systems [2]. System approach to the analysis of the evolutionary development of cyber physical systems can be divided into three strategic directions [3]. One of strategic directions is associated with the development of the principles of network organization and group control of individual intelligent systems that make up a distributed environment of artificial intelligence. This article describes a simple example of achieving group coordination. The considered system belongs to the class SEMS systems. Really, the behavior of this system at a whole looks like the behavior of a group of living systems, moreover the components of formation are electromechanical robots.

2 Formulation of the Problem

Use a group of mobile robots moving in the convoy type formation is one of the effective ways to solve a number of tasks, for example, transportation of a large number of cargos. A number of problems related to this task were solved in [4]. However, direct use of the developed approach may prove ineffective due to the specific nature of the task. Possible tasks of this type are:

1. Transportation of cargos from point A to point B, during the movement robot R_i leave convoy in point A, robot R_j insert into convoy in point B (see Fig. 1a).

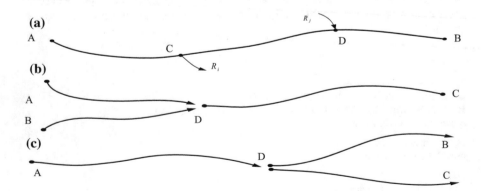

Fig. 1 Possible tasks

2. Transportation of cargos from point A and point B to point C using two group of mobile robots moving in the convoy type formation, in point D two convoys are combined into one (see Fig. 1b).
3. Transportation of cargos from point A to destinations B and C, while in point D the convoy is divided into two groups (see Fig. 1c).

The above tasks require the following convoy actions (see Fig. 2):

1. Formation of the convoy (Fig. 2a). This action consists in regrouping several "unorganized" robots from the schooling movement to the movement in convoy type formation.
2. Convoy movement (Fig. 2b). The essence of this movement is that the leading robot moves autonomously, and the remaining robots must follow the position of their own leading robot. Moreover, the group can stop and restart the convoy due to a dynamically changing environment; the group must continue its movement if one or more of the slave robots fails (except for leader of convoy).
3. Insert robot into convoy (Fig. 2c). The purpose of this action is that when the convoy is moving, it is necessary to provide a "gap" in the trajectory for the entry of an additional robot between the robots of the convoy or to designate the new robot as the final one in the convoy.
4. Detach a robot from convoy (Fig. 2d). Unlike the previous action, it is necessary to fill the "vacancy" after the robot leave from convoy.
5. Separation of the convoy (Fig. 2e). The numbers of robots in the new convoys are either determined in advance or are selected by some criterion.
6. Association of convoys (Fig. 2f). In this case, one of the two leading robots of convoys is appointed leader.
7. Dissolution of the convoy (Fig. 2a). After the task is completed, robots can move arbitrarily or stop.

The task to be solved in this work is to develop approaches to control a group of mobile robots when the topology of group is changed, and as a consequence, to develop a logical level of the control system that ensures the group's execution of a sequence of actions associated with change its structure.

There are several solutions to this problem, for example, in [5] a Petri net was used to describe the behavior of a group when the topology of the group is changed (bypass table in the room). Article [6] study formation changes for robot group in

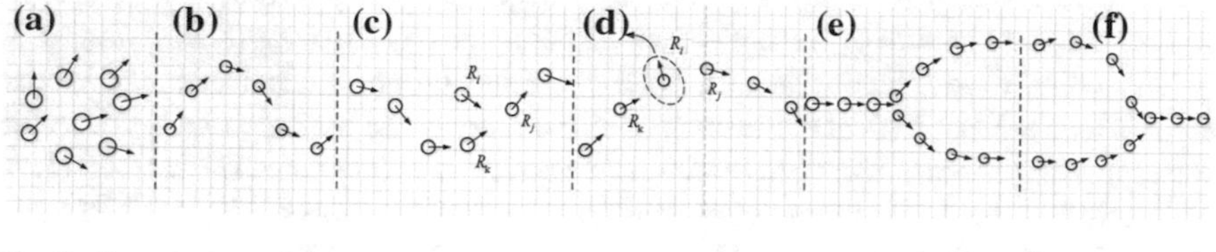

Fig. 2 Convoy behaviors

known environments. In [7] and [8] multi-robot collision avoidance algorithms were developed for reconfiguring the location of robot in the formation.

The article presents method for solving this problem using the theory of finite-state machine.

3 The Concept of a Functional Finite-State Machine

A functional finite-sate machine $A = [U,X,Z,Exec(),f,h]$ will be called a set of six objects: $U = \{u_1,...,u_r\}$—the set of input signals; $X = \{x_1,...,x_n\}$—a finite set of states; $Z = \{z_1,...,z_n\}$—a set of output signals; $Exec() = \{exec1(),...,execi(),...$ $execn()\}$—the set of operations that correspond to the state of the machine; f—functions of one-step transition to the next state $x(t + 1) = f(x(t),u(t))$; h—output function $z(t + 1) = h(x(t),u(t))$ (Mealy machines, which are characterized by producing outputs when a transition is taken) or $z(t + 1) = h(x(t + 1))$ (Moore machines, produce outputs when the machine is in a state, rather than when a transition is taken). Note that in our case, asynchronous automata are used, i.e., the parameter t characterizes not the time but the sign of the appearance of a signal from the set of input signals.

4 Planning and Coordination of the Behavior of an Individual Robot in Convoy–Convoy Coordinator (Control Automaton)

In order to solve the problem posed in this paper, a structural diagram of the logical level of the robot convoy control system is developed (see Fig. 3), which provides the implementation of different convoy modes, for example, insert robot into convoy, detach a robot from convoy, separation of convoy and others.

The control automaton—convoy coordinator performs the functions of controlling the behavior of each robot by the commands of operator, and coordinating their actions, depending on the result of the task execution by the corresponding robot. Automaton A_i is used to describe the logical model of individual robots. Switches allow control system to quickly rebuild the work of coordinators in accordance with the commands of the operator, for example, to perform the task of combining two convoys into one, the control system uses switches that provide the transfer of control of robots in convoy 2 from the coordinator convoy 2 to convoy coordinator 1.

The convoy coordinator is described by multi-input Mealy machines whose input is the result of logical expression, and the outputs are logical commands to the corresponding robots. The state diagram of convoy coordinator is shown in Fig. 4.

Specific values of input commands: u_1—FORM_CONVOY, u_2—PLAN_INSERT, u_3—PLAN_DETACH, u_4—PLAN_ASSOCIATION, u_5—PLAN_SEPARATION, u_6—ERROR, u_7—PLAN_PAUSE, u_8—SUCCESS, u_9—RESTART, p—isPrior_coordinator.

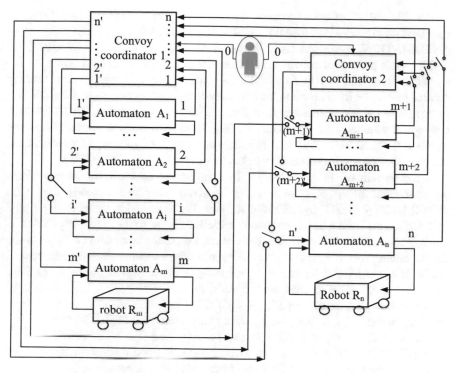

Fig. 3 Structural diagram of logic control of robots in group

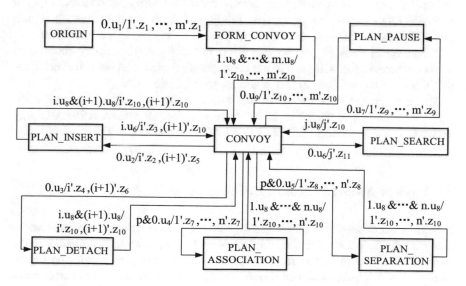

Fig. 4 State diagram of convoy coordinator

Specific values of output signals: z_1—READY, z_2 —INSERT, z_3—LEAVE, z_4—DETACH, z_5—AUXILIARY_INSERT, z_6—AUXILIARY_DETACH, z_7—ASSOCIATION, z_8—SEPARATION, z_9—PAUSE, z_{10}—GO, z_{11}—SEARCH.

The work process of convoy coordinator in each state looks as follows.

When command FORM_CONVOY (u_1) arrives from the top level of convoy control system or from operator, convoy coordinator sends command READY (z_1) to each robot at a certain interval of time. After receiving signal SUCCESS (u_8), convoy coordinator sends command GO (z_{10}) to the corresponding robot. When the number of signal SUCCESS (u_8) becomes equal to total number of robots in the convoy, the convoy coordinator switches to the basic state (CONCOY).

When command PLAN_INSERT (u_2) is issued or receiving request from an individual robot R_i (see Fig. 2c), convoy coordinator sends command AUXILIARY_INSERT (z_5) to robot R_k (see Fig. 2c), and after receiving signal SUCCESS (u_8) from it, sends command INSERT (z_2) to robot R_i (see Fig. 2c). The convoy coordinator switches to the basic state, i.e., sends command GO (z_{10}) to the corresponding robots R_i and R_k, which ensures the execution of the main control mode. If signal SUCCESS (u_8) from robot R_i does not appear during a specified time, then sends robot R_i command LEAVE (z_3).

When receiving command PLAN_DETACH (u_3) or request from a separate robot R_i (see Fig. 2d), convoy coordinator sends command DETACH (z_4) to robot R_i, and after taking signal SUCCESS (u_8) from it, sends command AUXILIARY_DETACH (z_6) to robot R_k (see Fig. 2d). The convoy coordinator switches to the basic state when signal SUCCESS (u_8) from robot R_k appears.

When command PLAN_ASSOCIATION (u_4) or PLAN_SEPARATION (u_5) arrives, the convoy coordinator sends command ASSOCIATION (z_7) or SEPARATION (z_8) to the appropriate robots and waits for the feedback signals SUCCESS from all the robots of convoy, and then switches to the basic state.

When signal ERROR (u_6) appears from the faulty robot, convoy coordinator sends command SEARCH (z_{11}) to the corresponding robot and waits for the signal SUCCESS to appear from it, then switches to the basic state.

When command PLAN_PAUSE (u_7) arrives, convoy coordinator sends command PAUSE (z_9) to all robots in the convoy. When command RESTART (u_9) appears, it sends command GO (z_{10}) to all robots in the convoy.

5   Logical Model of Robot in the Group

Moving in the convoy type formation, the robot R_i performs the basic control law—to repeat the trajectory of its leading robot with some time lag. Violation of the basic motion control law is required when it is necessary to ensure the implementation of the tasks described above. Such a situation arises in two cases: when the robot moves freely, namely, its belonging is not yet defined; or when moving in the convoy, robot receives appropriate commands from convoy coordinator.

The reasonable behavior of the robot in these cases, apparently, should be the following. In the first case, when command READY appears, the robot must implement the appropriate control law to reach the "Start" point during time T, at the same time, remember the trajectory of its leading robot. Upon arrival at the "Start" point (see Fig. 5), the robot sends signal SUCCESS to convoy coordinator.

When command INSERT appears, robot R_i must switch to implement the main control law, namely to follow robot R_{i-1} in the convoy and sends signal SUCCESS to convoy coordinator. The appearance of command GO means that the robot becomes a member of the convoy; otherwise, if command GO does not arrive at the specified time or command LEAVE appears, the robot switch to free movement mode.

In case moving in convoy type formation, when command DETACH arrives, the robot R_i switch to the mode of detaching from convoy, the strategy of which is as follows: the robot R_i must transmit to the robot R_{i+1} the trajectory of its leading robot R_{i-1} in the immediate period. After this, the robot R_i switches to free-motion mode and sends signal SUCCESS to convoy coordinator.

When command AUXILIARY_INSERT appears, the robot must continue to execute the basic control law for time T to save the trajectory of its leading robot in the immediate period 2*T. Then the robot executes another control law, repeating the trajectory of its leading robot with some time lag 2*T, and save the trajectory of its new leading robot in the near future T and sends signal SUCCESS to convoy coordinator. When command GO appears, robot again switches to the basic mode, but follows the new leader.

When command AUXILIARY_DETACH appears, robot begins to receive the trajectory of its future leader and does not follow the trajectory of its future leader until command GO appears.

When command ASSOCIATION/SEPARATION appears, robot must execute the control law to ensure relative location in relation to its leader, at the same to avoid collision. When command GO appears, robot switches to the basic mode. Note that when executing command SEPARATION, two robots move autonomously and both play the role of leader.

In case of malfunction, robot sends signal ERROR to convoy coordinator.

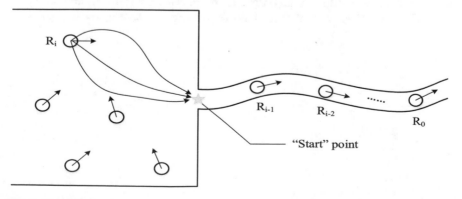

Fig. 5 The process of formation of convoy

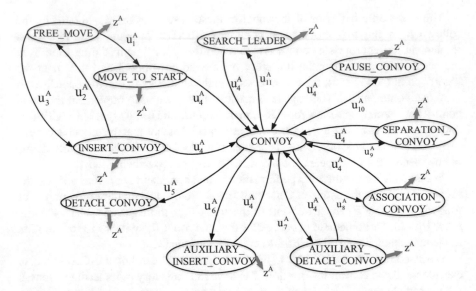

Fig. 6 Logical model of robot in convoy

When command SEARCH appears, robot switches to the mode of searching new leader. After receiving information about the new leader, robot sends signal SUCCESS to convoy coordinator.

The behavior of an individual robot in convoy at logical level is described by Moore machines, whose corresponding transition graph is shown in Fig. 6. The input signals of Moore machine come from convoy coordinator.

Specific values of input commands: u_1^A—READY, u_2^A—LEAVE, u_3^A—INSERT, u_4^A—GO, u_5^A—DETACH, u_6^A—AUXILIARY_INSERT, u_7^A—AUXILIARY_ DETACH, u_8^A—ASSOCIATION, u_9^A—SEPARATION, u_{10}^A—PAUSE, u_{11}^A— SEARCH.

Specific values of input commands: z^A—SUCCESS.

Thus, the control system of an individual robot has two-level: the lower level executes the actuator control according to the selected control law, and the upper (logical) level selects one of these laws depending on the convoy coordinator's command.

6 Example: The Process of Inserting a Robot $R_{k-1,\,k}$ into a Convoy Between Robots R_{k-1} and R_k

To describe convoy logic control process, in article detail investigate task of inserting a robot $R_{k-1,\,k}$ into a convoy between robots R_{k-1} and R_k. To implement these actions, convoy coordinator and controlled automata (model of individual robot) perform the corresponding transitions. The process of automata state transition is shown in Fig. 7, and the trajectories of the robots in convoy when carrying out the convoy's movement along a straight line are shown in Fig. 8.

7 The Simulation of Formation Behavior with a Change in Structure

The article presents formation behavior simulation in Stage_ros to verify correctness and efficiency of proposed approach using two convoys (convoy 1: robot 0 1 2, convoy 2: robot 3 4 5). Software architecture is shown in Fig. 9, which contains four levels: 1—convoy interface for operator; 2—strategic level, task of which is to determine the subtasks for each robot; 3—tactical level, whose outputs are motion control signals for robots; 4—the model of external environment and robots (executive level).

With consecutive commands (input signals of convoy coordinator) FORM_CONVOY (u_1), PLAN_DETACH (u_3), PLAN_INSERT (u_2), PLAN_ASSOCIATION (u_4) to convoy coordinators before association, PLAN_DETACH (u_3), PLAN_INSERT (u_2), PLAN_SEPARATION (u_5) to convoy coordinator before separation and PLAN_PAUSE (u_7) after separation, two convoys perform sequential appropriate actions. The simulation result is shown in Fig. 10. The video containing simulation results is presented in [9].

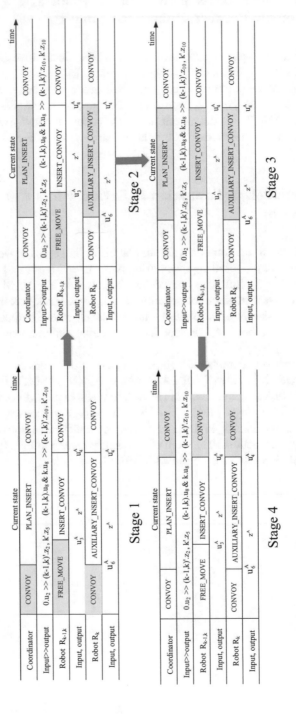

Fig. 7 Interaction of convoy coordinator and controlled automata

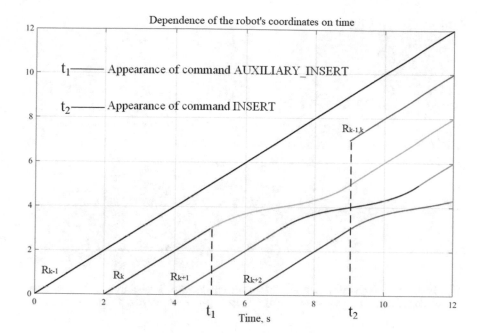

Fig. 8 The process of inserting robot into convoy

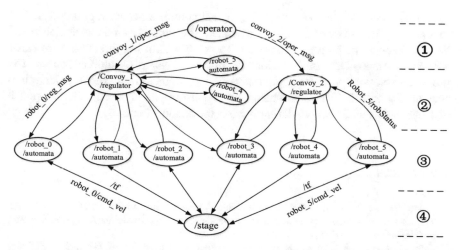

Fig. 9 Software architecture of control system

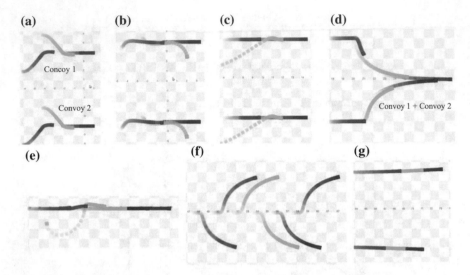

Fig. 10 Simulation results: **a** formation of the convoy, **b** detach a robot from convoy, **c**, **e** insert robot into convoy, **d** association of convoys, **f** separation of the convoy, **g** stop convoy

8 Conclusion

The article presents method for solving logical control tasks for a group of mobile robots using the theory of finite-state machine. The simulation results demonstrated the effectiveness of proposed approach. The possible further direction of research will focus on analysis of behavior stability of a group of mobile robots. The proposed solution provides high functionality of the group robot, taking into account the SEMS ideology, which allows perform tasks related to change the dislocation of mobile robots during the transportation of cargos.

References

1. Gorodetskiy, A.E. (ed.): Smart electromechanical systems. In: Studies in Systems, Decision and Control, vol. 49, 277 p. Springer, Switzerland (2016) http://dx.doi/org/, https://doi.org/10.1007/978-3-319-27547-5_4
2. Lee, E.: Cyber physical systems: design challenges. University of California, Berkeley Technical Report No. UCB/EECS-2008-8. Retrieved 7 June 2008
3. Shkodyrev, V.P.: Technical system control: from mechatronics to cyber-physical systems. In: Studies in Systems, Decision and Control, vol. 49, 277 p. Springer, Switzerland (2016) http://dx.doi/org/10.1007/978-3-319-27547-5_4
4. Zenkevich, S.L.: Zhu Hua Upravlenie dvizheniem gruppy robotov v stroju tipa "konvoj" (Control of a group of mobile robots moving in the convoy type formation). Mekhatronika, Avtomatizacia, Upravlenie (Mechatronics, Automation, Control), 18(1), 30–34 (2017) (in Russia)

5. Levi, P., Muscholl, M., Bräunl, T.: Cooperative mobile robots stuttgart: architecture and tasks. In: Proceeding of the 4th International Conference on Intelligent Autonomous Systems, IAS-4, Karlsruhe, pp. 310–317, March 1995
6. Hönig, W., Kumar, T.K.S., Ma, H., Koening, S., Ayanian, N.: Formation change for robot groups in occluded environments. Im: 2016IEEE/RSJ International Conference on Intelligent Robots and Systems (IROS), Daejeon, pp. 4836–4842 (2016)
7. van den Berg, J., Guy, S.J., Lin, M., Manocha, D.: Reciprocal n-body collision avoidance, In: Robotics Research: The 14th International Symposium ISRR, Springer Tracts in Advanced Robotics, vol. 70, pp. 3–19. Springer, Berlin, May 2011
8. Hennes, D., Claes, D., Meeussen, W., Tuyls, K.: Multi-robot collision avoidance with localization uncertainty. In: Proceedings of the 11th International Conference on Autonomous Agents and Multiagent Systems (AAMAS 2012), Valencia, Spain, June 2012
9. Video with simulation results. Available at https://drive.google.com/file/d/0B6UOi4Ja1li9 YVU4Y2FHNURLb1k/view?usp=sharing. Accessed 9 June 2017

Analysis of Use of Platonic Solids in Swarm Robotic Systems with Parallel Structure Based on SEMS

Sergey N. Sayapin

Abstract *Problem statement*: one of the most important directions in the development of modern mechatronics and robotics is associated with the development of a fundamentally new class of multi-link devices that can operate in extreme, a priori uncertain conditions and have an adaptive kinematic structure, automatically rebuilt depending on the specifics of the problem. The intelligent control system of such an object should have a distributed structure and provide the possibility of autonomous operation in conditions of uncertainty, which creates a number of problems associated with the creation of algorithms of functioning, self-learning and reconfiguration. Modules of parallel structure based on Platonic solids can be successfully used to create role-based systems. The attractiveness of using Platonic solids in the construction of mobile modular robots is due to the fact that of the huge variety of polyhedrons, only they are correct and therefore modules based on each of them have unified elements and the possibility of unlimited extension along each of the faces. *Purpose of research*: to choose the most effective type of mobile parallel robot of parallel structure based on Platonic bodies and SEMS for autonomous and collective application. *Results*: the review of known and perspective mobile robots of modular type on the basis of Platonic bodies is given. The comparative analysis of similar parallel robots on the basis of which the module in the form of an octahedron is chosen as a base sample. A new modular type of spatial mobile parallel robot with 12 dof based on octahedral structure, called Octahedral dodekapod (from Greek words dodeka meaning twelve and pod meaning foot or its counterpart leg). The analysis of its functional capabilities at its Autonomous and collective (swarm) application is carried out. *Practical significance*: the Octahedral dodekapod can be successfully applied for the solution of individual and collective

S. N. Sayapin (✉)
Mechanical Engineering Research named after A. A. Blagonravov
of the Russian Academy of Sciences, Moscow, Russia
e-mail: S.Sayapin@rambler.ru

S. N. Sayapin
Bauman Moscow State Technical University, Moscow, Russia

© Springer Nature Switzerland AG 2019
A. E. Gorodetskiy and I. L. Tarasova (eds.), *Smart Electromechanical Systems*,
Studies in Systems, Decision and Control 174,
https://doi.org/10.1007/978-3-319-99759-9_5

tasks in extreme, a priori undefined conditions. Its adaptive kinematic structure makes it possible to combine with similar modules to form multifunctional active intelligent robotic systems to solve a wide variety of problems.

Keywords Parallel robot · Platonic solids · Self-propelled robot
Self-reconfigurable robot · Swarm systems · Modular robot · SEMS

1 Introduction

One of the most important directions in the development of modern mechatronics and robotics is associated with the development of mobile robotic modules (MRM), which in combination with intelligent control systems become autonomous, and with information communication with each other—capable of self-organization in swarm robotic systems (SRS), which have collective intelligence. Thus application of MRM as a part of SRS allows them to unite mechanically in multimodule reconfigurable active structures. Such modules and structures are designed for operation in extreme, a priori uncertain environments and must have the adaptive kinematics, automatically tunable in accordance with the nature of the problem being solved. In this regard, there is a great interest all over the world in the creation of SRS for solving problems of various scientific, civil and military purposes [1–6]. For example, a collective study of the surfaces of planets, conducting chemical and radiation reconnaissance of contaminated areas and areas of industrial facilities, epidemiological survey of dangerous areas, as well as rescue operations in emergency situations. This interest is due to the possibility of individual solutions to numerous complex problems due to the autonomy of MRM, and due to the possibility of combining them into complex structures capable of solving problems that are impossible for autonomous modules. SRS is a new direction of robotics in the field of studying artificial collective intellectual systems like the study in the biology of insect behavior, which coordinate their actions in order to jointly solve problems that are impossible for the individual. Portability modular robots is provided in a variety of types of propulsion, e.g., wheeled, tracked, screw, push pull, snake-like etc. in combination with devices capable of ensuring their self-propelled in arbitrarily oriented in space, open and closed surfaces [7–11]. The ability to collectively solve problems that are impossible for an autonomous robot in SRS is borrowed from wildlife, which is the main biological inspiration for the creation of Autonomous robots and collective robotic systems [1–15]. For example, to overcome obstacles, groups of working ants are functionally combined into living bridges over which other ants move freely [1, 2, 13–15]. Figure 1 shows examples of overcoming obstacles by self-organized ants (a–f) and SRS chains consisting of identical Autonomous mobile robots s-bot (g, h) connected with each other [15, 16]. For offline solving complex tasks of chain can be formed of functional modules, such as, for example, a generic reconfigurable robotic system "Robotrain" (i) [18].

Fig. 1 Examples of overcoming obstacles by self-organized ants (**a–f**) and SRS chains consisting of identical Autonomous mobile robots s-bot (**g, h**) connected with each other

Such modules should be of the same type, mobile with the possibility of mechanical unification with each other in spatial or linear reconfigurable mobile robotic systems [2–7, 9–13, 16–18]. Currently, modules with a variety of geometric shapes, including polyhedron [2, 6–13, 16–18], are used to construct autonomous and multi-module mobile robotic systems. However, out of the variety, only five Platonic solids (Fig. 2) are correct and are capable of unlimited expansion along any of the faces, providing modularity of construction of a wide variety of structures and designs [19, 20]. This feature of the Platonic solids is observed in nature, for example in crystal lattices of minerals [21] and microorganisms [22].

In this regard, it is attractive to build autonomous mobile parallel robots (MPR) on basis of the Platonic solids (Fig. 2) [9, 10, 23–26], capable of unite in SRS.

TETRAHEDRON   OKTAHEDRON   HEXAHEDRON DODECAHEDRON ICOSAHEDRON

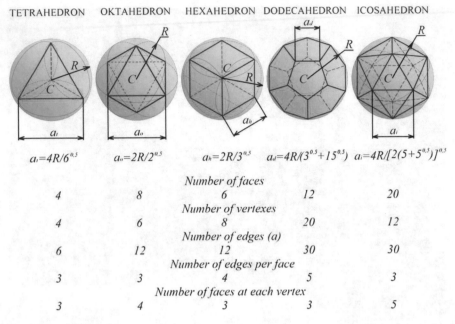

$$a_t=4R/6^{0.5} \qquad a_o=2R/2^{0.5} \qquad a_h=2R/3^{0.5} \quad a_d=4R/(3^{0.5}+15^{0.5}) \quad a_i=4R/[2(5+5^{0.5})]^{0.5}$$

		Number of faces		
4	8	6	12	20
		Number of vertexes		
4	6	8	20	12
		Number of edges (a)		
6	12	12	30	30
		Number of edges per face		
3	3	4	5	3
		Number of faces at each vertex		
3	4	3	3	5

Fig. 2 Five Platonic solids

2 The Analysis Use of the Platonic Solids as Universal Bases to Design for Adaptive Intelligent Spatial Parallel Mobile Modular Robots Based on SEMS

Of the variety polyhedrons, only the Platonic solids are correct and capable of unlimited expansion along any of the faces, providing modularity of construction of a wide variety of structures and constructions [19, 20]. However, despite numerous examples of successful application of the Platonic solids as the basic passive building modules of engineering and architectural structures, there are very few examples of such application in MPR. At the same time, there is growing interest in the creation of MPR based on the Platonic solids in the form of autonomous modules capable not only to move, but also to carry out manipulative operations, as well as to unite in reconfigurable active structures. MPR based on the Platonic solids are formed by the swivel of vertices with the ends of linear actuators, forming the right triangular (tetrahedron, octahedron, icosahedrons), quadrangular (hexahedron) and pentagonal (dodecahedron) faces, capable of compressing and decompressing the covered and covering objects of arbitrary shape. As a result, five types of robots are formed in the form of unified mobile parallel modules with adaptable faces and manipulation capabilities. Thus, the MPR based on the Platonic solids, in contrast to the well-known spatial parallel robots (SPR) [13], able not only to self-propelled, but also provide their vertices, edges and faces of the spacer adaptive compressive and impact on covered and covering the items of arbitrary shapes and manipulate them in space. In combination with intelligent control systems MPR is autonomous and when information communication with each other—capable of self-organization in swarm robot systems (swarm robotics) with the collective intelligence for solving tasks that would be overwhelming for self-contained units [1–11].

Figure 2 shows five Platonic solids and their basic parameters. Obviously, the frames constructed in the form of Plato and inscribed in a sphere of radius R, with the same design edge elements and their compounds have different mass, stiffness, the degree of approximation to the surface of a sphere of radius R. For example, the total length of the edges with radius R described the sphere for the tetrahedron, octahedron, hexahedron, dodecahedron and icosahedrons is: $S_t = 9.79R$; $S_o = 16.97R$; $S_h = 13.86R$; $S_d = 21.41R$; $S_i = 22.3R$. According to their mass m as follows: $m_t < m_h < m_o < m_d < m_i$. Let l denote the distance from the vertices of the Platonic solids to their connection to the linear actuators, and through $L_{max} = L$ and $L_{min} = l/2$ the maximum and minimum lengths of the non-telescopic linear actuators. Then $a_{max/min} = L_{max/min} + 2l$ and $R_{max} = R$, R_{min} for the Platonic solids will make (Fig. 2):

for tetrahedron

$$R_{tmax} = R = 6^{0.5}a_{tmax}/4 = 6^{0.5}(L_t + 2l)/4,$$
$$R_{tmin} = 6^{0.5}[(L_t/2) + 2l]/4 - 0.5R_{tmax} + 0.6l, \tag{1}$$

for octahedron

$$R_{omax} = R = 2^{0.5}a_{omax}/2 = 2^{0.5}(L_o + 2l)/2,$$
$$R_{omin} = 2^{0.5}[(L_o/2) + 2l]/2 = 0.5R_{omax} + 0.7l, \tag{2}$$

for hexahedron

$$R_{hmax} = R = 3^{0.5}a_{hmax}/2 = 3^{0.5}(L_h + 2l)/2,$$
$$R_{hmin} = 3^{0.5}[(L_h/2) + 2l]/2 = 0.5R_{hmax} + 0.87l, \tag{3}$$

for dodecahedron

$$R_{dmax} = R = \left(3^{0.5} + 5^{0.5}\right)a_{dmax}/4 = \left(3^{0.5} + 5^{0.5}\right)(L_h + 2l)/4,$$
$$R_{dmin} = \left(3^{0.5} + 5^{0.5}\right)[(L_d/2) + 2l]/4 = 0.5R_{dmax} + 1.4l, \tag{4}$$

for icosahedron

$$R_{imax} = R = \left[2\left(5 + 5^{0.5}\right)\right]^{0.5}a_{imax}/4 = \left[2\left(5 + 5^{0.5}\right)\right]^{0.5}(L_i + 2l)/4,$$
$$R_{imin} = \left[2\left(5 + 5^{0.5}\right)\right]^{0.5}[(L_i/2) + 2l]/4 = 0.5R_{imax} + 0.95l. \tag{5}$$

Thus, the maximum transformation coefficients ($k = R/R_{min}$) are those of the tetrahedron and octahedron-based MPR, and the minimum hexahedron, icosahedrons and dodecahedron ($k_t > k_o > k_h > k_i > k_d$).

To increase the stroke of the linear actuator MPR used paired (Fig. 3a), telescopic (Fig. 3b), gear-rack (Fig. 3), polypulley design (Fig. 3g) or their combinations (Fig. 3e) [27–32]. At the same time, in cases when the axis passing through

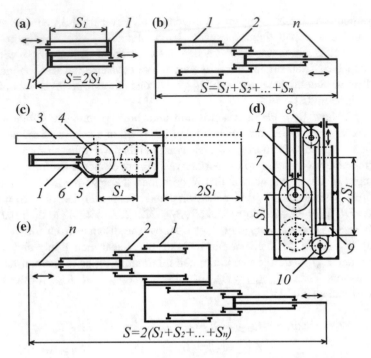

Fig. 3 Ways increase the stroke of the linear actuator MPR: pairing (**a**); telescopic connecting (**b**); gear-rack (**c**); polypulley design (**d**); combination connecting (**e**)

the connected vertices does not coincide with the axes of the pushed elements, the rods work not only for stretching and compression, but also for bending, which reduces their rigidity.

At each of the vertexes of MPR based on tetrahedron, hexahedron and dodec-ahedron hingedly connected to three edges, and in the octahedron and the s four and five respectively.

Use in these structures three, four and fife spherical joints with center in one vertex (Fig. 4a) allows to exclude the kinematic error in joint compounds and to increase the stiffness with the exception of hexahedron. In this case, all the links of the mechanism in this case work only for stretching-compression. However, in such joints because of the proximity of the ends of the rods of the ribs, with the actual size of cross sections, obtained by a small angular movement, which reduces the mobility of these structures. To eliminate this disadvantage, the diameter of the ball joint should be much larger than the diameter of the rod end, which will inevitably lead to an increase in the overall mass characteristics of MPR and significantly reduce the manufacturability of the swivel Assembly at the top of MPR. In this regard, in practice, the separation of the hinge joints of the top of the MPR with the ends of the rods from its center is used (Fig. 4b). As a result, the technological effectiveness of the swivel unit is increased, as well as the values of the angular displacements of the rods relative to each other at lower dimensional and mass

Fig. 4 Structures MPR with three, four and five spherical joints with center in one vertex (**a**) and near this vertex (**b**)

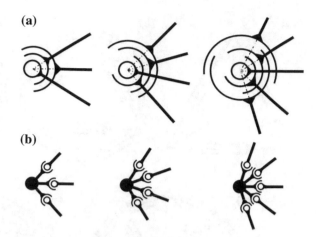

(**a**)

(**b**)

characteristics. At the same time, MPR rods work for tension-compression, and the elements of the swivel at the apex also for bending. The kinematic error of such MPR depends on the ratio of the distance between the joints of the rib rod and the joints of the vertex node: the more it is, the higher the kinematic accuracy of the mechanism. Also hinges can be made in the form of elastic elements. However, in this case, linear actuators need to overcome the elastic forces, which reduces their efficiency.

There are known the concepts of transformable modular structures based on the Platonic solids, in which the edges are made in the form of translational kinematic pairs with one degree of freedom. The ends of the edges are connected in the vertices (Fig. 5) [33, 34].

Such modules can be transformed and combined with similar spatial structures. Thus modules with tetrahedral (Fig. 5a), octahedral (Fig. 5b), dodecahedral (Fig. 5d) and icosahedral (Fig. 5e) structures have one degree of mobility, and the cubic module (Fig. 5c) has three degree of mobility. For self-propelled of such modules requires the use of special additional devices. Therefore, the establishment of MPR on their basis is not appropriate. However, they can be used to create multi-module transformable and reconfigurable active structures.

It should be noted that due to the three degrees of mobility, cubic modules can be combined into structures that can self-propelled by tilting. The article [35] presents a new type of robot consisting of three-step cubic modules. For rice 6 shows a sequence diagram of the move of the united eight angle structure by tilting. Tilting occurs as a result of displacement of the center of gravity of the eight-module structure beyond the support surface (Fig. 6a–d). After reducing the lengths of the edges to the initial length, the combined structure is transformed into the initial position, but in a new place (Fig. 6c). In this way, the combined structure can be rolled in two orthogonal directions and positioned on the surface with an accuracy of one step equal to two minimum lengths of edges of the cubic module (Fig. 6a, e). At the same time, it is possible to significantly improve the positioning accuracy

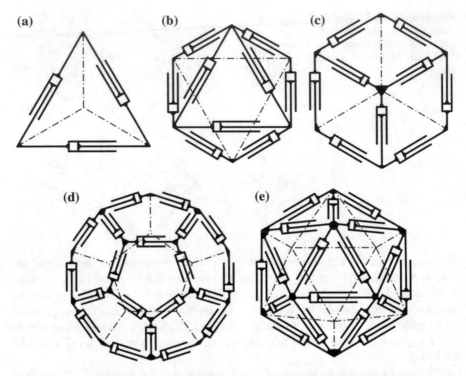

Fig. 5 Transformed modular structures with one degree of mobility: tetrahedral type (**a**); octahedral type (**b**); dodecahedral type (**d**); icosahedral type (**e**). Cubic transformed modular structure with three degree of mobility (**c**)

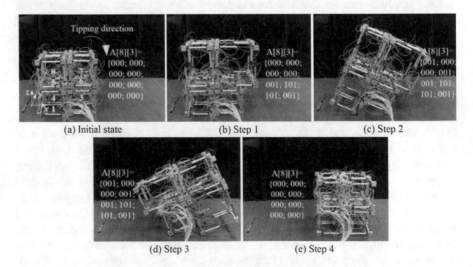

Fig. 6 Sequence diagram of the move of the united eight angle structure by tilting and of displacement of the center of gravity of the eight-module structure beyond the support surface (**a–d**)

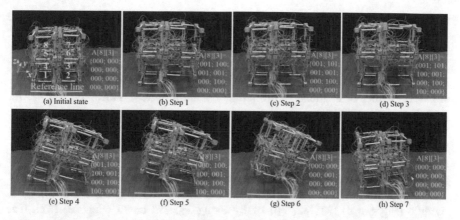

Fig. 7 The possibility of linear displacement (**a–d**) and rotation (**a–h**) of the combined eight-module structure

when moving on the surface from one point to another by changing the step in the motion cycles.

For this purpose, both forward and reverse motion of the combined structure can be organized in separate cycles. Figure 7 shows the possibility of linear displacement (a–d) and rotation (a–h) of the combined eight-module structure is shown. Thus, for the organization of self-propelled on the surface of the active structure formed from cubic modules with single-stage kinematic pairs, at least eight modules and 92 linear actuators, respectively, are required. Therefore, the use of such mobile eight modular robots in combat robotic systems is impractical.

The following are examples of building mobile robots of module type based on the Platonic solids and analyze their functionality.

3 Examples of Construction of Mobile Parallel Robots of Modular Type Based on Platonic Solids and the Analysis of Possibility of Their Application in SRS with Parallel Structure

As noted above, all the MPR based on Platonic solids have the following common functionality: self-propelled; self-assembling to self-reconfigurable structures; grasp-ability of the spacer faces; the spatial positioning of the vertices. Below are the individual properties of each type of MPR based on Platonic solids.

Tetrahedral MPR. Tetrahedral MPR was developed in NASA Goddard Space Flight Center for space research [31]. In General, MPR is a tetrahedral structure and contains telescopic linear actuators whose ends are pivotally connected to the vertices. In the article its description is given and evolution of its development is presented (Fig. 8): 2003—General concept; 2004—the first modification of the prototype with telescopic linear actuators; 2005—the second modification of the prototype (2005) with paired prismatic telescopic linear actuators; 2006—the third lightweight prototype with paired telescopic linear actuators.

The main advantages of tetrahedral MPR in comparison with other types of MPR based on Platonic bodies are higher rigidity and transformation coefficient, the smallest number of elements and a simpler control system. However, despite the grasping ability of the faces, they can not manipulate the captured objects and move with them. These abilities appear only in role-based robotic systems after integration into multi-module active structures. Therefore, in Autonomous mode, tetrahedral MPR can be used to monitor the terrain and various objects, as well as perform some 3D manipulation operations using a free vertex with 3 dof.

Self-propelled of similar The tetrahedral MPR can organized by tilting of module (Fig. 9a, c) [31, 32, 36] or by using leopard crawl (Fig. 9e) [37]. The advantage of the method of tilting is its independence from the irregularities of the surface on which the MPR moves. The ability of the movement on rocky terrain and to overcome the pits and fissures of comparable size faces the MNR in the initial position (Figs. 8 and 9a) [31, 36]. The disadvantages of this method include the inability to move on steep inclined surfaces, as well as the need for large working strokes of linear actuators. Therefore, linear actuators are telescopic with several knees (Figs. 8 and 9a) [31, 36], the coupled (Fig. 9b, c) [31, 32] or combined (Fig. 8) [31]. The cycle of this method includes only two stages: increasing the lengths of the upper linear actuators from the initial position to the moment of mixing the center of gravity of the tetrahedral MPR and its tilting; reducing the lengths of the rods to the initial state.

The advantage of the method move on their bellies (Fig. 9e) [37] is a small linear actuator stroke as well as the ability to move on steeper inclined planes. The disadvantages include the inability to move on a rocky, as well as uneven surfaces and overcoming recesses and cracks on the surface.

Compared to the previous method self-propelled of the tetrahedral MPR by using leopard crawl includes five stages (Fig. 9e): increase of the friction force in the supports B and C and reducing the friction forces in the support A by increasing the length of the linear actuator 1 and the displacement of the center of gravity in the direction of the supports B and C (2); A support moving ahead by increasing the lengths of the linear drives 2 and 3 (3); the increase of the friction force in the

Fig. 8 Evolution of development MPR with the tetrahedral structure

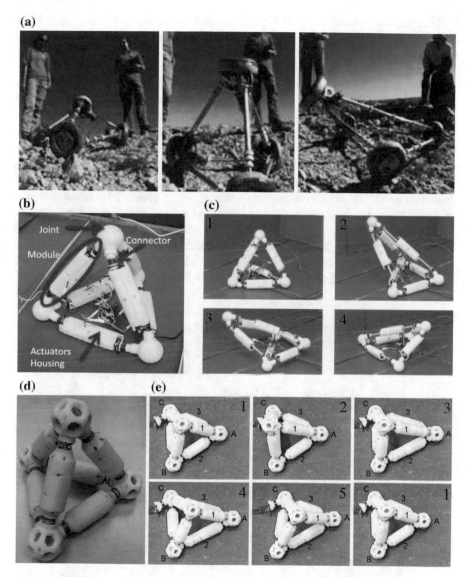

Fig. 9 Examples of self-propelled prototypes of tetrahedral MPR: by tilting (**a–c**); by using leopard crawl (**d, e**)

support A by reducing the length of the linear actuator 1 (4); moving the support forward by reducing the length of the linear actuator 3 (5); moving the support B by reducing the length of the linear actuator 2 to the initial state of tetrahedral MPR (1). As a result of a larger number of stages, the movement speed is plastically lower than that by tilting. However, the power consumption such method is three times less, because only one linear actuator works at a time, not three.

The common disadvantage of both methods is the inability to move such MPR on surfaces of bodies with microgravity (small satellites of planets, asteroids and comets).

Figure 10 shows a prototype of an adaptable tetrahedral MPR (a) [38, 39].

The vertex of each face contains a docking assembly in the form of pin 1 and slot 2, as well as a guide element 3. These vertices are made in the form of spherical hinges 4 connected to the ends of linear actuators 5. The presented MPR is autonomous and contains an individual control system 6, which includes a microcontroller, wireless communication and a power supply battery. The tetrahedral MPR group can exchange data with each other and with the operator through the wireless channel or a satellite communication system. If necessary, intelligent control in conditions of uncertainty and with maximum speed, the control system 6 of such MPR can be built on a coeducational neural network control system. Figure 10b–f shows as three modules are combined into an active linear truss (b) and active arc truss (c–f). Similarly, four tetrahedral MPR can be combined into an octahedral or hexsahedral MPR. However, such combined MPR would include 24 linear actuators rather than 12 as in octahedral or hexsahedral MPR.

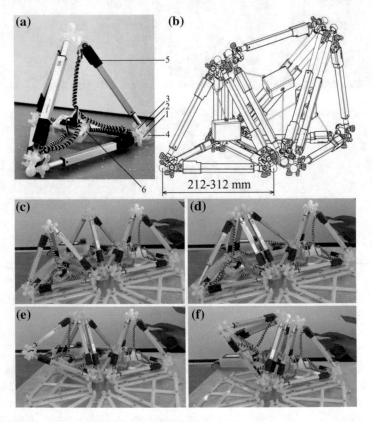

Fig. 10 Prototype of the adaptable tetrahedral MPR (**a**) and examples of reconfigurable structures from two (**b**) and three tetrahedral MPR (**c–f**)

Figure 11a shows 11 configurations for all the end positions of the linear actuators, which determine the angular criteria design tetrahedral MPR: Eleven different geometric shapes of a tetrahedral element with (a) 0 edges with maximum length L; (b) 1 L; (c, d) 2 L; (e, f, g) 3 L; (h, i) 4 L; (j) L 5; and (k) 6 L. Each of the faces tetrahedral or octahedral MPR can be used as a flat triangular 2D MPR "Triangle" with 3 DOF [40] (Fig. 12).

In MPR "Triangle" all the joints cylindrical, the axes of which pass through the point of intersection of the longitudinal axes of the linear actuators and is perpendicular to them. Therefore, the parallel structure of the flat mobile robot "Triangle" has the highest specific rigidity. The prototype MPR "Triangle" shows the ability of its self-propelled without vacuuming by using leopard crawl as in Fig. 9e, but without changing the friction force in the supports. In this case, the angle between the rods of the front support A should be $\alpha \leq 60°$ (Fig. 12a). Vacuuming at the contact points of the supports with the surface was not used. As a result, the number of stages in the cycle decreases from 5 to 3 and the speed of movement on the surface increases accordingly. The use of vacuuming at the contact points of the supports with the surface allows MPR "Triangle" to move itself on vertical surfaces, and perform manipulative operations with accurate positioning of the supports on the surface (Fig. 12b).

Figure 13a shows an example of reconfiguration of a plane active parallel structure formed by combining 20 triangular MPR of the "Triangle" type from one topology to another. Similarly, taking into account the possibilities shown in Fig. 11, the spatial active surfaces formed by the Union of tetrahedral MPR from one topology to another can be reconfigured (Fig. 13b, c–g) [38, 39]. Figure 13b shows an example of the steps in the process of reconfiguring the active structure from tetrahedral MPR. Tetrahedral MPR paths from the initial stage to the final position are marked in red when docking nodes are used to move as in Fig. 10. Figure 13c–g shows another example of the structure reconfiguration steps from tetrahedral MPR. The shortest path for the movement of the tetrahedral module along the surface is marked in red. As can be seen from the examples of the structure of tetrahedral MPR like a grid of finite elements and allow you to use the existing mathematical apparatus

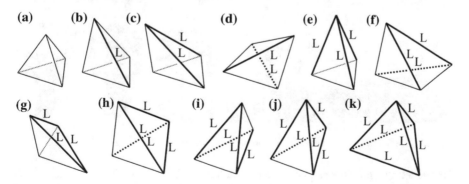

Fig. 11 Eleven different geometric shapes of a tetrahedral element: 0 edges with maximum length L (**a**); 1L (**b**); 2L (**c, d**); 3L (**e–g**) ; 4L (**h, i**); 5L (**j**); 6L (**k**)

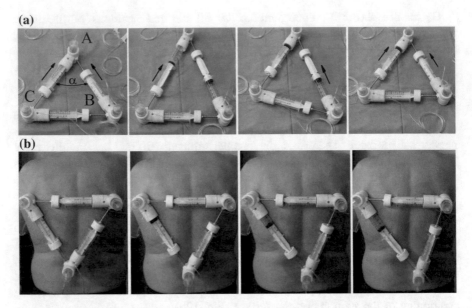

Fig. 12 Prototype MPR "Triangle': self-movement of MPR without vacuuming by using leopard crawl (**a**); vertical self-movement of MPR with use of vacuuming at the contact points of the supports with the surface (**b**)

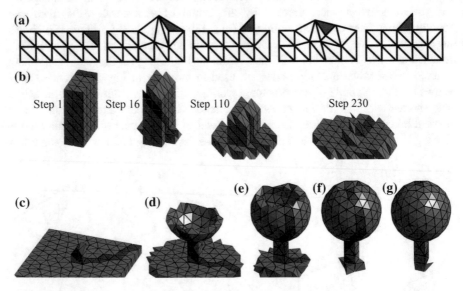

Fig. 13 Example of reconfiguration of triangular (**a**) and tetrahedral (**b–g**) MPR

for modeling and research of similar structures. Reconfigurable structures can be derived from the role SRS by combining tetrahedral MPR and can take a variety of forms to solve problems in various areas of the world.

Octahedral MPR. As noted above, octahedral MPR can be obtained by combining four tetrahedral MPR. The number of linear actuators will increase from 12 to 24. Thus, the formation of intelligent reconfigurable structures from role RTS based on tetrahedral MPR that do not require a high degree of surface approximation and rigidity, but require multi-functional manipulative abilities, is impractical. For these purposes, the most effective is the organization of role-based RTS based on intellectual octahedral MPR. The concept of intellectual octahedral MPR was developed at the Blagonravov Institute of machine science of the Russian Academy of Sciences (IMASH RAS). For the first time the concept of octahedral MPR called "Octahedral dodekapod of IMASH" (OD IMASH) (from the Greek words "dodeka"—twelve and "pod"—leg) and its functionality were presented in 2008 at the XXII International Congress of Theoretical and Applied Mechanics (ICTAM 2008) and other international conferences [41–45]. A number of papers have been published that reveal the functionality of IMASH RAS, including Patents of the Russian Federation for inventions in the part of IMASH RAS and ITS elements [9, 10, 23–26, 29, 41–60]. Figure 14 shows the kinematic (a) and structural (b) schemes of the OD IMASH with maximum equipment and its demonstration physical model (c). Depending on the functional tasks performed, the configuration can be reduced. The mechanism of the parallel structure is made in

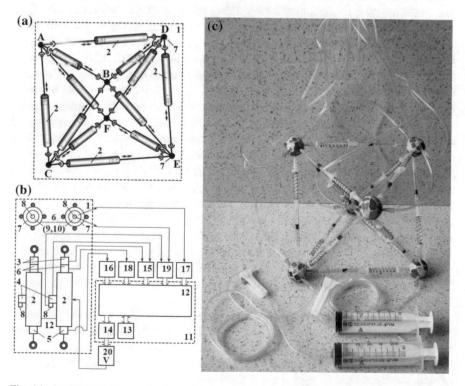

Fig. 14 OD IMASH: kinematic scheme (**a**); structural scheme (**b**); demonstration physical model (**c**)

the form of an octahedral module (OM) 1, which, when the linear actuators (LA) 2 are switched off, becomes a spatial farm and has a high specific rigidity. Each of the eight faces of OM 1 has grasping and spacer abilities. Methods of self-propelled of OD IMASH like of ways of self-propelled of tetrahedral MPR. Also, the possibility of self-propelled of OD IMASH by the method of tilting with the use of inertia forces and jumps is established. When self-propelled OM 1 on their bellies and push-grasping abilities with the ability to manipulate a clamped object and have the top three non-adjacent side faces. Also, in the process of self-propelled by using leopard crawl with help of coordinated interaction of linear drives of lateral faces, it is possible to provide active vibration isolation of the upper face with the clamped object from dynamic actions on the part of the lower face in real time. Tetrahedral MPR haven't such opportunities.

LA 2 are provided with axial 3 and median 4 force sensors and sensors of relative linear movement 5 and relative speed 6. Vertices 7 of OM 1 connected to the ends of LA 2 spherical hinges. If necessary, the vertices 7 and the median sections of LA 2 can be provided with radial stops and clutches with temperature sensors 8. At the same time, radial stops and tongs are used to transfer the spacer and compressive forces from the vertices 7 and the middle parts of LA 2 to the inner and outer surfaces to be contacted, respectively. In the vicinity of the radial end stops of the vertices 7 installed sensors of the spatial position of 9 acceleration and 10, which are used for operational monitoring of the spatial position of each of the vertices 7 and vibration acceleration along each of the axes of the rods 2 with LA 2. At the same time, sensors 9 and 10 can be made in the form of combined miniature three-axial blocks of gyroscopes-accelerometers. Control system (CS) 11 includes a neurocomputer 12 with software 13, which allows you to organize the work in real time, and digital-analog converters (DAC) 14. CS 11 inputs via data buses of analog-to-digital converters are connected respectively to outputs: 15 force sensors 3 and 4; 16 relative linear displacement sensors 5; 17 combined spatial position and acceleration sensors 9 and 10; 18 relative speed sensors 6 and 19 temperature sensors 8, and the CS 11 outputs via the output bus are connected to the corresponding inputs of software 13 and series-connected DAC 14, power amplifiers 20 and LA 2. Thus CS 11 and system of power supply can be autonomous execution as in Fig. 11 or remote execution. Operational control of the transmitted forces and temperatures at the points of contact of the stops of the vertices 7 and the middle parts of the LA 2 with the inner and outer surfaces is carried out with the help of force sensors 3, 4 and 8. Sensors of relative linear displacement 5 and relative velocity 6 (state observers) LA 2 record their relative displacements and velocities. Below are examples of the application of the functionality of OD IMASH in various fields.

Applications of OD IMASH for diagnose and repair the branched pipelines. Figure 15 shows examples of self-propelled of OD IMASH wounds inside pipes of various types and on the outer surface of the column. OD IMASH is able to self-adapt and perform manipulative operations inside the active branched pipelines of variable cross-section regardless of their spatial orientation.

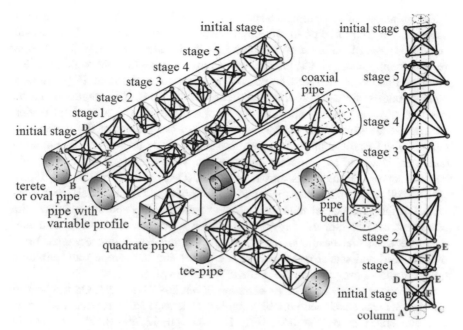

Fig. 15 Examples of self-propelled of OD IMASH wounds inside pipes of various types and on the outer surface of the column

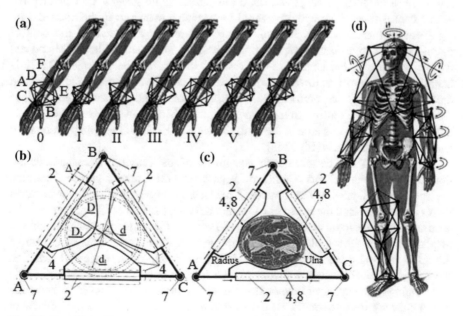

Fig. 16 Portable intelligent massage MPR: massage of elbow (**a**); change in the contact diameter of adaptive elastic tongs without finiteness (**b**) and with finiteness (**c**); massage of knee and neck (**d**)

The principle of self-propelled and manipulation of OD IMASH includes its installation in (on) the pipe in the initial position. Then, with the help of control commands from SU to the corresponding LA, the lengths of the ABCD OM edges are consistently changed and the necessary movement in the OM vertex space is provided relative to the base coordinate system. As a result, it is possible to organize both self-relocation and manipulative operations. The lengths of all the specified edges are monitored according to the readings of relative displacement sensors. At the same time, the spatial angular displacements of the vertices are judged by the readings of spatial position sensors made in the form of miniature three-axial blocks of gyroscopes-accelerometers, whose readings allow to increase the accuracy of measurements. In works [9, 10, 23–26, 41–48, 50–52, 54, 55, 57, 58] the following in-line manipulative capabilities of OD IMASH are shown: control of the geometry of the inner surface of the pipeline and its tracing; in-line movement of long-dimensional objects; in-line movement of objects with their simultaneous spatial positioning and active vibration protection; organization of rotation, feed and spatial movement of the processing tool; impact and vibration impact of the slotting tool on the end surface of the pipe.

Applications of OD IMASH for massage of the limbs and neck. On the basis of OD IMASH was developed portable intelligent massage MPR for massage elbow, knee and ankle, as well as neck (Fig. 16) [33, 41, 47, 49–51, 55, 56, 58–60]. Kinematic and structural schemes of massage OD IMASH are similar to those shown in Fig. 14. The difference is as follows: removed tongs LA 2 lateral faces with sensors 4 and 8; tongs LA 2 frontal and rear faces and median force sensors 4 are made in the form of adaptive elastic tongs with strain gauges. Due to adaptive elastic contractions, the range of covered transverse dimensions of the massed limb significantly increases (Fig. 16b, c). Figure 16b shows a change in the contact diameter of adaptive elastic tongs with finiteness (D and d) and without (D1 and d1) with a decrease in the lengths of the rods by Δ. During the massage, OM 1 is installed on the upper or lower limb, for example, on the forearm of the patient in the wrist (Fig. 16a) and perform the necessary movements and massage manipulations. Of massage on limb implemented agreed changes of the lengths of LA 2. After the massage in one area of the limb massage OD IMASH self-displacement to another area of the limb (Fig. 16a).

The vertices 7 can be made with possibility of installation of massage adaptations and ultrasonic sensors. Geometric invariance OM 1 allows you to determine the spatial coordinates of the vertices 7 measured lengths of all rods and control their movements. And the readings of the spatial position sensors 9 can improve the accuracy of measurement data. Figure 16a shows the sequence diagram of the displacement of OM 1 along the forearm and the radial circumference face of the ABC with adaptive elastic grippers 4. With the help of massage OD IMASH, the following rehabilitation regimens can be implemented for the patient: stroking and rubbing of the upper or lower extremities; longitudinal and transverse kneading of the upper or lower extremities; mobilization and manipulations on the elbow or knee joint of the extremities (Fig. 16d); manipulations of vibration; manipulations of forced motion of the free part of the limb relative to the adjacent one.

The organization of the required manipulations is carried out by means of an agreed change in the lengths of the rods OM 1.

In all modes, the patient has the opportunity to turn off the device, which CS 11 immediately sends a signal to the masseur. At the same time, all performed massage actions are registered in CS 11 for further analysis and appointment of subsequent procedures. According to the testimony of the axial force sensor 3 is judged on the resistance of the joint with his passive movements and to diagnose the degree of mobility (elaboration). It is also possible to use temperature sensors 8 to continuously measure the temperature of the limb at the point of contact and its values to judge the degree of its heating during massage procedures, for example, when rubbing. Dimension during the massage and the electrical resistance between the contact sections of the adaptive elastic tongs with the finality of the measured distances between them and the temperatures gives an indication of the dryness of the skin and the individual portability of the massage and clarify more favorable for the patient modes of influences.

Massage OD IMASH can be used at home for self-contained program self-massage. When connecting the CS 11 to the Internet can be organized on-line communication massage OD IMASH ran and the patient with the masseur for operational control of the massage and decision-making in the event of emergency situations, for example, to transfer the command to the CS 11 to stop the device and call medical help to the patient at home. As a result, patients who do not have the opportunity to visit the medical institution will be able to perform massage on-line under the control of the masseur without his physical presence in the room with the patient. As a result, the massage process will be automated, reduce physical fatigue and fatigue of the masseur.

Hexahedral MPR. The hexahedral structure can be applied in MPR, but only with elastic restraining bonds in spherical hinges or elastic joints of vertices. Due to the lowest stiffness among Platonic bodies, hexahedral MPR has not received distribution. Known only isolated cases of experimental work with the likes of MPR. These are hexahedral MPR with elastic constraints in spherical hinges (Fig. 17a) [61], and with the elastic connections of the vertices associated with any of the engines, for example, on the basis of an alloy with shape memory SMA (Shape Memory Alloy) (Fig. 17b) [62].

Dodecahedral and Icosahedral MPR. Dodecahedral and Icosahedral MPR can self-propelled. Their faces like octahedral parallel structure and spacer have prehensile abilities. However, the extremely small linear actuator stroke compared to the overall dimensions of the structure itself and the largest number of linear actuators make them unsuitable for efficient application in MPR during autonomous and collective operation.

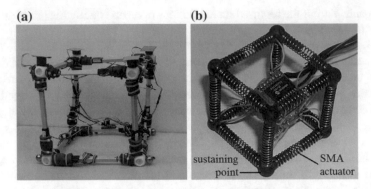

Fig. 17 Hexahedral MPR with elastic constraints in spherical hinges (**a**) and linear drives (**b**)

4 Conclusion

Analysis of parallel structures based on Platonic solids has shown that the MPR based on the tetrahedral structures have a maximum stiffness and ratio and are able to move on the surface of a roll-over, crawl on their bellies and jumping. Faces of a tetrahedral MPR does not have the spacer and grasping abilities. Therefore tetrahedral MPR is effective for individual and collective monitoring of the area. Tetrahedral MPR is able to unite with each other in multi-modular reconfigurable parallel structure and form of the active spatial structures of the most diverse forms, possess, grasp, push and manipulative abilities.

Octahedral MPR also has high rigidity and ratio and are able to move on the surface of a roll-over, crawl on their bellies and jumping. In addition, the facets of the octahedral MPR have grasping and spacer abilities and a variety of manipulative capabilities. Therefore, octahedral MPR has the greatest efficiency in individual application. In the case of collective application, octahedral IGR has the greatest efficiency in the individual solution of complex tasks, including not only monitoring the terrain, but also the use of its manipulation capabilities.

A hexahedral parallel structure with spherical or equivalent hinges is unsuitable for application in MPR.

Dodecahedral and icosahedral parallel structures inefficient for use in MPR.

The proposed octahedral and tetrahedral MPR and its unit may be used at the macroscopic level for use on land, underground, underwater, in medicine, and in the aerospace industry. It may also be used at the microscopic level. This research provides the basis for the development of up to date parallel spatial robots and for the expansion of their functional capabilities.

Further research will be continued to identify new functional capabilities of tetrahedral and octahedral MPR and their effective application in solving individual and collective problems.

References

1. Bonabeau, E., Dorigo, M., Theraulaz, G.: Swarm Intelligence: From Natural to Artificial Systems, p. 320. Oxford University Press, Oxford (1999)
2. Groß, R.: Self-assembling robots. Ph.D. thesis, Universite Libre de Bruxelles, Belgium (2007)
3. Yu, Ch.H.: Biologically-inspired control for self-adaptive multiagent systems. Ph.D. thesis, Harvard University Cambridge, Massachusetts (2010)
4. Kernbach, S., Meister, E., Schlachter, F., Kernbach, O.: Adaptation and Self-Adaptation Of Developmental Multi-Robot Systems. Int. J. Adv. Intell. Syst. 3(1 & 2), 121–140 (2010)
5. Yu, C.H., Werfel, J., Nagpal, R.: Collective decision-making in multi-agent systems by implicit leadership. In: van der Hoek, W., Kaminka, G.A., Lespérance, Y., Luck, M., Sen, S. (eds.) AAMAS '10 Proceedings of the 9th International Conference on Autonomous Agents and Multiagent Systems, vol. 3, Toronto, Canada, May 10–14, 2010, pp. 1189–1196. Association for Computing Machinery Press, New York, NY (2010)
6. Kadochnikov, M.: Mechatronno-modulnye roboty s adaptivnoy kinematicheskoy strukturoy: modely, algoritmy i programmnoe obespechenie system upravleniya (Mechatronic module robots with adaptive kinematic structure: models, algorithms and software of control systems), 196 p. Moscow, LAP LAMBERT Academic Publishing (2012) (in Russian)
7. Gradetskiy, V.G. et al.: Upravlyaemoe dvijenie mobilnych robotov po proizvolno orientirovannym v prostranstve poverxnostyam (Controlled movement of mobile robots in space on arbitrary orientation surfaces), 361 p. Moscow, Nauka Publ. (2001) (in Russian)
8. Gradetskiy V.G. et al.: Mechanika miniatyurnych robotov (Mechanics of miniature robots). Moscow, Nauka Publ., (2010) (in Russian)
9. Sayapin, S.N., Siniov, A.V.: Adaptive mobile 3D manipulator and method of organizing displacements and control over physical-mechanical properties, geometrical share of contact surface and displacement trajectory hereby. Russian Federation Patent 2424893, 11 Jan 2009
10. Sayapin, S.N., Karpenko, A.P., Dang, S.H., Kokushkin, V.V.: Application of mobile spatial parallel robot of "Dodekapod" type for diagnistics of branched out pipelines with variable cross-section. Eng. Autom Prob. 1, 26–40 (2016) (in Russian)
11. Vasiliev, A.V.: Printsipy postroeniya I classificatsiya shassi mobilnych robotov nazemnogo primeneniya I planetochodov (Development and Classification Principles of Ground Mobile Robots and Planet Rovers Chassis). Nauchno-technicheskie vedomosty of SPb. Polytechnical Institute, 1, 124–131 (2013) (in Russian)
12. Floreano, D., Mattiussi, C.: Bio-inspired artificial intelligence: theories, methods, and technologies, 659 p. The MIT Press, London, England, Cambridge, Massachusetts (2008)
13. Handbook of collective robotics: fundamentals and challenges, ed. Serge Kernbach. Singapore: Pan Stanford Publishing, Singapore (2013)
14. Anderson, C., Theraulaz, G., Deneubourg, J.-L.: Self-assemblages in insect societies. Insectes Sociaux 49, 99–110 (2002)
15. Thompson, K.: Swarm behavior and organizational teams. The Bumble Bee. http://www.bioteams.com/2006/03/21/swarm_behavior_and.html
16. Mondada, F., Gambardella, L.M., Floreano, D., et al.: The cooperation of swarm-bots: physical interactions in collective robotics. Robot. Autom. Mag., IEEE 12(2), 21–28 (2005)
17. Mondada, F., Bonani, M., Guignard, A., et al.: Superlinear physical performances in a SWARM-BOT. In: Capcarrèr, M.S., Freitas, A.A., Bentley, P.J., Johnson, C.G., Timmis, J. (eds.) Advances in Artificial Life. Proceedings of the 8th European Conference, pp. 282–291, ECAL 2005, Canterbury, UK, 5–9 Sept 2005. Springer, Berlin, Heidelberg (2005)
18. Ivanov, A.V., Yurevich, E.I., Иванов, А.В.: Mini- i Microrobototechnika (Mini- and Microrobotics), 96 p. SPb Publishing house of Polytechnical Institute. (2011) (in Russian)
19. Magnus, V.: Modely mnogogrannikov (Models of Polyhedrons), 236 p. Moscow, Mir Publ., (1974) (in Russian)
20. Martin, G.: Transformation Geometry: An Introduction to Symmetry. Springer, New York, NY (1982)

21. Wang, M., Wang, C., Hao, X.Q., Li, X., Vaughn, T.J., Zhang, Y.Y., Yu, Y., Li, Z.Y., Song, M.P., Yang, H.B., Li, X.: From trigonal bipyramidal to platonic solids: self-assembly and self-sorting study of terpyridine-based 3D architectures. J. Am. Chem. Soc. **136**(29), 10499–10507 (2014)

22. Solem, J.C.: Self-assembling micrites based on the platonic solids. Robot. Auton. Syst. **38**(2), 69–92 (2002)

23. Sayapin, S.N.: Parallel spatial robots of dodecapod type. J. Mach. Manuf. Reliab. Allerton Press, USA **41**(6), 457–466 (2012)

24. Sayapin, S., Karpenko, A., Hiep, D.X.: DODEKAPOD as universal intelligent structure for adaptive parallel spatial self-moving modular robots. In: Waldron, K.J., Tokhi, M.O., Virk, G. S. (eds.) Nature-Inspired Mobile Robotics, pp. 163–170. World Scientific, Singapore (2013)

25. Sayapin, S.N., Karpenko, A.P., Dang, S.H. Universal adaptive spatial parallel robots of module type based on the Platonic solids. Universal adaptive spatial parallel robots of module type based on the Platonic solids. Lecture Notes in Engineering and Computer Science, pp. 1365–1370 (2014)

26. Sayapin, S.N.: Octahedral dodekapod is novel universal adaptive spatial parallel mobile robot of module type based on the Platonic solids. In: Proceedings of the XI All-Russian Congress on Fundamental Problems of Theoretical and Applied Mechanical, pp. 3364–3365, Kazan, Russia, 20–24 Aug 2015, Kazan University Publ. (2015) (in Russian)

27. Sayapin, S.N.: Analysis and synthesis of flexible spaceborne precision large mechanisms and designs of space radiotelescopes of the petal type, Doctor Tech. Sci. Dissertation. M, IMash RAN, Moscow (2003) (in Russian)

28. Sayapin, S.N., Siniov, A.V.: The analysis and prospect of use of devices of increase of a course of a target link with reference to micro-and nanorosots. In: Proceedings of International Workshop on Micro- and Nano Production Technologies and Systems, pp. 151–155, 17–18 Oct 2007. Moscow, Russia (2007)

29. Sayapin, S.N., Sineov, A.V.: Linear drive. Russian Federation Patent 2373611, 20 Nov 2009

30. Merlet, J.-P.: Parallel Robots, 2nd edn. Springer, Dordrecht, Netherlands (2006)

31. Curtis, S., Brandt, M., Bowers, G., Brown, G., Cheung, C., Cooperider, C., Desch, M., Desch, N., Dorband, J., Gregory, K., Lee, K., Lunsford, A., Minetto, F., Truszkowski, W., Wesenberg, R., Vranish, J., Abrahantes, M., Clark. P., Capon, T., Weaker, M., Watson, R., Olivier, P., Rilee, M.L.: IEEE A&E Systems Magazine, pp. 22–30, June 2007

32. Belisle, R., Yu, C., Nagpal, R.: Mechanical design and locomotion of modular-expanding robots. In: Stoy, K., Nagpal, R., Shen, W.-M. (eds.) Proceedings of the IEEE 2010 International Conference on Robotics and Automation workshop "Modular Robots: State of the Art", pp. 17–23, 3 May 2010. Alaska, AK, USA (2010)

33. Agrawal, S.K., Kumar, S., Yim, M., Suh, J.W.: Polyhedral single degree-of-freedom expanding structures. In: Proceedings of the 2001 IEEE International Conference on Robotics & Automation, pp. 2228–2243, Seoul, Korea, 21–26 May 2001

34. Agrawal, S.K., Kumar, S., Yim, M.: Polyhedral single degree-of-freedom expanding structures: design and prototypes. J. Mech. Des. **124**, 473–478 (2002)

35. Ding, W., Wu, J., Yao, Y.A.: Three-dimensional construction and omni-directional rolling analysis of a novel frame-like lattice modular robot. Chin. J. Mech. Eng. **28**(4), 691–701 (2015)

36. Clarka, P.E., Curtis, S.A., Rilee, M.L.: A new paradigm for robotic rovers. Phys. Procedia **20**, 308–318 (2011)

37. Stoy, K., Lyder, A., Garzia, R.F.M., Christensen, D.J: Hierarchical robots. In: Proc., IEEE Int. Conf. on Intelligent Robots and Systems (IROS), Workshop on Self-Reconfigurable Robots, pp. 1–4, San Diego, USA (2007)

38. Pieber, M., Gerstmayr, J.: An adaptive robot with tetrahedral cells. In: Proceedings of the 4th Joint International Conference on Multibody System Dynamics, Montreal, Canada, May 29–June 2, 2016

39. Pieber, M., Gerstmayr, J.: A framework for cellular robots with tetrahedral structure. In: Roth, P.M., Vincze, M., Kubinger, W., Müller, A., Blaschitz, B., Stolc, S. (eds.) Proceedings of the

OAGM&ARW Joint Workshop Vision, Automation and Robotics on "Vision, Automation and Robotics", pp. 5–6, 10–12 May 2017. Vienna, Austria, (2017)

40. Sayapin, S.N.: Intelligence self-propelled planar parallel robot for sliding cupping-glass massage for back and chest. In: Hu, Z., Petoukhov, S., He, M. (eds.) Advances in Artificial Systems for Medicine and Education, Series "Advances in Intelligent Systems and Computing", vol. 658, pp. 166–175, 360 p, Springer, Cham, Switzerland (2018). https://doi.org/10.1007/978-3-319-67349-3_15

41. Sayapin, S., Siniov, A.: The adaptive spatial mobile robot - manipulator and way of diagnostics of physical and mechanical properties and the geometrical form of a surface of contact and trajectory of movement with his help. In: Abstracts book. XXII International Congress of Theoretical and Applied Mechanics, p. 217, ICTAM 2008. Adelaide, Australia, 25–29 Aug 2008

42. Sayapin, S., Siniov, A.: The adaptive spatial mobile robot - manipulator and way of diagnostics of physical and mechanical properties and the geometrical form of a surface of contact and trajectory of movement with his help. In: CD-ROM Proceedings XXII International Congress of Theoretical and Applied Mechanics, ICTAM 2008, 25–29 Aug 2008. Adelaide, Australia. ISBN: 978-0-9805 142-1-6, Index of author 11555

43. Sayapin, S.N., Sinev, A.V.: The adaptive spatial mobile robot manipulator and way of control of geometrical shape of the contacted surface and the trajectory of movement with his help. In: Proceedings of Symposium on Robotics and Mechatronics, 4–6 Nov 2008, Moscow, pp. 121–123, VVC, Pav. No. 69, Moscow, Russia (2008)

44. Sayapin, S.N., Sinev, A.V.: The adaptive spatial mobile robot - manipulator and way of diagnostics of physical and mechanical properties and the geometrical form of a surface of contact and trajectory of movement with his help. In: Proceedings of Conference "Problems of Mechanical Engineering Research", Москва, pp. 464–466, 13–14 Nov 2008, Moscow (2008)

45. Sayapin, S.N.: Mobile parallel robot-manipulator "Octahedral dodekapod": history, present and future. Eng. Autom. Prob., 3, 26–44 (2018) (in Russian)

46. Sayapin, S.N., Siniov, A.V. (2012) Dodekapod as universal intelligent structure for adaptive parallel spatial self-moving modular robots. In: Bai, Y., Wang, J., Fang, D. (eds.) Proceedings of the 23rd International Congress of Theoretical and Applied Mechanics, 19–24 Aug 2012, Beijing, China. The International Union of Theoretical and Applied Mechanics. The Chinese Society of Theoretical and Applied Mechanics. China Science Literature Publishing House. ISBN: 978-988-16022-3-7. Beijing, China. SM16-021 (2012)

47. Sayapin, S.N.: Dodekapod is novel module type of spatial parallel robot. In: Proceedings of Conference "Development of Extreme Robotics in Russia", pp. 58–60, 21–24 May 2013, Moscow, VVC, Pav. 75, Moscow (2013)

48. Karpenko, A.P., Dang, H.S., Sayapin, S.N.: The algorithm for the movement of dodekapod in the linear cylindrical pipe with variable cross-section. Nauka i obrazovanie, MGTU im. N.E. Baumana. Elektronniy zhurnal - Science and Education, the Bauman Moscow State Technical University. Electronic Journal, 8 (2013). https://doi.org/10.7463/0813.0587740 [in Russian]

49. Sayapin, S.N., Sokolov A.I.: Linear Drive. Russian Federation Patent 2499163, 20 Nov 2013

50. Sayapin, S.N., Karpenko, A.P., Dang, S.H. Universal adaptive spatial parallel robots of module type based on the platonic solids. In: Ao, S.I., Gelman. L., Hukins, D.W.L., Hunter, A., Korsunsky, A.M. (eds.) Proceedings of the World Congress on Engineering 2014, vol II, pp. 1365–1370, 2–4 July 2014, London, U.C. Newswood Limited, International Association of Engineering, Hong Kong. ISBN: 978-988-19253-5-0, ISSN: 2078-0958 (2014)

51. Sayapin, S.N., Sinev, A.V.: Adaptive spatial mobile robot manipulator and method of organization of movements and control of physical and mechanical properties and geometrical shape of the contacted surface and the path of movement with it. Inventors Mech. Eng. Mach. Tools, 8, 2–4 (2014) ISSN: 2074-6954

52. Sayapin, S.N., Karpenko, A.P., Dang, S.H.: Universal adaptive spatial parallel robots of module type based on the platonic solids. In: Physical and Mathematical Problems of

Advanced Technology Development: Abstract of International Scientific Conference, pp. 22–23, 17–19 Nov 2014, BMSTU, Moscow (2014). ISBN: 978-5-7038-4071-9

53. Sayapin, S.N.: An adaptive portable spatial rehabilitation robot manipulator and a means of organizing movements and its use for patient diagnosis. Russian Federation Patent 2564754, 10 Oct 2015

54. Sayapin, S., Karpenko, A., Dang, S.: Self-propelled intelligent robotic vehicle based on octahedral dodekapod to move in active branched pipelines with variable cross-sections", world academy of science, engineering and technology, international science index 108. Int. J. Mech. Aerosp. Ind. Mechatron. Manuf. Eng. 9(12), 2038–2044 (2015)

55. Dang, H.S., Karpenko, A.P., Sayapin, S.N.: In pipe parallel robot dodekapod type. In: Extreme Robotics, Proc. Int. Sc.-Tech. Conf. S.-Petersburg: "AP4Print", pp. 334–339 (2016)

56. Sayapin, S.N.: Novel principles of creation of portable self-propelled autonomous massage robots with triangular and octahedral parallel structures. Med. High Technol. 1, 64–72 (2017)

57. Sayapin, S.N.: Principles of the design of an adaptive mobile spatial rehabilitation manipulator robot based on an octahedral dodecapod. Biomed. Eng. 51(4), 296–299 (2017)

58. Sayapin, S.N., Karpenko, A.P., Dang, S.H.: Application of mobile parallel robots for diagnostics of active pipelines with variable cross-section. In: Proc. of III Sc.-Practical Seminar "Povyshenie nadegnosty magistralnych gazoprovodov, podvergennych korrozionnomu rastreskivaniyu pod napryageniem (KRN 2017)", p. 33, Moscow, 20–22 Sept 2017, "Gazprom VNIIGAZ" (2017)

59. Sayapin, S., Sayapina, E.: Novel approach to creation of portable self-propelled autonomous massage robots with triangular and octahedral parallel structures. In: WSEAS Transactions on Computers, ISSN/E-ISSN: 1109-2750/2224-2872, vol. 16, Art. #14, pp. 124–132 (2017)

60. Sayapin, S.N.: Novel principles of creation of robotics massager with parallel structure based on triangular and octahedral self-propelled modules. In: Fundamental and Applied Problems of Mechanics: Abstract of International Scientific Conference, p. 95, 24–27 November 2017, BMSTU, Moscow (2017). ISBN: 978-5-7038-4800-5

61. Yu, Ch.-H., Nagpal, R.: Self-adapting modular robotics: a generalized distributed consensus framework. In: Robotics and Automation, 2009. ICRA '09. IEEE International Conference on, 12–17 May 2009, Kobe, Japan, 1881-1888. Institute of Electrical and Electronics Engineers, Piscataway, NJ (2009)

62. Yu, H., Ma, P., Cao, Ch.: A novel in-pipe worming robot based on SMA. In: Proceedings of 2005 IEEE International Conference on Mechatronics and Automation, vol. IV, pp. 923–927, July 29–August 1, 2005. Niagara Fails, Ontario, Canada (2005)

Intellectualization of the Process of Designing Smart Electromechanical Systems

Aleksandr N. Shilin, Irina A. Koptelova and Dmitriy G. Snitsaruk

Abstract *Problem statement*: The main problem in the design of smart electromechanical systems (SEMS) is a reasonable choice of components and elements
that are characterized by a wide variety of functionality, technical characteristics,
and their cost. Currently, domestic and foreign firms produce a very large number
of devices, components and components. From this, it follows that design of SEMS
is a poorly formalized task. Therefore, for a reasonable choice of components and
elements of the SEMS of robotic complexes, the methods for solving multi-criteria
optimization problems with fuzzy representation of information are needed. In the
design of complex technical systems, in practice, heuristic methods are mainly
used. The purpose of the research: Formalization of the SEMS design process of
robotic complexes and the development of an algorithm for an automated design
system. Technical task of SEMS design should contain information about the
functions of the system being designed, a list of requirements and limitations to the
main characteristics. Moreover, the particular characteristics of the systems should
be combined with the general quality indicators, and the less significant characteristics should be excluded. In addition, to reduce the total number of operational
characteristics to the number of the most significant, it is necessary to investigate
their mutual correlation and compatibility. Results: The SEMS design methodology
was developed, based on the method of paired comparisons and prioritization. The
proposed method uses the technique of partitioning a complex problem into simpler
ones, which makes it possible to formalize the problem of multi-criteria of optimization of a technical solution choice from a set of alternatives with fuzzy
information representation. Based on the proposed methodology, an automated
SEMS design system was developed. The results obtained with the help of an
automated system coincide with the results obtained by teams and experienced

A. N. Shilin (✉) · I. A. Koptelova · D. G. Snitsaruk
Volgograd State Technical University, Volgograd, Russia
e-mail: eltech@vstu.ru

I. A. Koptelova
e-mail: shilina@yandex.ru

D. G. Snitsaruk
e-mail: norzes@mail.ru

© Springer Nature Switzerland AG 2019
A. E. Gorodetskiy and I. L. Tarasova (eds.), *Smart Electromechanical Systems*,
Studies in Systems, Decision and Control 174,
https://doi.org/10.1007/978-3-319-99759-9_6

specialists, but over a longer period. Practical meaning: The developed automated system allows significantly reducing the time and increasing the reliability of decision making when choosing alternative options. The system is versatile and has an intuitive interface that allows the expert not to have high skills in working with it. With the help of this system, an optic-electronic system for controlling geometric parameters of large-sized parts was developed.

Keywords System analysis · Decision theory · Hierarchy analysis method Morphological synthesis · Matrices of paired comparisons · Neural networks

1 Introduction

Evaluation The main problem of designing smart electromechanical systems (SEMS) is the reasonable selection of the already developed main components of the system being designed. The main components and elements produced by domestic and foreign companies are very diverse in terms of functionality, technical characteristics and their cost. At the initial conceptual stage of designing a technical system, a functional block diagram of the system is developed and its blocks and elements are selected. It should be noted that it is at this stage of the design that the potential capabilities of the system that determine its technical, operational and economic performance are laid. The choice of blocks and system elements from a wide variety of products is a poorly formalized task and it is very difficult to analyze the entire volume of the proposed products and make an informed choice. The task of choosing the most optimal variant from a large number of variants of alternative solutions is complicated by the fact that all these variants are characterized by fuzzy information. Therefore, for a reasonable choice of components and elements of the SEMS of robotic complexes, the methods for solving multi-criteria optimization problems with fuzzy representation of information are needed.

2 Method of Choosing an Electric Drive for a Control and Measuring Robot

Table 1 lists the criteria by which options for technical solutions are evaluated [1], in Table 2—technical solutions.

Evaluation in accordance with the above method is carried out using complex priorities P_{icom} according to the formula:

$$P_{icom} = \sum_{i=1}^{n} \beta'_j P'_{ij}, \tag{1}$$

Table 1 Criteria

Notation	Specifications
d_1	Range of speed rotation
d_2	Reliability
d_3	Starting torque
d_4	Integration with digital control systems
d_5	Dimensions
d_6	Investment costs

Table 2 Technical solutions

Notation	Variants of technical solution
X_1	DC motor with voltage regulator
X_2	Asynchronous motor with frequency control
X_3	Synchronous motor with frequency control
X_4	Stepper motor with digital control
X_5	Linear stepper motor with digital control

where β'_j is the relative priority (significance) of the j-th criterion; P_{ij}—the relative priority of the i-th option on the j-th criterion; n is the number of criteria ($n = 6$).

The first step determines the significance of the criteria (Table 3). For this, a square matrix of priorities is constructed. In the columns and rows, the criteria numbers are written, and at the intersection—the coefficients (1.5, 1.0, 0.5), indicating which option is preferable. The next step is to sequentially determine the absolute priorities of the Σ variants. For the calculation of β_j, each row (d_1 to d_6) in the matrix is multiplied by a column vector Σ. To obtain the normalized value, i.e. relative β'_j it is necessary to divide β_j by $\sum_{i=1}^{6} \beta'_j$. Criteria priorities was calculated by normalized value of β'_j (Table 4).

Table 3 The importance of criteria

Criteria index	d_1	d_2	d_3	d_4	d_5	d_6	Σ	β_j	β'_j
d_1	1	1.5	1	1.5	1.5	1.5	8	46	0.227
d_2	0.5	1	0.5	0.5	1.5	1.5	5.5	31.25	0.154
d_3	1	1.5	1	1.5	1.5	1.5	8	46	0.227
d_4	0.5	1.5	0.5	1	1.5	1.5	6.5	39.25	0.194
d_5	0.5	0.5	0.5	0.5	1	1.5	4.5	24	0.118
d_6	0.5	0.5	0.5	0.5	0.5	1	3.5	15.75	0.077
Total:								202.25	1

Table 4 Priority order of parameters

Priority	1	2	3	4	5
Criteria	d_1, d_3	d_4	d_2	d_5	d_6

It should be noted that the number of gradations can be increased, but with a fuzzy representation of information, it becomes difficult to set the coefficient of excellence.

From the analysis, it follows that for the control and measuring robot that the priority criteria are the speed range and the starting torque.

At the second stage, square matrices are constructed for each of the criteria. To do this, the method of prioritization is also applied, with the only difference being that the comparison objects are now not the evaluation criteria, but the decision options [2]. This step involves the sequential determination of the absolute priorities of the P_{ij} variants, and then the relative P'_{ij}, which are calculated in fractions of one. To calculate P_{ij}, each row (X_1 to X_5) in the matrix is multiplied by the column vector Σ. The normalized values, i.e. Relative P, are obtained by dividing P_{ij} by $\sum_{i=1}^{m} P_{ij}$. According to this rule, priorities P'_{ij} are calculated for each criterion table (Tables 5, 6, 7, 8, 9 and 10).

Then, the complex index (priority) for each of the P_{icom} variants (Table 11) is calculated, which is defined as the sum of the products of the relative priorities of the object by the relative priorities of the criteria. The last line shows the complex priorities that were obtained by summing the products of the elements of the first and subsequent columns of the table.

The variant that has received the greatest value β_{icom} can be considered the best of all. In the example under consideration, this is $P_{1com} = 0.263988$.

According to this method, the most complex indicator has a stepper motor.

Table 5 The adjacency matrix by the 1-st criterion

Index of the variant	X_1	X_2	X_3	X_4	X_5	Σ	P_{i1}	P'_{i1}
X_1	1	1.5	1.5	1	1.5	6.5	31.25	0.2637131
X_2	0.5	1	1	0.5	0.5	3.5	16.75	0.1413502
X_3	0.5	1	1	1	1.5	5	23.75	0.2004219
X_4	1	1.5	1	1	1.5	6	28.75	0.242616
X_5	0.5	1.5	0.5	0.5	1	4	18	0.1518987
Total:						25	118.5	1

Table 6 The adjacency matrix by the 2-nd criterion

Index of the variant	X_1	X_2	X_3	X_4	X_5	Σ	P_{i1}	P'_{i1}
X_1	1	0.5	0.5	0.5	0.5	3	14	0.1212121
X_2	1.5	1	1.5	1	1.5	6.5	31	0.2683983
X_3	1.5	0.5	1	0.5	0.5	4	17.5	0.1515152
X_4	1.5	1	1.5	1	1.5	6.5	31	0.2683983
X_5	1.5	0.5	1.5	0.5	1	5	22	0.1904762
Total:						25	115.5	1

Table 7 The adjacency matrix by the 3-rd criterion

Index of the variant	X_1	X_2	X_3	X_4	X_5	Σ	P_{i1}	P'_{i1}
X_1	1	1.5	0.5	0.5	0.5	4	18.5	0.1574468
X_2	0.5	1	1.5	0.5	0.5	4	18.5	0.1574468
X_3	1.5	0.5	1	0.5	0.5	4	18.5	0.1574468
X_4	1.5	1.5	1.5	1	1	6.5	31	0.2638298
X_5	1.5	1.5	1.5	1	1	6.5	31	0.2638298
Total:						25	117.5	1

Table 8 The adjacency matrix by the 4-th criterion

Index of the variant	X_1	X_2	X_3	X_4	X_5	Σ	P_{i1}	P'_{i1}
X_1	1	0.5	0.5	0.5	0.5	3	14	0.1206897
X_2	1.5	1	1	0.5	0.5	4.5	20	0.1724138
X_3	1.5	1	1	0.5	0.5	4.5	20	0.1724138
X_4	1.5	1.5	1.5	1	1	6.5	31	0.2672414
X_5	1.5	1.5	1.5	1	1	6.5	31	0.2672414
Total:						25	116	1

Table 9 The adjacency matrix by the 5-th criterion

Index of the variant	X_1	X_2	X_3	X_4	X_5	Σ	P_{i1}	P'_{i1}
X_1	1	0.5	1	0.5	0.5	3.5	16	0.1385281
X_2	1.5	1	1.5	0.5	0.5	5	22	0.1904762
X_3	1	0.5	1	0.5	0.5	3.5	16	0.1385281
X_4	1.5	1.5	1.5	1	1.5	7	34	0.2943723
X_5	1.5	1.5	1.5	0.5	1	6	27.5	0.2380952
Total:						25	115.5	1

Table 10 The adjacency matrix by the 6-th criterion

Index of the variant	X_1	X_2	X_3	X_4	X_5	Σ	P_{i1}	P'_{i1}
X_1	1	0.5	1.5	0.5	0.5	4	18	0.1518987
X_2	1.5	1	1.5	0.5	0.5	5	22.5	0.1898734
X_3	0.5	0.5	1	0.5	1	3.5	17.25	0.1455696
X_4	1.5	1.5	1.5	1	1	6.5	31.25	0.2637131
X_5	1.5	1.5	1	1	1	6	29.5	0.2489451
Total:						25	118.5	1

Table 11 Relative priority for each variant

Significance criterion of β_j'	X_1	X_2	X_3	X_4	X_5
0.2274413	0.263713	0.14135	0.200422	0.242616	0.151899
0.1545117	0.121212	0.268398	0.151515	0.268398	0.190476
0.2274413	0.157447	0.157447	0.157447	0.26383	0.26383
0.1940667	0.12069	0.172414	0.172414	0.267241	0.267241
0.118665	0.138528	0.190476	0.138528	0.294372	0.238095
0.0778739	0.151899	0.189873	0.14557	0.263713	0.248945
Complex priority	0.166207	0.180278	0.166039	**0.263988**	0.223487

3 Conclusions

The proposed methodology allows, in the process of multivariate search, already at the initial stage to work only with reliable, interdependent and consistent data and exclude non-significant results. In addition, to assist the expert in difficult situations related to the prioritization of priorities which indicating the preferences of alternatives and criteria, and, accordingly, it is more reasonable to choose the optimal variant of the technical solution.

The considered methods of individual expert assessments, even with their mathematical processing, still do not allow avoiding subjectivity. To overcome this disadvantage, especially in cases where it is necessary to obtain long-term values of certain indicators, for example, the significance of functions in typical functional models, resort to collective examination. The increase in the number of experts under collective expertise can be increased with the help of the Internet, and leading experts from different creative collectives and scientific schools can be involved.

The article presents the research results for the Erasmus + program № 573879-EPP-1-2016-1-FREPPKA2-CBHE-JP "InternationaliSation of master Programs in Russia and China in Electrical Engineering".

References

1. Shilin, A.N., Shilina, I.A., (Koptelova IA): Morphological synthesis of opto-electronic systems for measuring the dimensions of heated parts. Instruments and Systems. Management, Control, Diagnostics. No. 3 (2003)
2. Shilin, A.N., Budko, V.V., Koptelova, I.A.: Automation of the conceptual design of optoelectronic systems for measuring the dimensions of heated parts. Appliances. No. 4 (2006)

Part II
Methods and Principles of Designing of Automatic Control Systems

Reduction of Logical-Probabilistic and Logical-Linguistic Constraints to Interval Constraints in the Synthesis of Optimal SEMS

Andrey E. Gorodetskiy, Irina L. Tarasova and Vugar G. Kurbanov

Abstract *Problem statement*: modern dynamic robotic systems, such as SEMS, contain high-precision information-measuring and control systems (IMCS). The analysis and synthesis of such IMCS usually relies on structural identification techniques and computational optimization techniques. A feature of SEMS is the use of such systems in the contours of "non-rigid" models, whose structure, parameters and state vector are not specified at the design stage, but are in real time when they operate. This greatly complicates the synthesis of SEMS, especially when limiting the speed of microprocessor controls. In this case, an approach called the method of recurrent target inequalities (RTI) and based on the search for sub-optimal solutions and reduction of the original problem of synthesis to the search for the target set described by systems of inequalities, including dynamic ones, can be used. The need to select the best trajectories (scenarios of SEMS dynamics) in conditions of incomplete certainty requires taking into account the limitations of the environment of functioning which are introduced in the form of logical-probabilistic and/or logical-linguistic expressions. Therefore, this approach to the synthesis of SEMS to date has not been developed. However, the relevance of research in the direction of the development of the RTI method for the synthesis of SEMS is

A. E. Gorodetskiy (✉) · I. L. Tarasova · V. G. Kurbanov
Institute of Problems of Mechanical Engineering Russian Academy
of Sciences, V.O., Bolshoj pr., 61, Saint-Petersburg 199178, Russia
e-mail: g27764@yandex.ru

I. L. Tarasova
e-mail: g17265@yandex.ru

V. G. Kurbanov
e-mail: vugar_borchali@yahoo.com

I. L. Tarasova
Peter the Great St. Petersburg Polytechnic University, Polytechnicheskaya,
29, Saint-Petersburg 195251, Russia

V. G. Kurbanov
Saint-Petersburg State University of Aerospace Instrumentation,
B. Morskaia St., 67, Saint-Petersburg 190000, Russia

© Springer Nature Switzerland AG 2019
A. E. Gorodetskiy and I. L. Tarasova (eds.), *Smart Electromechanical Systems*,
Studies in Systems, Decision and Control 174,
https://doi.org/10.1007/978-3-319-99759-9_7

undeniable. In particular, the development of the RTSN method is closely related to the solution of the problem of reducing logical-probabilistic and logical-linguistic constraints to interval constraints. *Purpose of research:* solution of the problem of reduction of logical-probabilistic and logical-linguistic constraints to interval constraints and demonstration of decision-making efficiency in conditions of not complete certainty on the example of the algorithm for finding optimal trajectories of motion. *Results:* a brief overview of approaches to the synthesis of intelligent robots type SEMS. The peculiarities of the use of the RTI method in the presence of restrictions from the environment of functioning in the form of logical-probabilistic and/or logical-linguistic expressions are analyzed. Theorems of data of logical-probabilistic and logical-linguistic expressions to interval are given. The algorithm of decision-making at logical-probabilistic and logical-linguistic restrictions is shown on the example of the algorithm of search of optimum trajectories of movement of the SEMS type robot. *Practical significance:* the proven theorems for reducing logical-probabilistic and logical-linguistic expressions to interval expressions can be effectively used in robot control planning systems that provide the choice of the best scenarios of movements in conditions of not complete certainty, as evidenced by the example.

Keywords Robots · SEMS · Control planning · Dynamic models
The method of recurrent target inequalities · Decision-making in conditions of uncertainty · Logical-probabilistic and logical-linguistic constraints
Theorems reducing constraints to interval

1   Introduction

Structural identification methods and computational optimization methods are widely used in the analysis and synthesis of high-precision information-measuring and control systems (IMCS). A special feature of modern dynamic robotic systems, such as SEMS, is the use of such systems in the contours of "non-rigid" models in which the structure, parameters and state vector are not specified at the design stage, but are in real time when they operate [1, 2]. In this case, when setting the problem of analysis and synthesis of SEMS it is necessary to describe the control as a multi-criteria problem of optimal control with a large number of variables and constraints of different types. This significantly complicates the synthesis of cyber-physical systems such as SEMS with a constraint on the performance of microprocessor-based controls and the lack of fast methods of solving, use of force not only necessary but also sufficient optimality conditions, which allow to estimate the residence time of the solution. Adaptation of already existing optimization procedures for SEMS synthesis is one of the main classical approaches [3, 4]. An alternative approach is to find suboptimal solutions and reduce the initial problem to the search for the target set described by systems of inequalities, including dynamic ones [5]. This approach was developed by V. A. Yakubovich and his followers and was named the method of recurrent target inequalities (RTI) [6].

The peculiarity of the control of cyberphysical systems such as SEMS is the need to choose the best trajectories (scenarios) in conditions of not complete certainty with limitations on the part of the practical feasibility of algorithms for processing data about objects. This leads to the fact that the region determine the characteristics of the SEMS state the one way or the other is sampled based on the linear, nonlinear or ordinal scales [7, 8], and restrictions from the environment of functioning are introduced in the form of logical-probabilistic and logical-linguistic expressions [9, 10]. Therefore, the RTI method of further development, taking into account the logical-probabilistic and logical-linguistic limitations in the synthesis of IEMS to date has not received. However, the relevance of research in the direction of the development of the method of recurrent target inequalities for the synthesis of high-precision SEMS is undeniable.

The solution of the problem of reducing the initial problem of multidimensional multicriteria conditional optimization of large dimensions for the case when there are logical-probabilistic and logical-linguistic constraints to the equivalent with a given accuracy, the finite set of inequality systems describing the target set is closely related to the solution of the problem of reducing logical-probabilistic and logical-linguistic constraints to the interval.

2 Prerequisites for Reducing Logical-Probabilistic Constraints to Interval Constraints

Let [a, b] be the interval in which there was a random variable x with the normal distribution law during fuzzification (when a logical variable is obtained ξ), $\Phi(.)$— Gauss probability integral, m—the expectation of a random variable x, σ—the mean square deviation of the random variable x. Then the fair

Theorem 1 *If a Boolean variable with probability of truth* $P\{\xi = 1\}$ *was obtained by fuzzification from a random variable* x *having normal distribution law, then*:
$P\{\xi = 1\} = P\{a < \xi < b\} = \Phi((b - m)/\sigma) - \Phi((a - m)/\sigma)$ [11].
Consequence 1
If the random variable x has a standard normal distribution law (m = 0, σ = 1), then

$$P\{\xi = 1\} = \Phi(b) - \Phi(a).$$

Consequence 2
If m = 0, a = −b, then $\Phi(b/\sigma) = 0,5(1 + P\{\xi = 1\})$
Consequence 3
If $\quad m = (b + a)/2; \ \sigma = (m - a)/3; \ a^* = -b^*$, then $\Phi(b^*) = 0,5\,(1 + P\{\xi = 1\})$,

where: $b^* = (b\xi - m)/\sigma$, $a^* = (a\xi - m)/\sigma$,

$a\xi$—the lower boundary of the interval (quantum of fuzzification) of a Boolean variable ξ,
$b\xi$—he upper limit of the interval (quantum of fuzzification) of a Boolean variable ξ.

Consequence 4
If $a = kb, (m - 3\sigma)/b \leq k \leq 1$,   then $P\{\xi = 1\} = \Phi((b - m)/\sigma) - \Phi((kb - m)/\sigma)$. This condition allows for known k, m and σ to be selected b values using the function table Φ (.).

Let [a, b] be the interval in which the random variable x was located at the fuzzification (obtaining a Boolean variable ξ), $[x_{min}; x_{max}]$—the interval in which the random variable x has a uniform distribution law. Then the fair

Theorem 2 *If a Boolean variable ξ having the probability that $P\{\xi = 1\}$ was obtained by fuzzification of a random variable x having uniform distribution in the interval $[x_{min}; x_{max}]$, then*

$$P\{\xi = 1\} = P\{a < \xi < b\} = (b-a)/(x_{max} - x_{min})$$

Consequence 1
If $a = -b$, then $b = 0, 5P\{\xi = 1\}(x_{max} - x_{min})$
Consequence 2
If $a = kb, x_{min}/b \leq k \leq 1$, then $b = (P\{\xi = 1\}(x_{max} - x_{min}))/(1 - k)$

3   Prerequisites for Reducing Logical-Linguistic Constraints to Interval Constraints

Let [a, b] be the interval in which the fuzzy quantity x was located during fuzzification (obtaining the linguistic variable v), $[x_{min}; x_g]$ be the interval in which the membership function f(x) is set for the fuzzy quantity x. Then the fair

Theorem 3 *If the linguistic variable v having the value of the membership function $\mu(v)$ was obtained by fuzzification from a fuzzy variable x having the membership function   f(x)   on   the   interval $[x_{min}; x_{max}]$, then $a = x_{min} + \mu(v)f(x)$, $b = x_{max} + \mu(v)f(x)$*

Consequence 1
If the triangular appearance of the function f(x) is specified, that is,

$$f(x) = 0.5(x_{max} - x_{min}), \text{ at. } \quad x_{min} \leq x \leq 0.5(x_{max} - x_{min})$$

$$f(x) = -0.5(x_{max} - x_{min}), \text{ at. } \quad 0.5(x_{max} - x_{min}) < x \leq x_{max}$$

then we get the following intervals: $a = x_{min} + \mu(v)0.5(x_{max} - x_{min}), b = x_{max} - \mu(v)0.5(x_{max} - x_{min})$

4   Fuzzy Optimal Control Problem

The greatest number of optimal control methods considers such control processes, each of which can be described by a system of ordinary differential equations of the form:

$$dx_i/dt = f_i(x_1, x_2, \ldots, x_n, u_1, u_2, \ldots, u_m), i = 1, 2, \ldots, n, \qquad (1)$$

where: $x_1, x_2, \ldots, x_n$—values characterizing the process, i.e. the phase coordinates of the control object, determining its state at each time t, and $u_1, u_2, \ldots, u_m$—control parameters that determine the course of the process.

The typical optimal control problem is formulated as follows.

You want to find the control

$$u_j(t), j = 1, 2, \ldots, m \qquad (2)$$

which will transfer the control object from the state

$$x_i(t_0) = x_{i0}, i = 1, 2, \ldots, n \qquad (3)$$

into a state

$$x_i(t_1) = x_{i1}, i = 1, 2, \ldots, n \qquad (4)$$

so that the functionality

$$J = \int_{t_0}^{t_1} F^0(x_1, \ldots, x_n, u_1, \ldots, u_m) dt \qquad (5)$$

had a minimum value for constraints

$$(u_1, \ldots, u_m) \in U \qquad (6)$$

$$(x_1, \ldots, x_n) \in X \qquad (7)$$

where the set X reflects the specifics of the control object and the environment of its functioning, and the set U reflects the specifics of the control object.

The introduction of constraints often leads to nonclassical optimization problems that solve by computational methods. In particular, when forming control tasks, i.e. when displaying a real problem in some formal language, which is a professional language of the decision maker, i.e. in the language of the developer F: Q_Z Q_R, where F maps a set Q_Z objects of the displayed task (elements and their interrelations) to a set of QR objects of the developer language (concepts, relations, names, etc.), and $\Phi^{-1}(q_i)$—inverse mapping if at least one of the following conditions is met:

$$\exists q_j \in Q_3, /\Phi(q_i)/ > 1 \tag{8}$$

$$\exists q_j \in Q_p, /\Phi^{-1}(q_j)/ > 1 \tag{9}$$

$$\exists q_i \in Q_3, \Phi(q_i) = \emptyset \tag{10}$$

$$\exists q_j \in Q_p, \Phi^{-1}(q_j) = \emptyset \tag{11}$$

where: $/ \; /$ the cardinality of the set, and $\emptyset$—the empty set, q_i—element of Q_Z, qj-element of Q_R Will be the optimal control problem in conditions of not complete certainty or the fuzzy optimal control problem. In this case, there are logical-probabilistic and logical-linguistic limitations, such as:

$$x_i \otimes y_j \rightarrow v_i \otimes w_j, P(x_i = 1), P(y_j = 1), P(v_i = 1), P(w_j = 1), i = 1, 2. \ldots, n; j = 1, 2, \ldots, m$$

$$\tag{12}$$

$$x_i \otimes y_j \rightarrow v_i \otimes w_j, \mu(x_i), \mu(x_j), \mu(v_i), \mu(w_j), i = 1, 2. \ldots, n; j = 1, 2, \ldots, m \tag{13}$$

where: x_i, y_j, v_i, w_j—logical variable, $\rightarrow$ implication sign, $\otimes$—sign conjunction, $P(x_i = 1), P(y_j = 1), P(v_i = 1), P(w_j = 1)$—probabilities of logical variables, $\mu(x_i), \mu(y_j)\mu(v_i), \mu(w_j)$—functions of belonging of logical variables.

Knowing the intervals accepted at fuzzification, i.e. at reception of logical variables, it is possible, using the theorems, to pass from logical-probabilistic and logical-linguistic variables to interval variables and the fuzzy optimal control problem is reduced to the classical mathematical programming problem.

5   Example

5.1   Problem Statement

In the settlement having N * M intersecting streets, it is required to find the route of journey of the vehicle from point A with x_A, y_A coordinates in point B with x_B, y_B coordinates in the minimum time J.

The following logical variables characterizing the environment are introduced:

h_1—the logical variable "dry", which corresponds to humidity from 0 to 40%,
h_2—the logical variable "wet", which corresponds to 30–70% humidity%,
h_3—the logic variable "very wet", which corresponds to humidity from 60 to 100%,
t_1—the logical variable "hot", which corresponds to the temperature from 20 to 40 °C,
t_2—the logic variable "heat", which corresponds to the temperature from 5 to 20 °C,

t_3—the logic variable "cold", which corresponds to the temperature from -5 to 10 °C,
v_1—the logical variable "fast", which corresponds to the speed from 40 to 60 km/ h,
v_2—the logic variable "slow", which corresponds to a speed between 20 and 50 km/h,
w_1—the logical variable "sharply", which corresponds to the angular velocity from 4 to 8 g/s,
w_2—the logical variable "smoothly", which corresponds to an angular speed from 2 to 5 g/s.

The following logical-probabilistic and logical-linguistic limitations are known:

$$h_1 \otimes t_1 \to v_1 \otimes w_1, P(h_1 = 1) = 0.8, \mu(t_1) = 0.9, P(v_1 = 1) = 0.8, \mu(w_1) = 0,8 \quad (14)$$

$$h_1 \otimes t_2 \to v_1 \otimes w_2, P(h_1 = 1) = 0.9, \mu(t_2) = 0.8, P(v_1 = 1) = 0.7, \mu(w_2) = 0.7 \quad (15)$$

$$h_1 \otimes t_3 \to v_2 \otimes w_2, P(h_1 = 1) = 0.7, \mu(t_3) = 0.7, P(v_2 = 1) = 0.6, \mu(w_2) = 0.6 \quad (16)$$

$$h_2 \otimes t_1 \to v_1 \otimes w_1, P(h_2 = 1) = 0.8, \mu(t_1) = 0.9, P(v_1 = 1) = 0.75, \mu(w_1) = 0.75 \quad (17)$$

$$h_2 \otimes t_2 \to v_1 \otimes w_2, P(h_2 = 1) = 0.9, \mu(t_2) = 0.8, P(v_1 = 1) = 0.65, \mu(w_2) = 0.65 \quad (18)$$

$$h_2 \otimes t_3 \to v_2 \otimes w_2, P(h_2 = 1) = 0.7, \mu(t_3) = 0.7, P(v_2 = 1) = 0.7, \mu(w_2) = 0.7 \quad (19)$$

$$h_3 \otimes t_1 \to v_1 \otimes w_1, P(h_3 = 1) = 0.8, \mu(t_1) = 0.9, P(v_1 = 1) = 0.6, \mu(w_1) = 0.6 \quad (20)$$

$$h_3 \otimes t_2 \to v_1 \otimes w_2, P(h_3 = 1) = 0.9, \mu(t_2) = 0.8, P(v_1 = 1) = 0.9, \mu(w_2) = 0.9 \quad (21)$$

$$h_3 \otimes t_3 \to v_2 \otimes w_2, P(h_3 = 1) = 0.7, \mu(t_3) = 0.7, P(v_2 = 1) = 0.8, \mu(w_2) = 0.8 \quad (22)$$

where: h_i, t_j, v_i, w_j—logical variables, $\to$ implication sign, $\otimes$—sign conjunction, P $(h_i = 1)$, $P(v_i = 1)$—probabilities logical variables h_i and v_i, $\mu(t_i)$, $\mu(w_j)$—functions of belonging of logical variables t_i and w_j.

In addition, it is known that the view of each intersection configuration quadrilateral, φ_{ij}—rotation angles at i-th nodes (intersections) to j-th, l_{ij}—distances between i-th and j-th nodes and the waiting time at intersections τ_{ij}. In particular, if the coordinates of all nodes $s(x_s, y_s)$, $q(x_q, y_q)$ и $p(x_p, y_p)$ (intersections) are known, it is easy to calculate (see Fig. 1):

- distances between nodes using the following ratios:

$$l_{s,q.} = \left((x_s - x_q)^2 + (y_s - y_q)^2 \right)^{1/2} \quad (23)$$

$$l_{qp} = \left((x_q - x_p)^2 + (y_q - y_p)^2 \right)^{1/2} \quad (24)$$

- turning angles at intersections using ratios:

$$\varphi_{qp} = 2\pi - \text{arctg}((k_2 - k_1)/(1 + k_2 k_1)) \tag{25}$$

where: k_1—the angular coefficient of the straight line connecting the nodes q and p, k_2—angular coefficient of a straight line connecting s and q nodes.

The coefficients k_1 and k_2 are easily obtained from the equations of the specified straight lines:

$$k_1 = \left(y_q - y_s\right)/\left(x_q - x_s\right). \tag{26}$$

$$k_2 = \left(y_p - y_q\right)/\left(x_p - x_q\right). \tag{27}$$

In this case it is necessary to minimize the following functionality:

$$J(M_n) = \sum_{(i,j).} \left(a\left(l_{ij}/v_{ij}\right) + \sum_{(i,j).} b(\varphi_{ij}/w_{jj}) + \sum_{(i,j)} (c\tau_{ij}) \to \min_{(i,j)} \tag{28}$$

where: a, b, c—the coefficients of preference, v_{ij}—linear and w_{ij} is the angular velocity of movement associated with the h_i humidity and temperature t_j.

In Eq. (28) (i, j)—this is an element of an ordered set that characterizes the considered route from the start point to the end point. For example, if you look at the M_v route from node 1 to node 3, then from node 3 to node 7 and finally from node 7 to node 8 (see Fig. 1) then it means that the Eq. (28) is summed (i, j) ∈ {(1,3); (3,7); (7,8)}

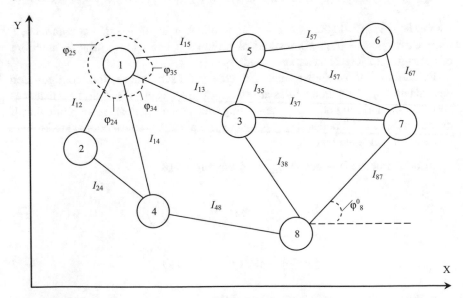

Fig. 1 Dynamic configuration space

5.2 Problem Solution

1. First on the terrain map the fragment is allocated, which contains the starting and ending points of the route. In the selected fragment (Fig. 1) there is a dynamic configuration space (DCS) containing 8 intersections (nodes), which are characterized by the distances between adjacent nodes l_{ij}, the rotation angle at the intersections φ_{ij} and the delay time at the intersection τ_{ij} (Fig. 2).
   In this case, the selected fragment should include all nodes adjacent to the start and end nodes. For example, from node 1 you need to get to node 7. Then, for node 1, it will be nodes 2,3,4 and 5, for node 7, it will be nodes 3,5,6 and 8.
2. All variants of hit from the start node to the end node of the selected fragment are constructed. For our example, these are the following route options: M_1: (1-2-4-8); M_2: (1-4-8); M_3: (1-3-8); M_4: (1-5-3-8); M_5: (1-3-7-8); M_6: (1-5-7-8); M_7: (1-5-6-7-8).
3. Intervals of logical variables characterizing the environment are calculated using the above theorems 1,3. The transition from logical-probabilistic and logical-linguistic constraints to logical-interval constraints is carried out and this problem is reduced to the classical problem of mathematical programming.
4. In any current dynamic configuration space (DCS) corresponding to the selected fragment, there will be 4 constraints at the same time (in our example - these are the conditions: 13, 14, 16 and 17), for each of which the minimum functional (28) is calculated from all variants of motion.
5. As an optimal route M_v^o is chosen, which will correspond to the minimum of the mean value of the functional $J(M_v)$

Let us consider paragraph 3 of the algorithm.

Select the condition 14. This condition describes h_1-the logical variable "dry", which corresponds to humidity from 0 to 40%, with a probability $P(h_1 = 1) = 0.8$

Fig. 2 The turn at the intersection

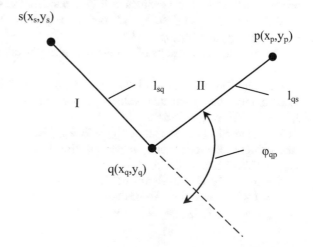

and t_1—the logical variable "hot", which corresponds to a temperature of 20–40 °C, with functions of belonging $\mu(t_1) = 0.9$.

Convert logical and probabilistic variable h_1 in logical-interval h_1^{H} To do this, we use the corollary 3 of theorem 1 and find the normal distribution function $\Phi(h_1^*)$ with the parameters: expectation $m = (b + a)/2$; standard deviation $\sigma = (m - a)/3$, where a—is the lower, b—is the upper bound of the quantum corresponding to the logical variable "dry" h_1 during fazzification.

For this case $a = 0\%$, $b = 40\%$, $\Phi(h_1^*) = 0.5$ $(1 + P\{\xi = 1\}) = 0.5$ $(0.8 + 1) = 0.9$.

Then we find the value of h_1^* in Table 1 of [11]. In this case $h_1^* = 1.29$ and for $(-h_1^*) = -1.29$. To find the intervals a_1^{H} and b_1^{H} h of the logical interval variable h_1^{H}, use the following expressions: $(-h_1^*) = ((a_1^{\text{H}} - m)/\sigma)$, $h_1^* = ((b_1^{\text{H}} - m)/\sigma)$.

Therefore, we obtain the following lower a_1^{H} and upper b^{H} intervals: $a_1^{\text{H}} = 11.5$ and $b^{\text{H}} = 28.5$. Thus, we moved from a logical-probabilistic constraint (13) to a logical-interval constraint:

$11.5 < h_1^{\text{H}} < 28.5$.

Now convert logical-linguistic variable t_1 in the logical-interval t_1^{H}. To do this, we use the consequence 1 of theorem 3. For the case under consideration $x_{\min} = 20$, $x_{\max} = 40$. Calculate the upper and lower at_1 bt_1 boundaries: $at_1 = x_{\min} + \mu(t_1)0.5$ $(x_{\max} - x_{\min}) = 29$,

$$bt_1 = x_{\max} - \mu(t_1)0.5(x_{\max} - x_{\min}) = 31$$

Thus, we moved from the logical-linguistic restriction (13) to the logical-interval constraint: $29 < t_1^{\text{H}} < 31$.

Table 1 shows the results of the transition from logical-probabilistic and logical-linguistic constraints to logical-interval constraints. At the same time, the linear speed is transferred from km/h to m/s.

Let us calculate the functional values (28) for the considered route variants by specifying the values of distances between intersections and angles of rotation at intersections. Delays at junctions will be considered is set to 10 s.

For route M_1: (1-2-4-8) will be: $l_{12} = 500$ m, $l_{24} = 400$ m, $l_{48} = 900$ m, $\varphi_{124} = 60°$, $\varphi_{248} = 30°$, $\varphi_{12} = 0°$.

Table 2 shows the results of calculation of the functional $J(M_v)$ for the route M_1: (1-2-4-8).

For route M_2: (1-4-8) will be: $l_{14} = 900$ m, $l_{48} = 900$ m, $\varphi_{14} = 0°$, $\varphi_{148} = 60°$. Accordingly, the calculated values of the functionals will be: $J_{\min} = 138.57$, $J_{\max} = 178.59$ и $J_v = 158.58$

For route M_3: (1-3-8) will be: $l_{13} = 700$ m, $l_{38} = 700$ m, $\varphi_{13} = 0°$, $\varphi_{138} = 30°$. Accordingly, the calculated values of the functionals will be: $J_{\min} = 117.41$, $J_{\max} = 137.77$ и $J_v = 127.59$.

For route M_4: (1-5-3-8) will be: $l_{15} = 700$ m, $l_{53} = 200$ $l_{38} = 700$ m, $\varphi_{15} = 0°$, $\varphi_{153} = 150°$, $\varphi_{538} = 60°$. Accordingly, the calculated values of the functionals will be: $J_{\min} = 168,77$, $J_{\max} = 223,55$ и $J_v = 196.16$.

Table 1 The results of the transition from logical-probabilistic and logical-linguistic constraints to logical-interval

№ restrictions	Intervals of quantization	Type of uncertainty	Logical-interval constraints
13	$0 < h_1 < 40$	$P(h_1 = 1) = 0.8$	$11.5 < h_1^{H} < 28.5$
13	$11.1 < v_1 < 16.7$	$P(v_1 = 1) = 0.8$	$12.7 < v_1^{H} < 15.1$
13	$20 < t_1 < 40$	$\mu(t_1) = 0.9$	$29 < t_1^{H} < 31$
13	$4 < w_1 < 8$	$\mu(w_1) = 0.8$	$5.6 < w_1^{H} < 6.4$
14	$0 < h_1 < 40$	$P(h_1 = 1) = 0.9$	$9.1 < h_1^{H} < 30.9$
14	$11.1 < v_1 < 16.7$	$P(v_1 = 1) = 0.7$	$12.9 < v_1 < 14.8$
14	$20 < t_1 < 40$	$\mu(t_1) = 0.9$	$29 < t_1^{H} < 31$
14	$4 < w_1 < 8$	$\mu(w_1) = 0.8$	$5.6 < w_1^{H} < 6.4$
16	$30 < h_2 < 70$	$P(h_2 = 1) = 0.8$	$41.5 < h_2^{H} < 58.5$
16	$11.1 < v_1 < 15.7$	$P(v_1 = 1) = 0.75$	$12.8 < v_1 < 14.9$
16	$20 < t_1 < 40$	$\mu(t_1) = 0.9$	$29 < t_1^{H} < 31$
16	$4 < w_1 < 8$	$\mu(w_1) = 0.75$	$5.4 < w_1^{H} < 6.6$
17	$30 < h_2 < 70$	$P(h_2 = 1) = 0.9$	$39.1 < h_2^{H} < 60.9$
17	$11.1 < v_1 < 15.7$	$P(v_1 = 1) = 0.65$	$13 < v_1 < 14.75$
17	$5 < t_2 < 15$	$\mu(t_2) = 0.8$	$9 < t_2^{H} < 11$
17	$2 < w_2 < 5$	$\mu(w_2) = 0.65$	$2.98 < w_2^{H} < 4.0$

Table 2 The results of calculating the functional J_i for the route M_1

№ constraints	v	w	J	J_{cp}	J_{min}	J_{max}	J_v
13	12.7	5.6	187.8	175.52	163.23	198.66	180.945
13	12.7	6.4	185.78				
13	15.1	5.6	165.27				
13	15.1	6.4	163.23				
14	12.9	5.6	185.6	175.66			
14	12.9	6.4	183.68				
14	14.8	5.6	167.69				
14	14.8	6.4	165.67				
16	12.8	5.4	187.29	176.615			
16	12.8	6.6	187.25				
16	14.9	5.4	167.48				
16	14.9	6.6	164.44				
17	13	2.98	198.66	186.725			
17	13	4.0	190.96				
17	14.75	2.98	182.84				
17	14.75	4.0	174.44				

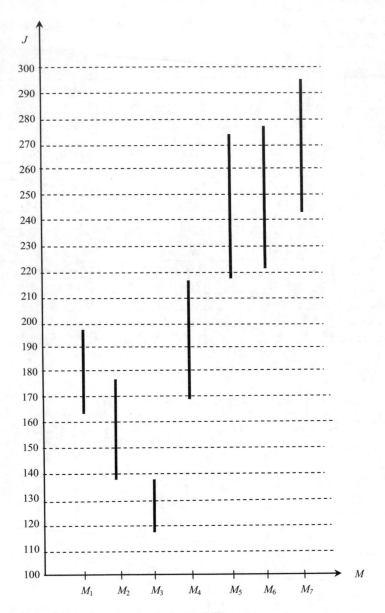

Fig. 3 Calculated intervals of functionals for M_1–M_7 routes

For route M_5: (1-3-7-8) will be: $l_{13} = 700$ m, $l_{37} = 900$ m, $l_{78} = 900$ m, $\varphi_{13} = 0$, $\varphi_{137} = 30°$, $\varphi_{378} = 120°$. Accordingly, the calculated values of the functionals will be: $J_{min} = 219.0$, $J_{max} = 272.65$ и $J_v = 245.82$.

For route M_6: (1-5-7-8) will be: $l_{15} = 700$ m, $l_{57} = 900$ м, $l_{78} = 900$ m, $\varphi_{15} = 0$, $\varphi_{157} = 45°$, $\varphi_{578} = 120°$. Accordingly, the calculated values of the functionals will be: $J_{min} = 221.68$, $J_{max} = 277.68$ и $J_v = 249.51$.

For route M_7: (1-5-6-7-8) will be: $l_{15} = 700$ m, $l_{56} = 700$ m, $l_{67} = 400$ m, $l_{78} = 900$ m, $\varphi_{15} = 0$, $\varphi_{156} = 10°$, $\varphi_{567} = 90°$, $\varphi_{678} = 45°$. Accordingly, the calculated values of the functionals will be: $J_{min} = 241.46$, $J_{max} = 296.35$ и $J_v = 268.91$.

Intervals of calculated functionals for all considered routes are shown in Fig. 3. From drawing it is visible that the M_3 route is the best, because it requires the least time (the smallest functional values (28)) to complete the route on average, as well as at worst and with a favorable forecast. Route M_7 is the worst, as it takes the most time. M5 and M6 routes are slightly different and their preference may depend on unaccounted for minor random factors, such as delays at pedestrian crossings.

6 Conclusion

The necessity to choose the best trajectories (development scenarios) in the group control of SENS in conditions of incomplete certainty requires taking into account nonlinear constraints in the form of logical-probabilistic and/or logical-linguistic expressions (rules). The solution of such problems can be significantly simplified by reducing the logical-probabilistic and logical-linguistic constraints to the interval and the use of the theorems described. Then knowing the spacing adopted in fuzzification, that is, when receiving the logical variables in the specified expression, using the theorems of reductions (1–3) go from fuzzy optimal control problem to a classical mathematical programming problem, as shown in the example.

Acknowledgements This work was financially supported by Russian Foundation for Basic Research, Grant 16-29-04424 and Grant 18-01-00076.

References

1. Gorodetskiy Andrey, E. (ed.): Smart electromechanical systems.. In: Studies in Systems, Decision and Control, Vol. 49, 277 p, Springer International Publishing, Switzerland (2016). https://doi.org/10.1007/978-3-319-27547-5
2. Kuchmin, A.Y.: Problems of Identification of Non-stationary dynamic objects. Informatsionno-upravliaiushchie sistemy [Information and Control Systems] (2), 18–25 (2018) (In Russian)
3. Gorodetsky, A.E., Tarasova, I.L.: Control and neural networks. SPb.: Publishing House of the Polytechnic. University Press, 312 p (2005)
4. Tobacco, D., Kuo, B.: Optimal control and mathematical programming. M.: Nauka, 280 p (1975) (In Russian)

5. Kuchmin, A.Y.: Analysis of polynomial constraints by the solution tree method. Informacionno-upravljajushhie sistemy **6**(91), 6–9 (2017) (In Russian). https://doi.org/10.15217/issn1684-8853.2017.6.9

6. Fomin, V.N., Fradkov, A.L., Yakubovich, V.A.: Adaptive management of dynamic objects. Nauka, Moscow. Ch. Ed. fiz-mat literature, 448 p (1981) (In Russian)

7. Gorodetskiy, A.E., Tarasova, I.L., Kurbanov, V.G.: Logical-mathematical model of decision making in central nervous system SEMS. In: Gorodetskiy, A.E., Kurbanov, V.G. (eds.) Smart Electromechanical Systems: The Central Nervous System, 270 p. Springer International Publishing AG, Berlin (2017). https://doi.org/10.1007/978-3-319-53327-8

8. Gorodetsky, A.E., Tarasova, I.L.: Fuzzy mathematical modeling of poorly formalized processes and systems [Nechetkoe matematicheskoe modelirovanie ploho formalizuemyh processov i sistem]. SPb. Publishing house of Polytechnic Institute, 336 p (2010) (In Russian)

9. Gorodetskiy, A.E., Kurbanov, V.G., Tarasova, I.L.: Methods of synthesis of optimal intelligent control systems SEMS. In: Gorodetskiy, A.E. (ed.) Smart Electromechanical Systems, 277 p. Springer International Publishing, Berlin (2016). https://doi.org/10.1007/978-3-319-27547-5

10. Gorodetsky, A.E., Tarasova, I.L., Kurbanov, V.G.: Logical and probabilistic methods of formation of a dynamic space configuration of the robot group. Materialy 10-j Vserossijskoj conference on governance] Divnomorskoye, Gelendzhik **2**, 262–265 (2017)

11. Venttsel, E.S.: Theory of probability: Proc. for high schools. Higher School, Moscow, 575 p (2001) (In Russian)

Nature-Inspired Algorithms for Global Optimization in Group Robotics Problems

Anatoliy P. Karpenko and Ilia A. Leshchev

Abstract Localization of plots of land which have the highest level of radiation, chemical or alternative contamination is one of the typical group robotics objectives. The research aim lies in development, software implementation and performance study of the original robotic group control algorithm based on nature-inspired Cat Swarm Optimization algorithm. This paper proposes Cat Swarm Optimization algorithm description, features of software realization and a vast computational experiment results. The practical value of this work lies in applicability of proposed *algorithm for decentralized robotic* group control systems synthesis.

Keywords Robot · Group robotics · SEMS · Decentralized robotic group control
Nature-inspired algorithms of global optimization · Contaminated plots of land
localization

1 Introduction

The main idea of group robotics lies in the fact that some tasks can be better solved by a large number of relatively small and simple robots that can act in coordination with one another rather than by one large and complex robot. In this case, failure of individual robots does not reduce or slightly reduces functionality of the group as a whole. As robots forming the group are relatively simple and low-cost, economic loss in this case is also insignificant. Due to a potentially small size of each of the robots in the group, they are difficult to destroy (which is important for military applications).

A group of robots can effectively solve such problems as: investigating a certain territory in order to detect a site with an extreme value of a certain substance; reaching a point on the terrain with given coordinates in the presence of various obstacles (to explore the vicinity of this point or to impact on the object(s) in its

A. P. Karpenko (✉) · I. A. Leshchev
Bauman Moscow State Technical University, Moscow, Russia
e-mail: apkarpenko@mail.ru

© Springer Nature Switzerland AG 2019
A. E. Gorodetskiy and I. L. Tarasova (eds.), *Smart Electromechanical Systems*,
Studies in Systems, Decision and Control 174,
https://doi.org/10.1007/978-3-319-99759-9_8

91

vicinity); determining the boundaries of the region all the points of which have some common characteristics (for example, the radiation level in which does not exceed a given value); drawing up a map of a given territory; patrolling the borders of a certain area to detect an intrusion.

From a formal point of view, in this paper we consider a class of problems that can be reduced to localization of extrema of an unknown scalar physical field. Detecting zones of radioactive, chemical, biological or other kinds of terrain contamination, damage caused by malignant algae, turbulence, temperature and salinity of the seas, and other similar problems can be posed in this form [1–4].

There are two main approaches that are known and used to solve the problem of group robotics: centralized and decentralized. The centralized control system for a group of robots has a large number of advantages which, however, level out its main drawback—a high vulnerability, because failure of the central control device results in performance loss of the entire system. The decentralized control system assumes that the overall goal of the functioning of a group of robots is achieved on the basis of an independent coordinated planning of their actions by each of its members, that is, on the basis of self-organization of a group of robots. The decentralized management system presupposes availability of channels of duplex communication between a part of the "closest" group members. The main advantage of such a system consists in keeping its operability when one or more robots fail; the main drawback consists of high requirements for "intelligence" of each of the robots.

The methodological basis of decentralized control in group robotics is bionic approaches, that is, approaches based on technical copying of effective solutions that were found by living species in the process of multimillion-year natural selection. So, to localize extrema of a scalar field of this or that physical nature in group robotics, such population algorithms as a particle swarm, a swarm of bees, an ant colony, etc. are widely used [5]. A large group of robots, the decentralized control of which is carried out on the basis of such algorithms, is called *swarm robotics* [6]. References to a significant number of works dedicated to swarm robotics (published before 2012 inclusive) are presented in [7].

It is well known that traditional Swarm Intelligence algorithms, such as Ant Colony Optimization (ACO), are poorly suited for the purposes of group robotics (for example, see [8]). The fact is that these algorithms require either global information about the group (positions of the globally better and/or worst of the robots), or presuppose that the robots change the environment in which they evolve. In real conditions implementation of both these processes is problematic. Therefore, as a rule, modified classical swarm algorithms are used to control robot navigation in group robotics problems.

We use the Cat Swarm Optimization (CSO) algorithm proposed by Shu-Chuan Chu, Pei-Wei Tsai in 2007 [9] as the basic swarm algorithm. To date, the CSO algorithm has developed significantly. In [10–14] a number of modifications of this algorithm were proposed, in [15] its parallel version is considered, in [15–18] the use of the algorithm for solving some applied problems is considered. Currently modifications of the CSO algorithm aimed at solving multicriteria optimization problems are also known [19]. Publications dedicated to group control of robots on the basis of the CSO algorithm are unknown to the authors.

The choice of the CSO algorithm is due to simplicity of its modification which takes into account peculiarities of hardware implementation of a group of robots. The deep modification of the CSO algorithm proposed in this paper was named ACSO (Advanced CSO). The ACSO algorithm includes a significant number of free parameters that allow for a wide variation in the diversification and intensification properties of the algorithm. Thus, the ACSO, in fact, is metaheuristic.

2　Problem Statement

We assume that the search space $R^{|X|}$ is two-dimensional: $|X| = 2$. We consider the problem of "strategic" control of robots,—that is, we exclude the "tactical" problem of overcoming obstacles from consideration.

Each of the devices s_i, $i \in [1 : |M|]$ is a mother robot (M-robot) carrying a fixed number $|C|$ of small non-returnable satellite devices $s_{i,j}$ (C-robots); $j \in [1 : |C|]$. These robots can *once* measure the values of the scalar field $\varphi(X) = \varphi(x_1, x_2)$

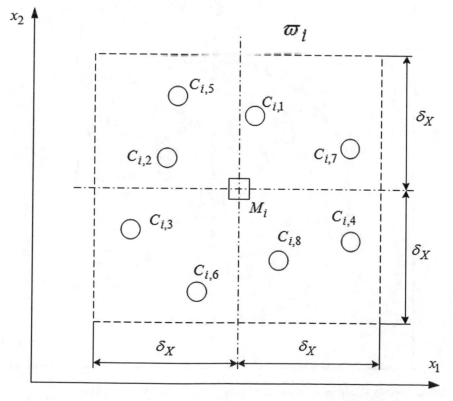

Fig. 1 Vicinity ϖ_i of the parent robot $s_i = M_i$: $s_{i,j} = C_{i,j}$—satellite robots; $j \in [1 : 8]$

(test) at distances from the robot s_i which do not exceed δ_X for each of the measurements x_1, x_2,—that is, in a square *vicinity* ϖ_i (Fig. 1). Thereby, the total number of tests that C-robots belonging to a given M-robot can perform is limited and equal to $|C| = n_C$. The coordinates of the tests made by C-robots and their results become known to the corresponding M-robot. Each of the M-robots can also perform tests (at their stop points), the total number of which does not exceed $n_M = \hat{t}$ (see below).

We assume that the results of the tests of M and C-robots are accurate, that is, the measured values of the function $\varphi(X)$ at the test points are not distorted by noise and/ or measurement errors. The robots can determine their coordinates x_1, x_2 with an error which is many times less than the value δ_X, so that these errors can be ignored.

The *vicinity* Ω_i of the robot s_i is a set of points in the search space, the Euclidean distance of which to the robot s_i does not exceed the value r_X, meaning the range of the communication channels between them. The set of the M-robots belonging to the vicinity Ω_i forms a set of N_i "neighbors" of this robot. We denote the cardinality of the set N_i as $|N_i|$ (Fig. 2).

Each of the M-robots s_i "remembers" its trace, that is, the coordinates of the points of the search space in which it performed the tests, as well as the results of

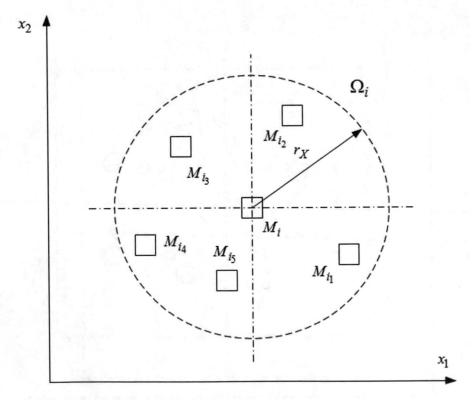

Fig. 2 Vicinity Ω_i and set N_i of the neighbours of the M-robot $s_i = M_i$: $|N_i| = 5$

these tests (fitness-values $\varphi(X)$). In addition, the robot s_i remembers the traces of all its satellite robots $s_{i,j}$, $j \in [1 : |C|]$. The aggregate of these traces together with the trace of the robot s_i itself forms an *extended* trace of this robot. To increase survivability of the entire system, at each iteration each of the M-robots receives extended traces of these robots from its neighbors and stores them in its memory.

If the M-robot localizes the extremum of the function $\varphi(X)$ with an accuracy δ_X, then it changes the search *region*—it moves to the vicinity of the center of one of the nearest rectangular subareas which were not explored by it and by its neighbors. In the event that the Euclidean distance between some two M-robots is less than the given value ρ_X, these robots "agree" with one another which of them will continue the research in the given region and which one will move to another region. The criterion for ending the search is achievement by each of the M-robots of a given number of iterations $\hat{\imath}$. In this case, all M-robots transmit information about all traces, that are stored in their memory, to the "center".

We introduce the following notations: $\Pi = \{X | X^- \leq X \leq X^+\} \in R^2$ is the investigated region (parallelepiped) of the search space, where the inequalities are understood component by component; $X^- = (x_1^-, x_2^-)$, $X^+ = (x_1^+, x_2^+)$ are the boundaries of the region; t is the number of the current iteration of the search; $X_i(t) = X_i = (x_{i,1}, x_{i,2})$ is the current position of the robot s_i; $X_i'(t) = X_i' = (x'_{i,1}, x'_{i,2})$ is the position of this robot at the next iteration; X^*—is the desired vector of optimal coordinates X, which delivers the maximum value of the fitness function $\varphi(X)$. The length of the diagonal of the parallelepiped Π (the diameter of the set Π) is denoted by d.

We consider the deterministic maximisation problem

$$\max_{X \in \Pi \subset R^2} \varphi(X) = \varphi(X^*) = \varphi^*.$$

3 The Original Cat Swarm Optimization Algorithm

As a basic algorithm, we use a simplified version of the CSO algorithm adapted to the needs of our modification of this algorithm. The essence of the algorithm comprises three stages: initialization of the population; seeking; tracing [9].

Initialization of the population is carried out according to the following scheme (here and below, the notations differ from the original ones).

(1) We initialize the values of all free parameters of the algorithm.
(2) In the search area Π we create a population $S = \{s_i, i \in [1 : |S|]\}$ of cats (M-robots) s_i that are uniformly randomly distributed in this area, that is, we assume that $X_i^0 \in \Pi$, $i \in [1 : |S|]$.
(3) We set the initial velocities ΔX_i^0, $i \in [1 : |S|]$ of the cats s_i that are uniformly randomly distributed in the region $\Pi_\Delta = \{\Delta X | \Delta X^- \leq \Delta X \leq \Delta X^+\} \in R^2$

where the components of two-dimensional vectors ΔX^-, ΔX^+ are free parameters of the algorithm; $\Delta X^- = (\Delta x_1^-, \Delta x_2^-)$, $\Delta X^+ = (\Delta x_1^+, \Delta x_2^+)$.

(4) We declare random cats in the population S in the number $0,8\,|S|$ in the state of *seeking*, and the rest—in the *tracing* mode.
(5) For each of the cats $s_i \in S$ we calculate the value $\varphi_i^0 = \varphi(X_i^0)$ of the fitness function and find the cats with the minimum φ^{worst} and maximum φ^{best} values of this function; $i \in [1 : |S|]$.

The *seeking mode* implements a local search and consists in exploring the cat's nearest vicinity in order to identify the direction of movement that leads to an improvement in the fitness function. The cat's behavior in this mode is determined by the following two free parameters:

- n_C (in the original—SMP)—the number of points under investigation in the vicinity of the cat, which is not variable, in contrast to item 1, as the iteration number increases;
- δ_X (in the original—SRD)—the size of the vicinity examined by the cat.

The scheme of the research process for a cat s_i, $i \in [1 : |S|]$ is as follows.

(1) We create n_C of copies of $X_{i,j}$, $j \in [1 : n_C]$ of the current coordinates X_i of the cat s_i.
(2) We change the position of each of the points n_C according to the expression

$$X'_{i,j} = (1 + U_2(-1;\ 1)\,\delta_X)\,X_{i,j},\tag{1}$$

where $U_2(-1;\ 1)$ is a two-dimensional vector of random variables uniformly distributed in the interval $[-1;\ 1]$

(3) We calculate the value of the fitness function $\varphi_{i,j} = \varphi(X'_{i,j})$ for each of the obtained points and find the maximum φ_i^{best} and minimum φ_i^{worst} of these values.
(4) In accordance with the formula

$$p_{i,j} = \frac{\varphi_{i,j} - \varphi_i^{worst}}{\varphi_i^{best} - \varphi_i^{worst}}, j \in [1 : n_C]\tag{2}$$

we calculate probability of choosing the point $X'_{i,j}$ as the next position of the cat s_i

(5) Basing on the probabilities (2) and using the roulette wheel method, we choose select the new position X'_i of the cat s_i.

The *tracing mode* is responsible for global search and for the cat s_i, $i \in [1 : |S|]$ is determined by the following sequence of steps.

(1) We update the speed (increment of coordinates) of the cat according to the formula

$$\Delta X_i' = \Delta X_i + c\, U_2(0;\ 1)\ \left(X^{best} - X_i\right),\qquad(3)$$

where c is a constant (free parameter); X^{best} is the position of the current globally best cat in the population.

(2) We check whether the components of the speed $\Delta X_i'$ are in the range of permissible values: $\Delta x_{i,k}' \in [\Delta x_k^-;\ \Delta x_k^+]$, $k = 1,\ 2$. If these limits are violated, then the value of the corresponding velocity component is set equal to the maximum permissible value.

(3) The position of the cat is updated in accordance with the expression

$$X_i' = X_i + \Delta X_i'.\qquad(4)$$

4   Modified ACSO Algorithm

4.1   New Entities Used by the ACSO Algorithm

Neighborhood of M-robots. The ACSO algorithm uses the concept of the neighborhood of M-robots in the search space R^2. The neighborhood *topology* is defined by the Euclidean norm $\left\|X_i - X_j\right\|_E$, $i,\ j \in [1:\ |S|]$, so that the set $N_i(t) = N_i$ of neighbors of the robot $s_i \in S$ is formed by M-robots whose positions satisfy the condition

$$\{X_j | \left\|X_i - X_j\right\|_E \le r_X, s_j \in S\}, i \in [1:|S|]$$

where r_X is the radius of the vicinity Ω_i (Fig. 2).

Trace and extended trace of the M-robot. Trace $Tr_i(t)$ of the M-robot s_i, $i \in [1:\ |S|]$ is a collection of all its positions in the search space on iterations $\tau \in [0:\ t]$, as well as the corresponding values of the fitness function: $Tr_i(t) = Tr_i = \{(X_i(\tau), \varphi_i(\tau)), \tau \in [0:t]\}$. Similarly, the *extended tracing* $\mathbf{Tr}_i(t)$ of this robot is defined by the expression

$$\mathbf{Tr}_i(t) = \left\{Tr_i(t) \bigcup (X_{i,j}, \varphi_{i,j}), j \in [1:n_C(t)]\right\},$$

where $\varphi_{i,j} = \varphi(X_{i,j})$; $n_C(t) \in [0:\ n_C]$ is part of the C-robots $s_{i,j}$ used (despatched) by the robot s_i by the end of the iteration t.

Perspective search direction. Allow $(X_{i_1},\ \varphi_{i_1})$ $(X_{i_2},\ \varphi_{i_2})$ to be two points that are "close" to the region Π_i, and $\varphi_{i_1} > \varphi_{i_2}$. The direction defined by these points is called *perspective* for the region Π_i (Π_i-perspective), if the ray $(X_{i_2}, X_{i_1}, \infty)$

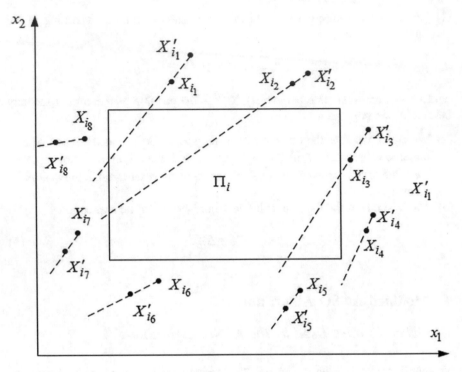

Fig. 3 Determination of perspective directions: $m = 8$; $m_i = 4$; $\varphi_{i_k} > \varphi'_{i_k}$, $k \in [1 : 4]$; $\varphi_{i_k} < \varphi'_{i_k}$, $k \in [5 : 8]$

crosses the region Π_i (Fig. 3). The meaning of this definition is as follows. Due to the fact that $\varphi_{i_1} > \varphi_{i_2}$, in the direction of the ray $(X_{i_2}, X_{i_1}, \infty)$ the values of the function $\varphi(X)$ increase, and there is a chance that the best values of this function can be achieved in the region Π_i which the specified ray crosses. The number Π_i of perspective directions m_i is determined by the following scheme (Fig. 3).

(1) We find $m = \min(\tilde{m}, |\mathbf{Tr}_i(t)|)$ of the points (X_{i_1}, φ_{i_1}), $(X_{i_2}, \varphi_{i_2}), \ldots, (X_{i_m}, \varphi_{i_m})$ which belong to the tracing $\mathbf{Tr}_i(t)$ and whose distances to the center X_i^c of the region Π_i are minimal. Here $\tilde{m}$ is the free parameter of the algorithm.
(2) For each of these points X_{i_j}, in the trace $\mathbf{Tr}_i(t)$ we find the nearest point $(X'_{i_j}, \varphi'_{i_j})$; $j \in [1 : m]$.
(3) We define all Π_i-perspective directions determined by points (X_{i_j}, φ_{i_j}), $(X'_{i_j}, \varphi'_{i_j})$; $j \in [1 : m]$. The number of these directions is denoted by m_i.

We note that the inequality $\min(\tilde{m}, |\mathbf{Tr}_i(t)|) < \tilde{m}$ may turn out to be valid at the initial iterations of the search, when the number of points in the trace $\mathbf{Tr}_i(t)$ is small.

Attractiveness of the region. Each of the regions Π_i, as well as a *candidate* for the region (see below), has a characteristic called *attractiveness* a_i. We use the two-criteria rule for determining this value, which has the form of

$$a_i = \lambda \frac{d_i}{d} + (1 - \lambda) \frac{m_i}{m}, \lambda \in [0; 1], \tag{5}$$

where d_i is the region's diameter equal to the length of the diagonal of the rectangle Π_i; d is the diameter of the search area Π; m_i, m is the number of Π_i-perspective directions and the total number of directions, respectively; λ is the weight coefficient (free parameter of the algorithm).

The formula (5) means that attractiveness of the region Π_i is proportional to its weighted relative diameter d_i and the relative number of perspective directions. It is easy to see that the values λ that are close to 1 diversify the search, and those that are close to zero, on the contrary, intensify it. The coefficient λ can be adaptively or programmatically decreased in the search process, providing a transition from the primary exploration of the search space (diversification) to the primary localization of the desired by the maximum of the function $\varphi(X)$ (intensification).

Region and candidates in the search region. The M-robot s_i, $i \in [1 : |S|]$ in its current position $X_i(t)$ can use the *region* Π_i—one of the nearest unexplored subregions of the search area Π. The M-robot s_i selects the region Π_i if the set of extended traces of this robot has sufficient power. The selection is performed in three stages (we omit some details).

Stage 1—definition of *candidates* for the region Π_i

(1) We sort the coordinate x_1 of all extended traces in the memory of this robot, in ascending order: $x_1^-, x_1^1, x_1^2, \ldots, x_1^+$. Similarly, we sort the coordinate x_2: $x_2^-, x_2^1, x_2^2, \ldots, x_2^+$. For simplicity, here and below, we omit the indication that these lists are associated with the robot s_i.

(2) We find in the first list a pair of numbers (x_1^k, x_1^{k+1}) for which the value $x_1^{k+1} - x_1^k$ is the maximum. Similarly, we find a pair of numbers (x_2^l, x_2^{l+1}). As a candidate $\widetilde{\Pi}_{i,1}$, we take a rectangle whose vertices are the indicated points. We calculate the diameter $\tilde{d}_{i,1}$ of the rectangle, as well as the coordinates $\tilde{X}_{i,1}^c$ of its center.

(3) Similarly, we select the regions $\widetilde{\Pi}_{i,2}, \widetilde{\Pi}_{i,3}, \ldots, \widetilde{\Pi}_{i,e}$ and determine their diameters and the coordinates of the centers. The criterion for the end of this process is the condition $\tilde{d}_{i,e} = d(\widetilde{\Pi}_{i,e}) < 2\delta_X$.

Stage 2—narrowing the set of candidates for the region Π_i.

From the set of candidates $\widetilde{\Pi}_{i,1}, \widetilde{\Pi}_{i,2}, \ldots, \widetilde{\Pi}_{i,e}$ we exclude subregions that do not satisfy the restrictions on the speed of the robot's movement: $\left(\tilde{X}_{i,j}^c - X_i\right) \in \Pi_\Delta$. The remaining regions are denoted by $\Pi'_{i,1}, \Pi'_{i,2}, \ldots, \Pi'_{i,r}$; $r \leq e$.

Stage 3—selection of the region Π_i from the candidates $\Pi'_{i,1}$, $\Pi'_{i,2}$,...,$\Pi'_{i,r}$.

(1) If the specified set of candidates for the regions is empty (that is, if $r = 0$), then we skip this iteration (leaving the M-robot in its current position).
(2) If not ($r > 0$), we determine the attractiveness $a'_{i,j}$ of each of the regions $\Pi'_{i,j}$, and on the basis of these values, according to the roulette wheel rule, we select the region Π_i; $j \in [1 : r]$.

Inadmissible approach of M-robots. In the process of evolution of M-robots s_{i_1}, $s_{i_2} \in S$ in the search space the situation of their inadmissible approach is possible when the Euclidean distance between these robots is less than ρ_X: $\|X_{i_1} - X_{i_2}\| < \rho_X$. The rule of behavior of the robots s_{i_1}, s_{i_2} in this situation has the following form.

    (1) M-robots s_{i_1}, s_{i_2} determine the regions Π_{i_1}, Π_{i_2} and, thereby, their attractiveness a_{i_1}, a_{i_2}, and exchange the values of these quantities among themselves.

(3) If $a_{i_1} > a_{i_2}$, then we move the robot s_{i_1} to the center of the region Π_{i_1} (tracing mode), and the robot s_{i_2} continues to evolve in the seeking mode. If not, the robot s_{i_2}, being in the tracing mode, moves to the center of the region Π_{i_2}, and the robot s_{i_1} continues to evolve in the seeking mode.

Landing of C-robots. The operation algorithm of the M-robot $s_i \in S$ includes the rules for selecting the points of time $t_{i,1}$, $t_{i,2}$, ..., $t_{i,e}$ and values $n_C(t_{i,1})$, $n_C(t_{i,2})$, ..., $n_C(t_{i,e})$. We use the mediated choice of values $t_{i,1}$, $t_{i,2}$, ..., $t_{i,e}$: at the point of time $t_{i,j}$ the C-robots land, if the value

$$n_C(t_{i,j}) = \begin{cases} \lceil a_i \rceil, & \lceil a_i \rceil \geq n_C^{rem}(t_{i,j}), \\ n_C^{rem}(t_{i,j}), & \text{else}, \end{cases} \tag{6}$$

is different from zero. Here $n_C^{rem}(t_{i,j})$ is the number of C-robots remaining with the M-robot s_i by the point time $t_{i,j}$; $\lceil \cdot \rceil$ is the symbol of the nearest integer. Thus, the landing number of C-robots $n_C(t_{i,j})$ is proportional to the attractiveness a_i of the region Π_i. The formula (5) means that if the given M-robot runs out of C-robots, then only the M-robot continues the investigation.

Inter-robot communication. At each iteration t, neighboring M-robots exchange information with one another. In this case, each of the robots $s_i \in S$ performs the following actions:

(1) it transmits its extended trace $\mathbf{Tr}_i(t)$ to all robots in the set N_i, as well as extended traces of all the robots which it stores in its memory;
(2) it receives similar information from each of the robots in the set N_i.

Obviously, in technical implementation, one can significantly reduce the amount of information exchanged by robots, by eliminating duplicate transfers.

4.2 The ACSO Algorithm Scheme

Initialization of the population is carried out according to the scheme of the original CSO algorithm (i. 2), with the difference that we declare all $|S|$ robots in the state of investigation. We assign the meaning of the required localization accuracy of the maximum of the fitness function $\varphi(X)$ to the value δ_X.

The investigation mode of the ACSO algorithm differs from the same mode of the CSO algorithm because in this case not all n_C of the satellite robots participate on the iteration t in the local search but only a part $n_C(t)$ of them, which, in general, is different for each of the M-robots. If the robot s_i drops C-robots landing on iterations $t_{i,1}, t_{i,2}, \ldots, t_{i,e}$, then, as the C-robots are non-returnable, the following equality is true:

$$n_C(t_{i,1}) \bigcup n_C(t_{i,2}) \bigcup \ldots \bigcup n_C(t_{i,e}) = n_C.$$

The sequence of the ACSO algorithm changes cardinally in comparison with the CSO algorithm.

Any of the M-robots can switch to the tracing mode in the following two situations:

- in the seeking mode on the iteration t, it turned out that the values of the fitness function at all points $n_C(t)$ differ by no more than by the value δ_φ (the situation is interpreted as localization of one of the contamination sources with accuracy δ_X);
- the Euclidean distance between some two M-robots turned out to be less than the value ρ_X (Fig. 2).

The M-robot s_i in the tracing mode moves to the point with coordinates

$$X_i(t+1) = X_i^c + d_i N_2(0, \sigma), \tag{7}$$

where X_i^c is the center of the region Π_i; $N_2(0, \sigma)$ is a two-dimensional vector of normal random numbers with zero mean and standard deviation which is equal to σ. Movement takes place only if the region Π_i is large enough, that is, if the inequality $d_i > 2\delta_X$ is true.

We should note that rule (7) ensures that in all cases the M-robot $s_i \in S$ does not violate the boundaries of the region Π. In contrast, the C-robots $s_{i,j}, j \in [1 : n_C]$ of a given M-robot can violate the constraint $X_{i,j} \in \Pi$ (obviously, by no more than the value δ_X).

5 Software Implementation and Computational Experiment

The software implementation of the ACSO algorithm was performed in the integrated development environment of Microsoft Visual Studio, which allows to connect third-party extensions at various levels. The object-oriented programming language C # is used, which allows to quickly create a graphical user interface and graphics of varying complexity.

The research was carried out in the multi-start mode (the number of starts is 100) with the following values of the free parameters of the algorithm: $|S| = 10$; $c = 1,0$; $\tilde{m} = 8$; $\delta_X = \rho_X = \Delta x_i = \frac{d}{100}, i = 1, 2$; $r_X \approx \frac{d}{10}$; $n_C = 25$; $\hat{t} = 100$; $\delta_\phi = 0,01\,(\varphi_{\max} - \varphi_{\min})$, where the minimum and maximum possible values of the fitness function $\varphi_{\max}$, $\varphi_{\min}$ should be evaluated before the beginning of the solution of the problem; $\sigma \approx \frac{d}{1000}$; restrictions ΔX^-, ΔX^+ on the speed of the M-robot s_i are determined from the conditions $\Delta x^-_{i,k} = 0,5\,x^-_k$, $\Delta x^+_{i,k} = 0,5\,x^+_k$; $i \in [1 : |S|]$, $k = 1, 2$.

Objective functions:

- function that is inverse to Himmelblau's function;

$$x^-_k = -6, x^+_k = 6, k = 1, 2;$$

- function that is inverse to Rastrigin's function

$$x^-_k = -0,5, x^+_k = 0,5, k = 1, 2.$$

Figure 4 presents estimated probability of localizing the maxima of a function inverse to the four extreme Himmelblau's function for the number of iterations $\hat{t} = 50, 75, 100$ and parameter values $\lambda = 0.25, 0.50$. Proximity of the computed value of the objective function to its exact maximum value with an error not exceeding the value δ_φ was used as the criterion of localization. We remember that small values of the value λ intensify search, and values close to one diversify it.

Figure 4 shows that the *ACSO* algorithm in all its 100 starts provides, with a specified accuracy, localization of the global maximum of the objective function. Estimated probability of localization of all four maxima reaches 0.8.

Figure 5 shows similar results for the function that is inverse to Rastrigin's function which has one global and eight local maxima in the search space. We used two criteria for localizing the maximum of the objective function—based on the value δ_φ and value δ_X.

Figure 5 shows that even for a complex multimodal function, the ACSO algorithm can provide probability of localizing the global maximum of the objective function close to one at relatively small iteration numbers (50, 75).

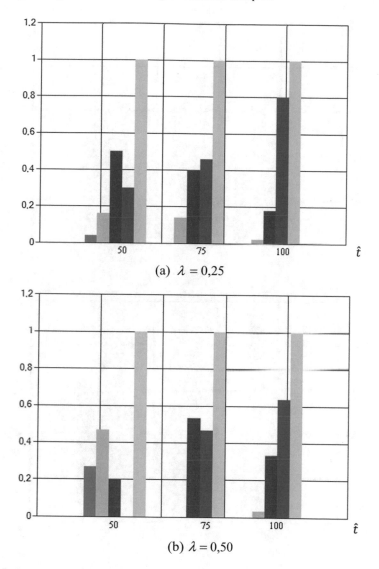

(a) $\lambda = 0{,}25$

(b) $\lambda = 0{,}50$

Fig. 4 Estimated probability of localization the maxima of the function that is inverse to Himmelblau's function: ▮, ▮, ▮, ▮—one, two, three, four maxima, respectively; ▮—at least one maximum

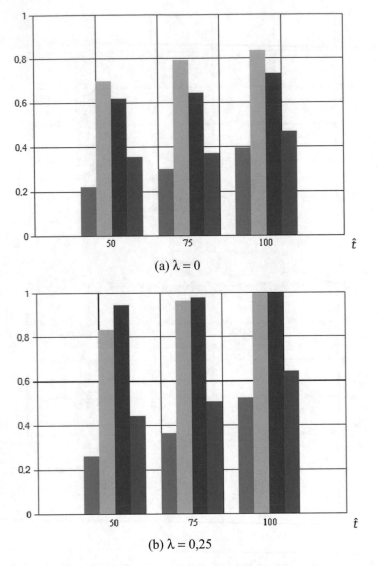

(a) $\lambda = 0$

(b) $\lambda = 0,25$

Fig. 5 Estimated probability of localizing the maxima of a function that is inverse to Rastrigin's function: ■, ▨—the global maximum and at least one local or global maximum, respectively (the localization criterion based on δ_φ); ■, ■—one maximum and two maximums respectively (the localization criterion based on δ_X)

6 Conclusion

The choice of the *CSO* algorithm is due to the simplicity of its modification which takes into account the peculiarities of the hardware implementation of the group of the robots that we use. The deep modification of the *CSO* algorithm proposed in this paper was named *ACSO* (*Advanced CSO*). The *ACSO* algorithm includes a significant number of free parameters that allow a broad range of balancing diversification and intensification properties of the algorithm. Thus, *ACSO*, is, in fact, metaheuristic.

The computational experiment showed rather high efficiency of the ACSO algorithm even in the case when the scalar physical field under test is determined by the complex Rastrigin's multimodal objective function.

We believed that the results of tests of the M- and C-robots are accurate. In the development of the work, the authors propose to investigate the situation when the measurement results are distorted by additive noise or measurement errors with known statistical characteristics. It is also proposed to expand the algorithm for the dimension of the search space equal to three. In addition, it is important to study the effectiveness of the asynchronous variant of the ACSO algorithm.

References

1. Pettersson, L.M., Durand, D., Johannessen, O.M., Pozdnyakov, D.: Monitoring of Harmful Algal Blooms. Praxis Publishing, London (2012)
2. White, B.A., Tsourdos, A., Ashokoraj, I., Subchan, S., Zbikowski, R.: Contaminant cloud boundary monitoring using UAV sensor swarms. In: Proceedings of the AIAA Guidance, Navigation, and Control Conference, pp. 1037–1043, San Francisco, USA (2005)
3. Hayes, A.T., Martinoli, A., Goodman, R.M.: Distributed odor source localization. IEEE Sens. J. 2(3), 260–271 (2002)
4. Lilienthal, A., Duckett, T.: Creating gas concentration grid maps with a mobile robot. In: Proceedings of the IEEE/RSJ Conference on Intelligent Robots and Systems, pp. 118–123, Las Vegas, USA (2003)
5. Xing, B., Gao, W.-J.: Innovative Computational Intelligence: A Rough Guide to 134 Clever Algorithms, 451 p. Springer International Publishing Switzerland (2014)
6. Sharkey, A.J.C.: Swarm robotics and minimalism. Connection Sci. 19(3), 245–260 (2007)
7. Swarmanoid: Towards Humanoid Robotic Swarms. http://www.swarmanoid.org/publications_byyear.php
8. Vaishak, N.L., Shilpa, B.: A review of swarm robotics: a different approach to service robot. Int. J. Sci. Eng. Technol. Res. (IJSETR) 2(8), 1560–1565 (2013)
9. Chu, S.C., Tsai, P.W., Pan, J.S.: Cat Swarm Optimization, LNAI 4099, 3(1), 854–858 (Berlin, Heidelberg: Springer-Verlag)
10. Orouskhani, M., Mansouri, M., Teshnehlab, M.: Average-Inertia weighted Cat Swarm Optimization. LNCS, pp. 321–328. Springer, Heidelberg
11. Orouskhani, Meysam, Orouskhani, Yasin, Mansouri, Mohammad, Teshnehlab, Mohammad: A novel cat swarm optimization algorithm for unconstrained optimization problems. I.J. Inf. Technol. Comput. Sci. 11, 32–41 (2013)

12. Crawford, B., Soto, R., Berríos, N., Johnson, F., Paredes, F., Castro, C., Norero, E.: A binary cat swarm optimization algorithm for the non-unicost set covering problem. Math. Probl. Eng. **2015**, 8 (2015)
13. Kanwar, N., Gupta, N., Swarnkar, A., Gupta, N.: Improved cat swarm optimization for simultaneous allocation of DSTATCOM and DGs in distribution systems. J. Renew. Energy **2015**, 10 (2015)
14. Nie, X., Wang, W., Nie, H.: Chaos quantum-behaved cat swarm optimization algorithm and its application in the PV MPPT. Comput. Intell. Neurosci. **2017**, 11 (2017)
15. Tsai, P.-T., Pan, J.-S., Chen, S.-M., Liao, B.-Y., Hao, S.-P.: Parallel cat swarm optimization. In: Proceedings of the Seventh International Conference on Machine Learning and Cybernetics, Kunming, vol. 6, pp. 3328–3333, 12–15 July 2008
16. Saha, S.K., Ghoshal, S.P., Kar, R., Mandal, D.: Cat swarm optimization algorithm for optimal linear phase FIR filter design. ISA Trans. **52**(6), 781–794 (2013)
17. Kotekar, S., Kamath, S.S.: Enhancing service discovery using cat swarm optimization based web service clustering. Perspect Sci **8**, 715–717 (2016)
18. Khalaf, A.H., El-Bakry, H.M., Sabbeh, S.F.: University courses scheduling using cat swarm optimization algorithm. Int. J. Adv. Res. Comput. Sci. Technol. (IJARCST 2016) **4**(1), 2347–8446 (2016)
19. Orouskhani, M., Teshnehlab, M., Nekoui, M.A.: Integration of cat swarm optimization and Borda ranking method for solving dynamic multi-objective problems. Int. J. Comput. Intell. Appl. **15**(03), 18 (2016)

Distributed Solution of Problems in Multi Agent Robotic Systems

Anaid V. Nazarova and Meixin Zhai

Abstract Over the past decade, multi-agent systems (MAS) have become widespread, especially in the context of advances in smart electromechanical systems (SEMS) and the solution of distributed problem in swarm control. The properties of agents, such as autonomy and reactivity, as well as the possibility of dynamic interaction, make it possible to implement their cooperative actions in achieving common goals, and motivate SEMS a matter of serious scientific interest. Multi-agent robotic systems, combining different specialization of robot-agents, are able to accomplish assignments without any external interference, which ensure high reliability and adaptability of such systems. In order to successfully complete a common task, robot-agents must conduct complex negotiations, cooperate and coordinate their actions with each other. Each part of multi-agent robotic systems is impossible achievement common goal without the dynamic redistribution of tasks between agents in changing environmental. *Purpose of research*: The main approaches to the construction of models for the distribution of tasks among the MAC were analyzed. The various algorithms were compared and their mathematical modeling were established. *Results*: The centralized and decentralized methods of distribution tasks among agents were investigated with the aim of achieving an optimal result in minimal time and no conflicts in SEMS. *Practical significance*: The presented algorithms can be used to control a multi-agent system, considered as SEMS, especially to complete tasks, which are critical to the execution time, such as search and rescue operations in the case of natural or man-made disasters.

Keywords Multi-agent systems · Robot · SEMS · Static and dynamic distribution of tasks · Dynamic models · Swarm intelligence · Ant colony algorithm and genetic algorithm · Auction algorithm

A. V. Nazarova (✉) · M. Zhai
Bauman Moscow State Technical University, Moscow, Russia
e-mail: avn@bmstu.ru

M. Zhai
e-mail: 982696853@163.com

© Springer Nature Switzerland AG 2019
A. E. Gorodetskiy and I. L. Tarasova (eds.), *Smart Electromechanical Systems*,
Studies in Systems, Decision and Control 174,
https://doi.org/10.1007/978-3-319-99759-9_9

107

1 Introduction

The agents are based on SEMS [1] are a kind of cyber physical systems [2–4]. A distinctive feature of such complexes is parallelism in obtaining and processing information, in calculating and forming control actions and in performing various movements similar to the central nervous system of human. From the point of view of information control systems, a group of such robots can be considered as a group of "cognitive information control systems built into their environment" [5]. Consequently, such devices should have two-way communication with the outside world and other agents involved in-group interaction [6], which is inherent to SEMS. The main feature of agents on based SEMS in solving complex tasks of group control is the ability of intelligent processing of information available to them for making decisions about preferred actions in a changing situation.

In many publications, team of agents is regarded as a system, and each agent has its own behavior, the group of such agents is called multi-agent systems (MAS). Often the execution of complex tasks requires group of agents of autonomous operation without any outside interference, which causes high reliability and adaptability of such systems. To accomplish any task, it must have to be allocated to the agents. Therefore, we consider various approaches to the problem of distribution of tasks among agents with the goal of achieving an optimal result with minimal time and no conflicts in the system, such as swarm intelligence, evolutionary method, market economy principles and others. At the same times, to perform the task correctly, agents must conduct complex negotiations, cooperate and coordinate their actions with each other. The functioning of team members in a changing environment is impossible without the dynamic redistribution of tasks between agents.

We consider the problem of the distribution of a certain set of tasks; the purpose of distribution is to ensure that there is no conflict between agents and to minimize expenditure of their resources. In order to describe the problem, we introduce the following notation: $M = \{m_1, m_2, \ldots, m_k\}$,—the set of tasks (can change during the process of execute task), $N = \{n_1, n_2, \ldots, n_p\}$—the set of robot agents (also can change). In addition, the matrices $Q = \{q_{ij}\}$, $R = \{r_{ij}\}$ and $L = \{l_{ij}\}$ have the dimension of $k \times p$, where q_{ij} is the reward for the fulfillment of the subtask $i \in \{1, \ldots, k\}$ by the agent $j \in \{1, \ldots, p\}$, r_{ij}—the resources which have spended, and l_{ij}—the possibility of performing task i by the robot j, $l_{ij} = \{0,1\}$. It is necessary to distribute the tasks M between the agents N to make $C = \sum_{j=1}^{p} \sum_{i=1}^{k} (l_{ij} * (q_{ij} - r_{ij}))$ maximal.

After establishing the "task-agent" correspondences, new tasks can appear or the number of agents maybe changed during the process of fulfilment task, which leads to the need for redistribution of dynamic tasks. Thus, in the MAS there are two types of tasks distribution: static and dynamic. In addition, according to the types of organization of system, they are divided into centralized [7], decentralized and hybrid [8]. Centralized distribution supposed the presence of a control center in the team whose role can be played either by a team member or a separate control section that commands the processes and agents and also collects global information. When a new task appears, the control center forms a price array containing the estimated costs of the agent-robots (for example, time or distance), and then, depending on the functionality, distributes the tasks among the robots in such a way as to minimize costs [9]. At the same time, such approaches as search algorithms, linear programming, swarm intelligence [10, 11] and others. With a method of decentralized distribution, both the single control center and the leader among the objects are absent. In the distribution of tasks involved all agents who, in the process of interaction, coordinate their actions, compete and jointly solve complex problems. For this type of distribution, game theory, Markov solutions, auction [11–15] etc. are used. Hybrid type is a combination of the above two types and breaking up the team into groups taking into account the specifics of the tasks.

With the increasing complexity of tasks that are fulfilment by large teams of robots, bionic and intelligent distribution algorithms become more and more important. Therefore, this article discusses the methods of swarm intelligence, at the same time, assesses market based approaches in the problem of task allocation.

2   Distribution the Tasks Based on Method of Swarm Intelligence

Swarm intelligence (Swarm Intelligence) was introduced by Gerardo Beni and Jing Wang in 1989 [16]. Swarm intelligence is a self-organizing multi-agent system in which the interaction of agents with the environment is governed by simple rules of behavior. Examples of such interactions are the ant algorithm, particle swarm optimization (PSO) and the genetic algorithm. These principles are applied to the centralized task allocation.

2.1   The Ant Algorithm

The ant algorithm [17, 18], which is based on the behavior of the colony of ants, was proposed by Marco Dorigo in the mid-90s of the twentieth century. This algorithm is one of the variants of solving the problem of traveling salesman and remains relevant for more than 20 years.

The colony of ants is able to find the shortest path from an anthill to a source of food (see Fig. 1), while the team adapts to changing environmental conditions and is able to find a new optimal path.

Due to the fine interaction between the individuals in the group, which is carried out by a special chemical substance—pheromone. The choose a path of ants is guided not only by their desires, but also by the experience of other individuals, the

A. V. Nazarova and M. Zhai

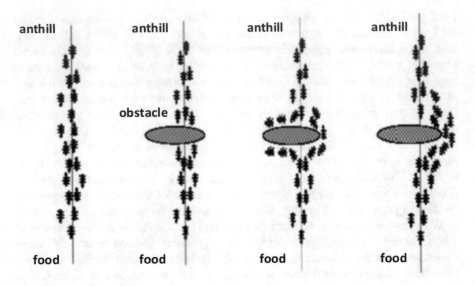

Fig. 1 The shortest path from an anthill to food

knowledge of which it receives, evaluating the "paths" of pheromones on each path. Since the concentration of pheromone differs in each way, it is possible to calculate the most "attractive" variant. The movement to the goal and back takes relatively little time, the trace of the pheromone along the short path remains brighter, and makes all new and new individuals choose it, which ultimately determines the unambiguous priority of the particular path, which is optimal.

To allocation task by method, the following parameters are taken into account: the food search space, the colony number (a set of effective solutions in the search space), the pheromone concentration, all paths from the anthill to the food source, the only optimal path.

In technical systems, the primary path is formed by "greedy" algorithm that ensures the adoption of locally optimal solutions at each stage of the search, and assuming that the final solution is also optimal. Then a shorter path is chosen in process of iteration. The advantage of this method is highly stable, globally searchable, using parallel and distributed computation. In addition, it is well combined with other algorithms and fast and easy convergence, which speeds up the optimization process. However, in some cases, this algorithm can only obtain local optimal solutions and the time of convergence becomes uncertain in many global problems. For this reason, the ant algorithm is often used in combination with other algorithms.

2.2 Particle Swarm Optimization

The algorithm of the particle swarm optimization (POS) is one of the evolutionary optimization methods which was proposed by James Kennedy and Russell Eberhart [19] in 1995 based on the research and modeling of bird behavior in the pack (see Fig. 2).

The method is based on the principle of the random distribution of a flock of birds in space. Each bird knows its own distance to food and the distances to the food of all other birds in the pack. At first, the birds have a random speed (module and direction) then, they adjust the speed, moving to the bird closest to the food.

The observation of birds foraging inspired Craig Reynolds to create a computer model in 1986, which he named Boids. In order to simulate the behavior of a flock of birds, Reynolds programmed the behavior of each bird separately, as well as their interaction. At the same, he used three simple principles: first, every bird in his model tries their best to avoid collisions with other birds; second, each bird moves in the same direction as the nearby birds; third, the birds try to move to keep the same distance from each other (see Fig. 3).

The result of simulation surprised the author: although rules of algorithm are very simple, the flock on the screen of the monitor looked extremely plausible. Birds were huddled together in groups, escaped from collisions and rushed chaotically just like real ones. As a specialist in computer graphics, Reynolds was interested in the results of the program he created. However, in the Boids article, he noted that the behavioral model developed could be expanded by the additional factors such as searching for food.

The optimal solution obtained after a certain number of iterations, at each step of the iteration the particle (the bird) updates its position, striving for its own best solution (pbest)—the local best solution of the particle and a better solution among all the swarm particles (gbest)—a global best solution. Then the correction of the

Fig. 2 The behavior of the birds in the flock

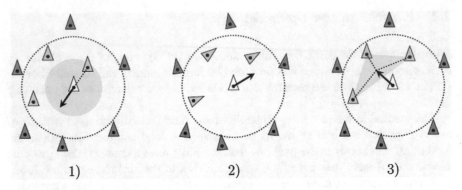

Fig. 3 Rules of behavior of birds in the Reynolds model

velocity and position of the particle will be gotten according to the following formulas [14]:

$$v_{i+1} = v_i + a_1 \cdot rnd() \cdot (pbest_i - x_i) + a_2 \cdot rnd() \cdot (gbest_i - x_i)$$

$$x_{i+1} = x_i + v_{i+1}$$

where v_i is the velocity of the current particle (bird i), x_i is the position of the current particle, a_1 and a_2 are the acceleration, $pbest_i$ is the best position of particle i, $gbest_i$ is the best position by all particles, $rnd()$ is random number from 0 to 1.

In consideration of task allocation in the MAS, this method is implemented in the following sequence of steps:

(1) Stochastic initialization of the "particle swarm"—agents are distributed randomly in the entire search area and each has a random speed and direction;
(2) Calculation the value of the objective function- it is the distance between agents, and then determine the local and global optimal solutions based on objective function;
(3) Adjusting the position of the agent so that it does not go beyond the area of the search;
(4) Checking the conditions of termination—all agents are at equal distances from each other and move at the same speeds. If the conditions are met, the search ends, otherwise return to step 2.

This algorithm is very simple, relatively small number of parameters, as well as high accuracy and fast convergence. However, with the increase in the number of iterations, the rate of adjustment becomes lower and lower, as a result, all the particles are collected either near the "optimal" particle or in some local extremum, which is caused by the premature convergence of this algorithm.

2.3 The Genetic Algorithm

In 1975, John Henry Holland published his most famous work "Adaptation in Natural and Artificial Systems", in which, he first introduced "genetic algorithm" [20]. The algorithm based on a principle borrowed from natural biological evolution. Currently, the genetic algorithm is one of the most popular method; it is effectively used to solve global optimization problems.

At the beginning of the algorithm, the initial decision space is defined. Then, in the process of evolution new generations appear, the search of best solution is happened in the possible space of solutions. In the genetic algorithm of Holland, each parameter of the fitness function is encoded by a string of bits (see Fig. 4).

After that, selections and crossings are performed (see Fig. 5a), for which a single-point crossover (1-point crossover) and mutations (Fig. 5b) are used. At the same time, evolution is designed to preserve the positive characteristics of the parents in subsequent generations.

The process of applying the genetic algorithm for task allocation among the MAC through the following stages (see Fig. 6):

(1) Creation of the initial population - formation of a set of characteristic parameters;
(2) Setting the objective function (fitness) for the "individual" of the population— in the task allocation problem, fitness can be viewed as the expenditure of time or resources by agents, the distance between agents and goals, or the total time or material costs to perform the task;
(3) Selection, crossing and mutation for the creation of new "generations";
(4) Checking the conditions of termination, either a given number of "generations" or a convergence of the population process can be considered.

Fig. 4 Coding parameters   1010   10110   101   ...   10101

$$| \ x_1 \ | \ x_2 \ | \ x_3 \ | \ ... \ | \ x_n \ |$$

$$\left. \begin{array}{l} 011010.01010001101 \\ 111100.10011101001 \end{array} \right\} \Rightarrow \left\{ \begin{array}{l} 111100.01010001101 \\ 011010.10011101001 \end{array} \right.$$

(a) The application of a single-point crossover operator

10110011000101101 -> 10110011010101101

(b) Inverting the bit of each individual in the population with a certain probability

Fig. 5 Operations on genes

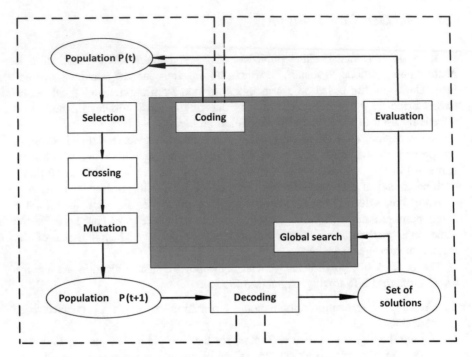

Fig. 6 Process of genetic algorithm

The genetic algorithm is used to find universal optimal solutions and optimize multiparameter functions, but it has very low efficiency in finding exact solutions. Adaptive genetic algorithm to the MAS allows to gain global optimization in the centralized task allocation and to find the best solution through iterations. However, for dynamic task allocation this algorithm is not suitable, which forces us to turn to the market based approaches.

3 Task Allocation with Market Based Approaches

Recently, a new mechanism for distributed control "market algorithm" based on the principles of market economy, has emerged. The essence of the mechanism is that agents in the MAS dynamically distribute tasks among themselves through negotiation, harmonization and competition. These processes are regulated by rules and protocols, and their goal is to acquire the maximum "profit" for the team as a whole or for each participant personally. These methods include the contract network protocol, auction algorithms, game theory and others.

3.1 The Contract Network Protocol

The contract network protocol [21] is an algorithm that allows to dynamically distribute tasks in the MAC. Agents are nodes that form a contract network. Each node, depending on a particular task or within a single subtask, can perform the function of a manager or contractor (performer).

The operation of the algorithm is described as follows:

(1) Announcement: if a node for some reason cannot cope with a particular task, if possible, breaks the task into subtasks and announces them to other agents, at the time of the announcement, this node becomes a "manager", and other agents "contractors", that is, executors;
(2) Bidding: "contractors" decide for themselves whether they should offer a price for the task;
(3) The choice of the performer: the "manager" evaluates the received proposals and decides on whom to contract with, the decision is communicated to all "contractors" who proposed the price;
(4) Sign the contract: the selected "contractor" concludes the contract and execution the task.

The described process is shown in Fig. 7.

If the agent—the "contractor" who concluded the contract, for some reason cannot fulfill the task (for example, in case of a breakdown), then he becomes a "manager" and repeats the process of above. Thus, any agent can be either a manager or a contractor, thus forming a hierarchical network of the contract network.

This algorithm is effective in the dynamic task allocation, but its effectiveness is significantly reduced when the network is expanded, that is, increase in the number of agents. In addition, in cases where several agents simultaneously announce tasks, the process of decision making in the system is significantly complicated. Another problem of this method is unilateral decision making, that is, the "free" contractor cannot stimulate its services to the "manager". Elimination of these shortcomings requires the development of new, improved protocol algorithms.

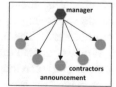 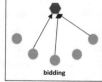 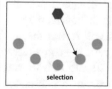 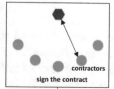

Fig. 7 Working process of the contract network protocol

3.2 Auction Algorithm

The auction is a competitive sale of goods according to pre-established rules. From the point of tasks allocation, this method can be attributed to centralized algorithms. Nowadays, there is a wide variety of auctions: English and Dutch auctions, auctions of the first and second prices and others.

In the case of the MAS, the process of task allocation using the auction in the following steps: the team of robots is divided into auctioneers, which are responsible for announcing tasks and concluding contracts, and participants, reacting to the tasks reported by auctioneers. After receiving information about the task, participants analyze it, assess their capabilities and usefulness, and decide whether to participate in the competition for the task. Then, the auctioneer chooses the best bidder for the specific task. In cases where the task is too complex or too large for one participant, it becomes necessary to form coalitions.

The auction algorithm is an effective method for tasks allocation in the MAS. It is distinguished by great flexibility, as well as the possibility of dynamic task redistribution. However, it does not always provide a global optimal solution, and with the increase of the number of agents, its effectiveness drastically decreases.

In addition to the approaches discussed above, there are also contract theory, the Markov decision algorithm, and others. They are distinguished by high flexibility and the ability to combine with other methods to achieve the best result.

3.3 Game Theory

Game theory is a branch of mathematics related to the study of optimal strategies in games and prediction of the behavior of participants in games. A game is understood as a process in which two or more parties participate in the struggle for the realization of their interests. Each of the parties pursues a certain goal and has a certain strategy leading to a win or lose, depending on the behavior of the other players. The optimal strategy is Nash's equilibrium, which means that the player always chooses the action for gain the best utility in response to the opponent's action.

In the case of a multi-agents robotic system, when distributing tasks among agents, we will consider robots as players, methods for task allocation as strategies, and payment matrices will reflect the utility of robots in performing a task. There are two main types of games—cooperative and non-cooperative. With a cooperative game, the goal of task allocation is to maximize the overall interest or minimize the total expense. In the case of a non-cooperative game, each agent plays "for himself", striving for a single goal—maximizing his own interest. For the latter case, the optimal strategy is determined by the Nash equilibrium, this means that if the player decides to deviate from such a strategy, he will minimize the gain. For this reason, it is necessary to find the Nash equilibrium in the problem of task allocation.

For this, all possible strategies and their usefulness are determined, and then they select a strategy that will bring maximum utility.

In [13] Chapman used game theory to distribute tasks between robots. Chapman viewed this process as a Markov process, then, led this continuous process to a potential statistical game for which he calculated the global utility functions and payment functions of each robot. Consistently finding equilibrium processes in similar games, which is possible to ensure an optimal task allocation. The use of this method on RoboCup Rescue by Chapman is an excellent confirmation of its success.

4 Investigation the Efficiency of Algorithms

In the previous sections, theoretical approaches to the problem of task allocation in multi-agent robotic systems were discussed, now we turn to the examples of the implementation of static and dynamic task allocation. In order to compare the efficiency of algorithms, the computer simulation was performed to study the rate of convergence of various methods, as well as to determine the time for the MAC to distribute a certain number of tasks between a given numbers of robots.

4.1 Static Distribution of Problems Using the Swarm Algorithm and the Genetic Algorithm

As follows from the statement of the problem, the aim of our research is to ensure the optimal distribution of the set of problems M between the available number of robot-agents N taking into account the benefit or reward of each robot and the resources spent by it. It is necessary to distribute the tasks in such a way that each robot, in the performance of the chosen task, aspires to achieve the least possible expense and greater utility (gain), taking into account the capabilities and "interests" of other robots, that is, the total utility of the group of robots should be maximum.

Let six robots $P = 6$ together perform k tasks, and each task can be performed by any robot independently.

A. Simulation of the particle swarm optimization (POS) with different number of tasks and iterations:

Table 1 The number of tasks $k = 20$

Number of iterations characteristic	50	100	300	500	1000	2000	3000
Utility(expend)	487.9	365.4	328.1	313.5	298.2	286.9	279.1
Efficiency (%)	–	25.1	10.2	4.45	4.88	3.79	2.78
Modeling time/s	0.99	1.35	3.26	4.99	10.35	17.2	24.06

Table 2 The number of iterations = 2000

Number of tasks characteristic	5	10	20	30	50	100	200
Modeling time/s	11.09	12.9	15.2	19.6	28.7	48.5	85.3
Rate of increase in time (%)	–	16.3	17.8	28.9	46.4	69.0	75.9

Table 3 The number of tasks $k = 20$

Number of iterations characteristic	50	100	300	500	1000	2000	3000
Utility(expend)	669.3	659	625.5	615.1	600.74	589.1	572.6
Efficiency (%)	–	1.5	5.1	1.66	2.3	1.9	2.8
Modeling time/s	1.5	3.1	7.1	11.2	19.7	28.3	37.8

B. Simulation of the genetic algorithm (GA) with a different number of tasks and iterations:

Comparative analysis: when comparing Tables 1 and 3 in (Fig. 8a, b), it becomes obvious that for the same number of iterations, the POS converges faster and takes less time than the GA. Comparison of Tables 2 and 4 in (Fig. 9) shows that the speed of the algorithms decreases sharply with the increase in the number of tasks. Thus, with the static distribution of tasks between the robots, the POS is better than GA.

4.2 Dynamic Task Allocation Based on the Auction Algorithm and Game Theory

In order to compare the efficiency of dynamic task allocation algorithms in MAS, we consider the following model problem. Let several cargoes are placed on a limited area, the locations of which are unknown. On the same area, there is a certain number of robot-agents, who need to transport the goods to the specified places. It is assumed that the MAS control center provides a continuous search for cargo. Once the load is found, the information from the center is transferred to the agents, they cooperate, interact with each other to collect and transport the goods. In this case, robots can perform tasks individually, if the weight of the cargo allows it or form a coalition, if the cargo is too large (Table 5).

As noted in the auction method, robots are participants, and the detected cargo is a "commodity" for which they compete. When a cargo is detected, the control center sends information to all free agents about the location of the cargo, and simultaneously one or more tasks can be offered. Participants in the auction analyze the features of the task, assess their capabilities and the time of the task takes, and

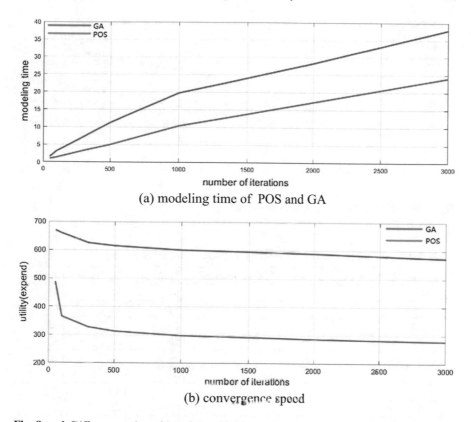

(a) modeling time of POS and GA

(b) convergence speed

Fig. 8 a, b Different number of iterations of POS and GA

Table 4 The number of iterations = 2000

Number of tasks characteristic	5	10	20	30	50	100	200
Modeling time/s	21.0	23.2	28.3	35.8	47.8	84.3	122.3
Rate of increase in time (%)	–	10.4	21.98	26.5	33.5	76.3	45.1

send requests to the managing center, which decides which agent to complete this task.

The process of task allocation is as follows. Each agent Ai has two states "free" and "busy", agent always knows its coordinates and its capacity vector C_i^r, whose dimension is r: $C_i^r = \left(c_i^1, c_i^2, \ldots, c_i^r\right)$, whose elements $c_i^k \geq 0, k = 1, 2, \ldots, r$, correspond to the ability to perform tasks of robot-agents. Each task T_j is connected with the vector C_i^r, whose dimension is r: $C_j^r = \left(c_j^1, c_j^2, \ldots, c_j^r\right)$, $c_j^k \geq 0, k = 1, 2, \ldots, r$, characterizing the needs to complete the task. For example, in order to transport the load from the point (4, 6) to the point (0, 0), and for this task need two

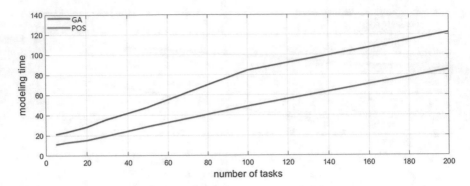

Fig. 9 Different number of tasks of POS and GA

Table 5 Cost matrix (squares of distances between cargo and robot-agents)

Tasks, coordinates of cargoes agents, initial coordinates	T_1, (2,5)	T_2, (4,10)	T_3, (3,7)
A_1, (10,5)	64	61	53
A_2, (6,7)	20	13	9
A_3, (3,12)	50	5	25
A_4, (7,0)	50	109	65
A_5, (0,10)	29	16	18
A_6, (4,5)	4	25	5
A_7, (3,9)	17	2	4
A_8, (7,7)	29	18	16

robots $\left(c_j^k = 2 \right)$, then the parameters of the task j will have the form T_j = <(4, 6), (2)>. Each agent can move in any direction at a constant rate, assumed that agents do not collide with each other.

If there are several tasks, the control center sorts the tasks in order of increasing needs. If the task can be performed by one robot, then the agent receiving the greatest reward in performing this task is chosen, otherwise agents form coalitions so that the coalition's "ability" vector is greater than the "needs" vector of the task.

For example, ten isomorphic robot-agents A_1, A_2 ... A_8 have the same ability vectors: $C_1^1 = C_2^1 = \ldots = C_8^1 = (1)$. The control center detected three loads (three tasks) with the following parameters: T_1 = <(2,5), (1)>, T_2 = <(4,10), (4)>, T_3 = < (3,7), (3)>, that is, *1* robots are required to move the 1st load, *4* robot is required to move the 2nd load, *3* robots are required for the 3rd load. The task of the agents is to transport the loads to the point with coordinates (0, 0).

A. Application of the auction algorithm

The auction procedure is iterative and includes the next steps:

① The control center reports the information of cargos to the robot-agents;
② Agents send their information to the center—state (free/busy) and coordinates;
③ The control center starts to deal with the data like this: (a) forms a cost matrix—squares of distances from robots to cargo, (Table 6). (b) Sorts all tasks in order of increasing needs: T_1, T_3, T_2. (c) Since one agent can perform task T_1, agent A_6 is selected for a minimum of costs, its status becomes "busy", and in the future, it does not participate. For task T_3, three agents are needed, agents A_8 and A_9 are selected, their statuses are changed to "busy". For task T_4, four agents are required, and Agents A_1, A_2, A_6 and A_{10} are selected from the remaining "free". The result of the task allocation is presented in Table 6.
④ The control center reports the results to agents, and agents begin to perform tasks.

In this example, the process of tasks allocation based on auction algorithm. However, with the increasing number of tasks, the computational speed will reduce, so the algorithm must be updated. In addition, the algorithm usually can only get a local optimal solution.

B. The application of game theory

Given the previous problem, we suggest using the methods of game theory instead of step 3 in the auction algorithm.

Assume that all agents are "reasonable", they know the decisions of other agents, and all agents have one goal, which is to reduce costs. In this example, this is the sum of the distances $U = \sum_{j=1}^{3} u_j$, where U is the global payment function, u_j is the payment function for the task j. Thus, we have a reasonable game with known information and our goal is to find the "Nash equilibrium", that is, the smallest of all possible solutions $U^* : U^*(u_1^*, u_2^*, u_3^*) \leq U(u_1, u_2, u_3)$.

In the example above, each task has a set of pure strategies:

$$T_1 = \{(A_1), (A_2), (A_3)\ldots\}$$

$$T_2 = \{(A_1, A_2, A_3, A_4), (A_1, A_2, A_3, A_5), (A_1, A_2, A_3, A_6)\ldots\}$$

$$T_3 = \{(A_1, A_2, A_3), (A_1, A_3, A_4), (A_1, A_4, A_5)\ldots\}$$

Table 6 The result of the task allocation

Tasks	Coordinates of cargoes	Needs (number of robots)	Results—robots' numbers
T_1	(2,5)	1	A_6
T_3	(3,7)	3	A_2, A_7, A_8
T_2	(4,10)	4	A_1, A_3, A_4, A_5

Table 7 The results of the task allocation

Tasks	Coordinates of cargoes	Needs (number of robots)	Results—robots' numbers
T_1	(2,5)	1	A_6
T_2	(4,10)	4	A_2, A_3, A_5, A_7
T_3	(3,7)	3	A_1, A_6, A_8

To find the best group strategy, need to go over all possible group strategies, which is too much work. However, in our "game" agents choose tasks that require the least expenditure of their resources, so the best solution is:

$$U_1 = \{u_1(A_6), u_2(A_2, A_3, A_5, A_7), u_3(A_2, A_6, A_7)\} = 58.$$

However, since A_2, A_6, A_7 agent can perform only one task at one time, it is necessary to look for other options:

$$U_2 = \{u_1(A_6), u_2(A_2, A_3, A_5, A_7), u_3(A_1, A_6, A_8)\} = 114$$
$$U_3 = \{u_1(A_6), u_2(A_3, A_4, A_5, A_6), u_3(A_2, A_7, A_8)\} = 188$$
$$U_4 = \{u_1(A_4), u_2(A_2, A_3, A_5, A_7), u_3(A_1, A_6, A_8)\} = 160$$
$$U_5 = \{u_1(A_8), u_2(A_1, A_3, A_4, A_5), u_3(A_2, A_6, A_7)\} = 238$$
$$U_6 = \{u_1(A_4), u_2(A_1, A_3, A_5, A_8), u_3(A_2, A_6, A_7)\} = 168$$

It is easy to see that the variant U_2 is better, which is the "Nash equilibrium" (Table 7). Thus, the result of task allocation for our example is:

$$U^*\left(u_1^*, u_2^*, u_3^*\right) = \{u_1(A_6), u_2(A_2, A_3, A_5, A_7), u_3(A_1, A_6, A_8)\}$$

Comparing Tables 6 and 7 we see that the results are different, game theory is better than auction algorithm. Using game theory can get global optimal solution, but it takes a long time to negotiate between agents. Therefore, when the numbers of tasks and agents are not too many, game theory should be used to obtain global optimal solution, in other cases, auction algorithm should be used.

5   Conclusion

In this article, we consider and compare various algorithms and methods for task allocation among agents. The advantages and disadvantages of centralized and decentralized distribution methods are noted. These methods of distribution tasks among agents were investigated with the aim of achieving an optimal result in minimal time and no conflicts in SEMS. Examples of the distribution of tasks among agents are given. The results of simulation showed the capabilities of each method and revealed the greater efficiency of decentralized methods in the work of

MAS in a dynamically changing environment. The presented algorithms can be used to control a multi-agent system, considered as SEMS, especially to complete tasks, which are critical to the execution time, such as search and rescue operations in the case of natural or man-made disasters.

References

1. Gorodetskiy Andrey, E. (ed.): Smart electromechanical systems. In: Studies in Systems, Decision and Control, vol. 49, 277 p. Springer International Publishing Switzerland (2016). https://doi.org/10.1007/978-3-319-27547-5_4
2. Lee, E.: Cyber Physical Systems: Design Challenges. University of California, Berkeley Technical Report No. UCB/EECS-2008-8. Retrieved 07 June 2008
3. NSF Cyber-Physical Systems Summit. Retrieved 01 Aug 2008
4. NSF Workshop On Cyber-Physical Systems. Retrieved 09 June 2008
5. Fridman, A.Y.: SEMS-Based control in locally organized hierarchical structures of robots collectives. In: Gorodetskiy, A.E., Kurbanov, V.G. (eds.) Studies in Systems, Decision and Control, vol. 95, pp. 31–47, 270p. Smart Electromechanical Systems: The Central Nervous System. Springer International Publishing Switzerland (2017). https://doi.org/10.1007/978-3-319-53327-8_3
6. Gorodetsky, A.E., Tarasova, I.L., Kurbanov, V.G.: Safe Control of SEMS at Group Interaction of Robots. Materialy 10-j Vserossijskoj mul'tikonferencii po problemam upravlenija [Proceedings of the 10th All-Russian Multi-Conference on Governance], vol. 2, pp. 259–262. Divnomorskoye, Gelendzhik (2017)
7. Shehory, O., Kraus, S.: Methods for task allocation via agent coalition formation. Artif. Intell. **101**, 165–200 (1998)
8. Юревич И.Е. и др. Интеллектуальные роботы // Машиностроение. С.360 (2007)
9. 唐苏妍.朱一凡.李群. 雷永林. 多Agent系统任务分配方法综述[期刊论文]//系统工程与电子技术 (10) (2010)
10. Berman, S., Halasz, A., Hsieh, M.A., Kumar, V.: Optimized stochastic policies for task allocation in swarms of robots. IEEE Trans. Rob. **25**(4), 927–937 (2009)
11. Meng Y., Gan J.: Self-adaptive distributed multi-task allocation in a multi-robot system. In: IEEE Congress on Evolutionary Computation, pp. 398–404 (2008)
12. Dias, M., Stentz, A.: Opportunistic optimization for market-based multirobot control. In: International Conference on Intelligent Robots and Systems, vol. 3, 2002, pp. 2714–2720
13. Chapman, A.C., Micillo, R.A., Kota, R., Jennings, N.R.: Decentralized dynamic task allocation: a practical game-theoretic approach. In: Proceedings of 8th International Conference on Autonomous Agents and Multiagent Systems, AAMAS 2009, pp. 915–922, Budapest, Hungary, 10–15 May 2009
14. Shaheen Fatime, S., Wooldridge, M.: Adaptive task resource allocation in multi-agent systems. In: Proceedings of 5th International Conference on Autonomous Agents, AGENTS' 01, pp. 537–544. ACM, New York, NY, USA (2001)
15. Kong, Y., Zhang, M., Ye, D.: A group task allocation strategy in open and dynamic grid environments. Presented at 7th International Workshop on Agent based Complex Automated Negotiations, ACAN 2014, Paris, France, 5–6 May 2014
16. Beni, G., Wang, J.: Swarm intelligence in cellular robotic systems. In: Proceedings NATO Advanced Workshop on Robots and Biological Systems, Tuscany. Italy, 26–30 June 1989
17. Bonavear, F., Dorigo, M.: Swarm Intelligence: From Natural to Artificial Systems. Oxford university Press, Oxford (1999)
18. Corne, D., Dorigo, M., Glover, F.: New Ideas in Optimization. McGrav-Hill (1999)

19. Kennedy, J., Eberhart, R.: Particle swarm optimization. In: IEEE International Conference on Neural Networks of 1995 Proceedings, vol. 11, issue 4, pp. 1942–1948 (1995)
20. Holland, J.H.: Adaptation in Natural and Artificial Systems, 1st edn, 1975. 2nd edn, MIT press, Cambridge, MA (1992)
21. https://en.wikipedia.org/wiki/Contract_Net_Protocol

Principles of Docking of Modular Mobile Robots Based on Sems in Their Group Interaction

Sergey N. Sayapin

Abstract *State of the problem*: one of the important tasks in the organization of group (swarm) modular robotic systems is the development of simple and reliable reusable systems of docking/undocking modules. Depending on the purpose, the docking devices can be one-time and reusable and provide automatic docking/undocking of the command connection from the control system. Also, the docking/undocking units can be semi-automatic, when the mobility of the modules themselves is additionally used. In necessary cases, in order to simplify the robotic systems, the operations of docking and undocking of modules can be performed manually by the operator or manipulator. In all cases, after docking, the connection Assembly must be highly rigid and exclude uncontrolled mobility of the joined elements. *The purpose of the study*: the Choice of the basic types of systems of docking/undocking for mobile robots modular type that are grouped in the active group structure. *Results*: an overview of the known principles and devices that can be used for docking modular mobile robots is given, and their classification is given. A comparative analysis of docking devices and recommended basic samples for solving group problems are shown. New original devices and docking joints of connected modules with the formation of group structures of various applications are presented. *Practical value*: of the Presented device docking/undocking of the mobile robots module type allow you to create a group (swarm) of active multifunctional robotic structures with based on SEMS. Such group structures will be able to solve a wide variety of problems in extreme and a priori uncertain conditions.

Keywords Devices of docking/undocking of modular mobile robots
Types of connections of modules · Group connections of modules in swarm
robotic systems based on SEMS

S. N. Sayapin (✉)
Mechanical Engineering Research Named After A. A. Blagonravov
of the Russian, Academy of Sciences, Moscow, Russia
e-mail: S.Sayapin@rambler.ru

S. N. Sayapin
Bauman Moscow State Technical University, Moscow, Russia

© Springer Nature Switzerland AG 2019
A. E. Gorodetskiy and I. L. Tarasova (eds.), *Smart Electromechanical Systems*,
Studies in Systems, Decision and Control 174,
https://doi.org/10.1007/978-3-319-99759-9_10

1 Introduction

Research of Self-assembly systems called Swarm Robotic Systems have become since 1980s. Now also it has a wide interest for Researchers and Designers of such systems due to the following advantages: Parallelism, Robustness, Scalability, Heterogeneousness, Flexibility, Complex Tasks, Cheap Alternative [1]. Swarm Robotic Systems or self-reconfigurable modular robots (SRMR) include several automatically mobile robots with variable morphology, which have wireless communication with each other. One of fundamental applications of SRMR is creating of intelligence reconfigurable linear and spatial structures by automatic mechanical connection between such robots. Most of the known mobile robots for Swarm Robotic Systems, that move on wheels, tracks or with the help of legs [1–7]. Therefore, the basic principles of docking devices are developed for these types of mobile robots. The principles of construction of docking devices are based on the principles of construction of numerous types of gripping mechanisms and devices for industrial robots [8–11]. Figure 1 shows the based grasping principles of last decades which can be applied at macro-, meso- and micro-scale [11]. From this variety in [3] four main principles of docking applied in SRMR are distinguished (Fig. 2). It is also necessary to add well-known compounds with the help of adhesion [2, 11], compounds of the "Velcro fastener" type (Fig. 3a) and "snap fastener" (Fig. 3b), as well as non—pump (c—bellows, d—diaphragm, e—piston) and pump vacuum connections [9].

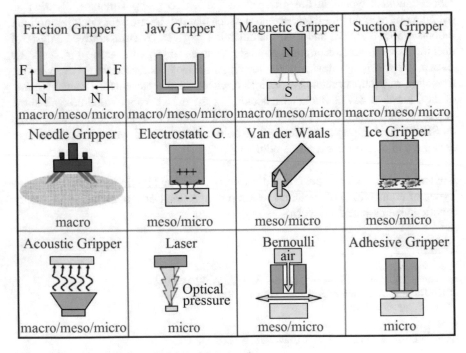

Fig. 1 The based grasping principles of last decades

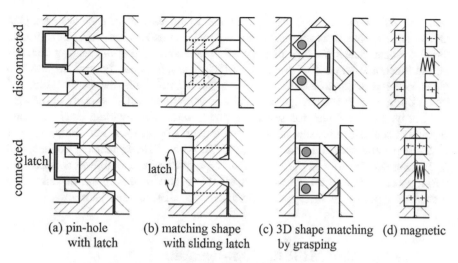

(a) pin-hole with latch   (b) matching shape with sliding latch   (c) 3D shape matching by grasping   (d) magnetic

Fig. 2 Four main principles of docking applied in SRMR

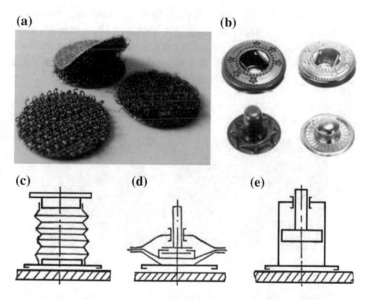

Fig. 3 Principles of docking applied in SRMR: "Velcro fastener" type (**a**); "Snap fastener" type (**b**); non-pump and pump vacuum connections types (**c**-bellows, **d**-diaphragm, **e**-piston)

Currently, there is a promising class of intelligent spatial truss self-reconfigurable structures (SRS), which are built of the same type of intelligent autonomous mobile parallel robots (MPR) with tetrahedral and octahedral structures (see Chap. 5 in this book). It has also a wide interest for Researchers and Designers of SRS. Such SRS are characterized by versatility and versatility and are able to solve a variety of tasks. Therefore, now there is the wide interest to the research of such systems. One of the most problems is automatic mechanical connection between such mobile parallel robots. Select the type of connection and construction of docking depends on the functional purpose of the SRS. In this regard, SRS can be divided into two types: non-separable self-assembled SRS, including structures that are disassembled by an operator or robot manipulator and self-assembled (self-disassembled) SRS.

First type of SRS use non-separable nodes automatic docking. Such SRS are capable of self-relocating and perform manipulative operations. For second type of SRS components mechanical docking of MPR, is able to automatically connect and disconnect two or more MPR. The second type of SRS is also able to be separated into a group of intelligent autonomous MPR for individual solution of tasks. The main base types of docking joint and devices for MPR are shown below.

2   Nodes Joining MPR for Non-separable Self-assembled SRS

Nodes joining the MNR for this type of SRS may be based on the use of adhesive bonding or foam and other application of self-hardening polymeric materials. Such compounds, except for some adhesives, do not require any effort to dock.

Figure 4 shows a prototype of a self-centering conical joint of automatic mechanical docking with increased load-bearing capacity in the axial direction [12].

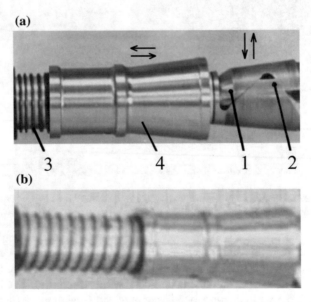

Fig. 4 The prototype of the self-centering conical joint of automatic mechanical docking with increased load-bearing capacity in the axial direction: initial position (**a**); work position (**b**)

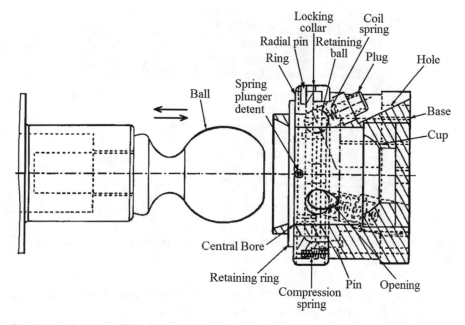

Fig. 5 Three degrees-of-freedom capability of the passive capture joint

The docking joint allows to completely eliminate the backlash between the rods in the axial and radial directions and provide reliable automatic fixation with high bearing capacity in the axial direction.

The joined elements are made in the form of semi-conical shanks 1 and 2 with teethes. One of semi-conic liner installed spring-loaded by a spring 3 a conical sleeve 4. At the radial convergence of the semi-conical shanks 1 and 2 at the time of coupling, the tooth of the semi-conical shank 1 presses the protruding element of the conical sleeve 4 lock and releases it. Under the action of the spring 3, the conical sleeve 4 moves axially, centers and fixes the semi-conical shanks. To fix the docking assembly, it is necessary to move the conical sleeve 4 from the semi-conical shanks 1 and 2 until it stops and separate them from each other. The protruding element of the lock will return to its original position and the conical sleeve will be fixed in the axial direction. The operation of decoupling of MPR may be made by an operator or by a robot manipulator.

Figure 5 shows a simple three degrees-of-freedom capability of the passive capture joint provides for quick connect and disconnect operations [10].

This joint allows for quick connection between any two structural elements where it is desirable to have rotation in all three axes. The joint can be fastened by moving the two halves into position. The joint is then connected by inserting the ball into the bore of the base. When the joint ball is fully inserted, the joint will lock with full strength. Release of this joint involves only a simple movement and rotation of one part. The joint can then be easily separated. This joint can be useful

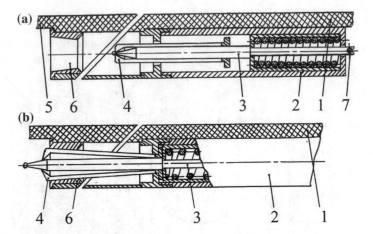

Fig. 6 Elastic connection of the free faces of the MPR by collet clamp: initial position (**a**); work
position (**b**)

in a variety of applications, including assembly of structures where simple rapid
deployment is essential, such as in space, undersea etc.

If necessary, the elastic connection of the free faces of the MPR in a continuous
active reconfigurable surface, the junction node faces can be made in the form of
collet clamps (Fig. 6) [12]. The retainer contains a housing 2 fixed on the free face 1
of one of the MPR, in which a spring-loaded rod 3 with a conical end with a collet 4
is placed with the possibility of axial movement. On the free face 5 of the MPR to
be joined, a sleeve 6 with a conical hole is fixed. In the fixed position, the rod 3 is
held from the axial movement by the check 7 (Fig. 6a). After pulling the check 7,
for example, with the help of a linear drive MPR, from the rod 3, its axial movement occurs along with the collet 4 in the conical hole of the sleeve 6. Further
movement of the rod 3 leads to the expansion of the collet 4 inside the conical hole
of the sleeve 6 and fixing the surfaces of the free faces 1 and 5 with each other
(Fig. 6b). Actuation of abutment does not require the axial force after mutual
rapprochement between the abutting MNR. To increase the stiffness of the connection, it is necessary to use two docking nodes located at the vertices of the faces.
As an MPR it is advisable to use an octahedral parallel structure with an inactive
docked face and nine linear actuators.

3   Docking Joints of MPR for Self-assembled
(Self-disassembled) SRS

The docking joints of MPR for self-assembled (self-disassembled) SRS include the
following types: docking joint, of MPR, that requires an effort for docking and
undocking; docking joint, that do not require from MPR effort to docking, but it

requires effort on undocking; docking joint, that do not require from MPR effort on docking and undocking.

Docking joint of MPR, that requires the effort for docking and undocking. Most of the docking units of this type include elastic elements of the latches, which must be pressed during fixing and fixing like "snap fastener" type connections (Fig. 3b). The design of the node connections of this type for tetrahedral MPR is shown in Fig. 10 in Chap. 5 in this book.

Also, this type of compounds include non-pump vacuum docking units (Fig. 3c–e).

Docking joint of MPR that does not require the effort for docking, but it requires effort for undocking. First of all, such devices include the docking joints with permanent magnets (Fig. 2d).

Also this type includes the docking joints of MPR that requires minimal effort to dock. A well-known example of such a connection is the "Velcro fastener" type connection (Fig. 3a).

Another example of this type of docking assembly is shown in Fig. 7 [12]. The docking assembly does not require high-precision mutual movement of the socket 1 and pin 2. In the socket 1 two sleeves are coaxially installed with the possibility of axial movement: the first sleeve 3, and in it the second 4. The first sleeve 3 is spring —loaded away from pin 2, and the second-to the pin. In the walls of the first sleeve there are holes, the axes of which are perpendicular to the longitudinal axis of the socket and a ball 5 is installed in each hole. On the inner surface of the socket 1, an annular groove 6 with a trapezoidal profile is made, and on the pin, the profile of the annular groove 7 corresponds to the ball 5. The inclined surfaces of the circular groove of the trapezoidal profile form an angle from 0° to 45° with the longitudinal axis of the nest. The lock works as follows. After contact with pin 2 in the conical socket 1, the sleeve 4 is pressed and under the action of the spring, the balls 5 are fixed through the groove 7 pin 2. After that, the conical part of the pin moves to the

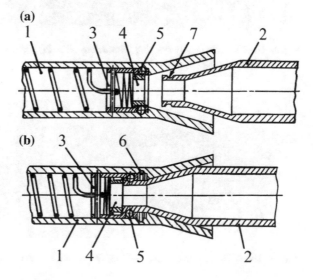

Fig. 7 The docking assembly does not require high-precision mutual movement of the socket 1 and pin 2: initial position (**a**); work position (**b**)

mate conical socket 1 until they are fully paired. Undocking MPR occurs in reverse order after the application of the required axial force.

Docking joint of MPR that does not require the effort for docking and undocking. To this type belong the device is equipped with independent drives or combined fixing of the abutting elements. This type of docking devices is the most common and diverse. As the actuators used electromechanical, electromagnetic, vacuum, pneumatic and many other devices.

Figure 8 shows Electromechanically Docking joint: (a) Male Piece; (b) Female Piece; (c) pulling of the Pieces; (d) Prototype of Docking Joint [13]. The cone has an annular groove and a transverse groove. When the cone is fixed by the interaction of the lock with the response cylindrical part and the annular groove, the cone can be rotated relative to the mate around the longitudinal axis. If it is necessary to exclude the rotation of the cone relative to the mate, the cone is fixed through the transverse groove.

Notice that the MPR with wheels or tracks are needed that has some devices for spatial interaction pieces of docking joints, for example, as Fig. 9 shows [14]. The tetrahedral and octahedral MPR does not require any additional devices for spatial

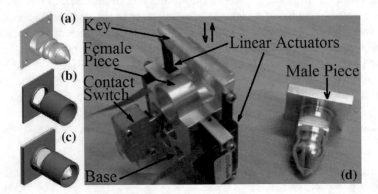

Fig. 8 Electromechanically docking joint: Male piece (**a**); Female piece (**b**); Pulling of the pieces (**c**); Prototype of docking joint (**d**)

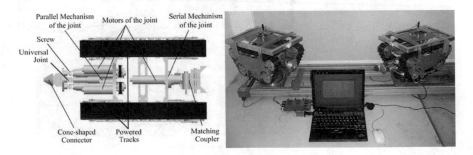

Fig. 9 The MPR with tracks and devices for spatial interaction pieces of docking joints

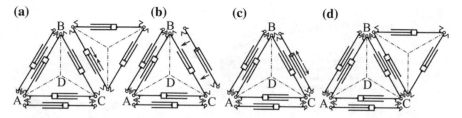

Fig. 10 Basis stages of connecting of tetrahedral MPR with each other by linear drives: initial position of faces (**a**); rapprochement of faces (**b**); connection of faces (**c**); end position (**d**)

interaction pieces of docking joints due to 3D mobility them vertices with these docking joints.

For the docking of tetrahedral and octahedral MPR b can be used their linear drive (Fig. 8). Figure 10 shows Basis stages of connecting of tetrahedral MPR with each other by linear drives.

The MPR in the form of Octahedral dodekapod IMASH (see Chap. 5 in this book) can simultaneously with the movement on their bellies to grab the free faces of objects of arbitrary shape. This ability allows to dock and undock such MPR through intermediate objects, such as rods, pipes, beams, etc. Figure 11 shows the possibility of connections through a third object, not only of Octahedral dodekapod IMASH (a), but their connection to tetrahedral MPR (b), and flat-robots type "Triangle" (see Chap. 5 in this book). As a result, it becomes possible to create a combined SRS including the combination of tetrahedral and octahedral MPR and flat MPR of the "Triangle" type both directly with each other and with the help of a third object.

If necessary, the elastic connection of the free faces of the MPR in a continuous active reconfigurable surface, the junction of the faces can be made in the form of elastic band lock 1, stretched on the drive and tension drums 2 (Fig. 12) [12]. Belt lock 1 automatically locking and unlock free face 3 abutting the MNR. Figure 10 shows the jointed free faces 3 before (a) and after (b, c). The edges of the joined faces 3 are made with longitudinal grooves 4, which interact with the elastic ribs 5 of the elastic band lock 1. The design of the elastic band lock 1 (d) is made in the form of a closed tape 1, which is half the length is made with a reduced width with one row of ribs 5. The wide part of the closed belt is made with two rows of elastic ribs 5. The tape lock works in the following way. After the convergence of the free

Fig. 11 The possibility of connections through by some third object (beam) for the Octahedral dodekapods IMASH (**a**) and the tetrahedral MPR

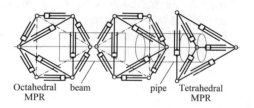

Octahedral   beam   pipe   Tetrahedral
MPR   MPR

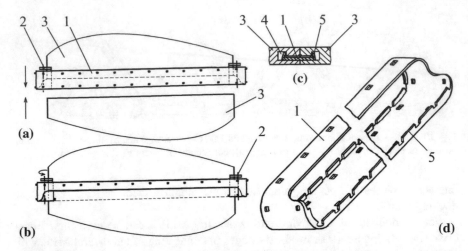

Fig. 12 The elastic connection of the free faces of the MPR in a continuous active reconfigurable surface: initial position (**a**); rapprochement of faces (**b**) and its connection (**c**); the general view of elastic band lock with the elastic ribs (**d**)

faces joined 3 and the coupling of the longitudinal grooves 4, at the command of the control system, the drive drum 2 is activated. As a result, a wide area of elastic band lock 1 with elastic ribs 5 is dragged through the longitudinal grooves 4 and the mutual fixation of the joined free faces 3. Unfixing of the free faces 3 is carried out in reverse order.

4　Conclusion

The analysis of docking joints and devices showed that for the tetrahedral and octahedral MPR the most effective are docking joints using the kinematics of their linear actuators (Fig. 11).

The tetrahedral and octahedral MPR does not require any additional devices for spatial interaction pieces of docking joints due to 3D mobility them vertices with these docking joints.

The organization of docking of octahedral MPR and in combination with tetrahedral MPR and flat MPR of type "Triangle" through the third subject is of interest for future research. This is particularly relevant for the group use of the SEMS-based MPR in rescue and recovery operations in the emergency area. In this case, any of the objects located in the immediate vicinity of the work site can be used for joining the MPR.

References

1. Yogeswaran, M., Ponnambalam, S.G.: Swarm robotics: an extensive research review. In: Fuerstner I (ed.) The Book: Advanced Knowledge Application in Practice, Chapter 14, pp. 259–278. InTech (2010)
2. Gradetskiy, V.G., et al.: Upravlyaemoe dvijenie mobilnych robotov po proizvolno orientirovannym v prostranstve poverxnostyam [*Controlled movement of mobile robots in space on arbitrary orientation surfaces*], 361 p. Nauka Publ., Moscow (2001) (In Russian)
3. Mondada, F., Bonani, M., Magnenat, S., Guignard, A., Floreano, D., Groen, F., Amato, N, Bonari, A., Yoshida, E., Kröse, B.: Physical connections and cooperation in swarm robotics. In: Groen, F., Amato, N., Bonarini, A., Yoshida, E., Kröse, B. (eds.) Proceedings of the 8th Conference on Intelligent Autonomous Systems (IAS8), Amsterdam, NL, 10–14 March 2004, pp. 53–60
4. Groß, R.: Self-assembling robots. Ph.D. thesis, Universite Libre de Bruxelles, Belgium, 201 p (2007)
5. Bruno, S., Oussama, K. (eds.): Springer Handbook of Robotics, 1611 p. Springer-Verlag Berlin Heidelberg (2008)
6. Ivanov, A.V., Yurevich, E.I., Ивановм, А.В.: Mini- i Microrobototechnika [Mini- and Microrobotics], 96 p. SPb. Publishing house of Polytechnical Institute (2011) (In Russian)
7. Kernbach, S. (ed.): Handbook of Collective Robotics: Fundamentals and Challenges, 93 p. Pan Stanford Publishing Pte. Ltd. Singapure (2013)
8. Chen, F.Y.: Gripping mechanisms for industrial robots. Mech. Mach. Theory **17**(5), 299–311 (1982)
9. Oteniy, Ya.N., Olshtynskiy, P.V.: Vybor I raschet zachvatnych ustroystv promyshlenych robotov, 65 p. RPK "Politichnik"SPb, Volgograd (2000) (In Russian)
10. Sandin, P.E.: Robot Mechanisms and Mechanical Devices Illustrated, 300 p. McGraw-Hill Companies, Inc (2003)
11. Fantoni, G., Santochi, M., Dini, G., Tracht, K., Scholz-Reiter, B., Fleischer, J., Lien, T.K., Seliger, G., Rinhart, G., Franke, J., Hansen, H.N., Verl, A.: Grasping devices and methods in automated production processes. CIRP Ann Manuf Technol **63**, 679–701 (2014)
12. Sayapin, S.N.: Analysis and synthesis of flexible spaceborne precision large mechanisms and designs of space radiotelescopes of the petal type, Doctor Tech. Sci. Dissertation. M.: IMash RAN, Moscow (2003) (In Russian)
13. Delrobaei, M., McIsaac, K.A.: Connection mechanism for autonomous self-assembly in mobile robots. IEEE Trans. Rob. **25**(6), 1413–1419 (2009)
14. Li, d., Fu, H., Wang, W.: Ultrasonic based autonomous docking on plane for mobile robot. In: Proceedings of the IEEE International Conference on Automation and Logistics, Qingdao, China, September 2008, pp. 1396–1401

The Control Complex Robotic System on Parallel Mechanism

V. A. Glazunov, S. V. Kheylo and A. V. Tsarkov

Abstract *Problem statement*: Parallel mechanisms analysis and design are the main trends of the SEMS. These mechanisms as universal modules are widely used in different industrial measuring, handling, and orienting mechanisms as well as in the biological structural system and crystalline structures simulation. These modules provide the translational motions and represents as 3-DOF mechanisms. SEMS elements of intelligent robotic systems provide the maximum precision of the motion, structural rigidity, and high specific lifting capacity. These systems are similar to space truss. Control of these systems is a major challenge of the design process. Translational parallel mechanisms are well understood. New translational motion mechanism is considered in this paper. The inverse dynamic problem is presented on the basis of virtual displacement principle. *Purpose of research*: Object of the article is to suggest the control procedure of parallel mechanisms that is universal module of SEMS. *Results*: Control algorithm is developed. This algorithm is based on the deviation minimization in coordinate, velocity, and acceleration as well as in inverse dynamic problem solution. The problem of dynamic accuracy has been examined. Particularly the influence of mechanism deviations was examined. The control procedure efficiency was checked under offsetting conditions of input and output links masses. The influence of mechanism dimensional deviation on positional accuracy was considered too. The algorithm efficiency has been examined using numerical simulation of mechanism motion on the specified trajectory. *Practical significance*: SEMS elements with suggested

V. A. Glazunov (✉)
The Blagonravov Institute for Machine, Russian Academy of Sciences,
Moscow, Russia
e-mail: vaglznv@mail.ru

S. V. Kheylo
The Kosygin State University of Russia, Moscow, Russia
e-mail: sheilo@yandex.ru

A. V. Tsarkov
Bauman Moscow State Technical University Kaluga Branch, Kaluga, Russia
e-mail: andrey.tsarkov@mail.ru

© Springer Nature Switzerland AG 2019
A. E. Gorodetskiy and I. L. Tarasova (eds.), *Smart Electromechanical Systems*,
Studies in Systems, Decision and Control 174,
https://doi.org/10.1007/978-3-319-99759-9_11

137

control procedure can be used for the various intelligent robotic systems of different applications (technical and manufacturing systems, transport industry, medical equipment, testing systems).

Keywords Parallel mechanisms · Robot · SEMS · Control · Accuracy Dynamic

1 Introduction

Robotics emerged as the copy of parts of the human body or the animal world (for example, a hand or musculoskeletal system). In these mechanisms the links arranged in series, each joint has an actuator.

Parallel structure mechanisms became an alternative to "inventions" of nature. Their appearance has become one of the trend of robotics and can have a large industrial application [1–5]. In such mechanisms the output link connected to the base by several kinematic chains or leg in parallel. The actuators are situated on the base and the mass of the links are lights. Parallel mechanisms have high stiffness, payload capacity, high speed, high accuracy, high stiffness and low inertia. Light links made of carbon fibre composite with good damping properties.

These parallel handling mechanisms connected in parallel or sequential order may form SEMS. This allows to extend operational and technological capabilities of the systems being created. Such SEMS have great potential applications to their characteristics. However, these systems possess more complex kinematics, dynamic behavior and control. Besides, mass of the structure increases in these systems, as the driving motors of each following module are located on the output arms of the previous one.

Sophisticated robotic system built in different architectures permits to solve various technological problems (processing tasks) in case of spatial motion performance. Parallel mechanisms with different DOF number can be SEMS elements. If DOF exceeds three, such property of parallel mechanisms as kinematic and dynamic decomposition may be applied. It helps to analyze translation and rotation motions separately. It gives us possibility to simplify the control task.

The subject of control of 3-DOF translational motion mechanism was considered in the paper.

2 Control of Translation Parallel Mechanism

The control system is an important functional part of the manipulator. The main objectives of the control system are:

- implementation of the movement of the output link according with prescribed trajectories;
- ensuring the set of laws of change coordinates.

The control system must be stable, the transients must correspond to the set parameters of quality.

The control system parallel mechanisms have a difficulties, caused by changing moments of inertia, nonlinearity of the equations, the changeable gear ratio. For the proposed control algorithm based on minimizing deviations from a set law of motion [6–10]. The actuators are servo. They have feedback in the coordinate, velocity and acceleration of output link. Presented translational parallel manipulators 3-PPaPa have three degrees of freedom and three actuators [7, 11] (Fig. 1).

This mechanism has three P-Pa–Pa legs and provides pure translational motion to its moving platform in three dimensions. The first P-joint in each leg is actuated. Robot workspace is a cube (Fig. 2).

The constrain equations represent a relationship between Cartesian coordinates x_0, y_0, z_0 and generalized coordinates q_1, q_2, q_3 and can be represented by the following system:

$$
\begin{cases}
F_1 = q_1 - x_0 - l_3 - l_2 \cos\left(-\arcsin\dfrac{y_0}{l_2}\right) - l_1 \cos\left(\arcsin\dfrac{z_0}{l_1}\right) = 0; \\[2mm]
F_2 = q_2 - y_0 - l_3 - l_2 \cos\left(-\arcsin\dfrac{z_0}{l_2}\right) - l_1 \cos\left(\arcsin\dfrac{x_0}{l_1}\right) = 0; \quad (1)\\[2mm]
F_3 = q_3 - z_0 - l_3 - l_2 \cos\left(-\arcsin\dfrac{x_0}{l_2}\right) - l_1 \cos\left(\arcsin\dfrac{y_0}{l_1}\right) = 0.
\end{cases}
$$

The desired laws of the coordinates of the mobile platform are described by equations $x_T(t)$, $y_T(t)$, $z_T(t)$. After differentiation we get the required velocity $\dot{x}_T(t)$, $\dot{y}_T(t)$, $\dot{z}_T(t)$ and acceleration $\ddot{x}_T(t)$, $\ddot{y}_T(t)$, $\ddot{z}_T(t)$. The task is to minimize the error in the coordinate:

$$\Delta_1(t) = x_T(t) - x(t), \Delta_2(t) = y_T(t) - y(t), \Delta_3(t) = z_T(t) - z(t),$$

speed: $\dot{\Delta}_1(t) = \dot{x}_T(t) - \dot{x}(t), \dot{\Delta}_2(t) = \dot{y}_T(t) - \dot{y}(t), \dot{\Delta}_3(t) = \dot{z}_T(t) - \dot{z}(t),$

acceleration: $\ddot{\Delta}_1(t) = \ddot{x}_T(t) - \ddot{x}(t), \ddot{\Delta}_2(t) = \ddot{y}_T(t) - \ddot{y}(t), \ddot{\Delta}_3(t) = \ddot{z}_T(t) - \ddot{z}(t),$

where $x(t)$, $y(t)$, $z(t)$, $\dot{x}(t)$, $\dot{y}(t)$, $\dot{z}(t)$, $\ddot{x}(t)$, $\ddot{y}(t)$, $\ddot{z}(t)$—the actual values of the coordinates, velocities, and accelerations of the output link.

Presented the algorithm of control transfers the system from the initial stage to the set neighborhood above mentioned trajectory in finite time, minimizing functional J built in the deviation $\Delta(t)$:

$$J = \int_{t_0}^{T} (\Delta^2 + k_1 \cdot \dot{\Delta}^2 + k_2 \cdot \ddot{\Delta}^2)dt \qquad (2)$$

where k_1, k_2—feedback coefficients.

V. A. Glazunov et al.

Fig. 1 Kinematic scheme of
translational manipulator

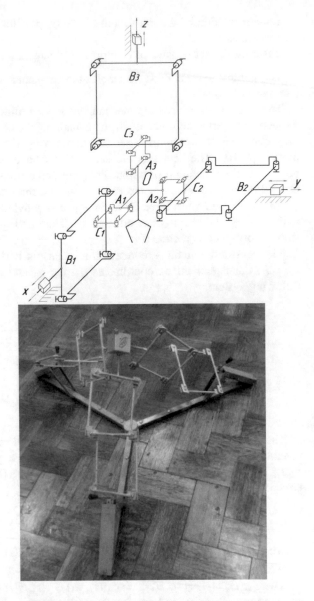

The value feedback gains coefficients are determined according the theory of robotic control [7–9]. The value of this integral should be of minimal value.

The integral can be represented as:

$$\int_{t_0}^{T} (\Delta_i^2 + k_{1i} \cdot \dot{\Delta}_i^2 + k_{2i} \cdot \ddot{\Delta}_i^2) \mathrm{d}t = \int \left(\ddot{\Delta}_i + \gamma_{1i} \cdot \dot{\Delta}_i + \gamma_{0i} \cdot \Delta_i \right)^2 \mathrm{d}t + C_i\left(\Delta_0, \dot{\Delta}_0\right)$$

Fig. 2 Robot workspace

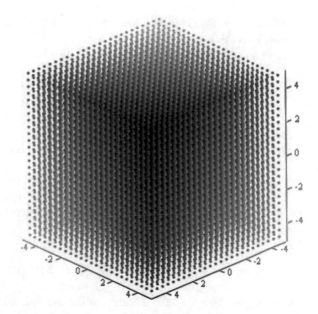

where $C_i(\Lambda_0, \dot{\Lambda}_0)$—some constant depending on the initial state of the controlled mechanism.

Functional J minimum is in trajectories realized:

$$\ddot{\Delta}_i + \gamma_{i1} \; \dot{\Delta}_i + \gamma_{i0} \cdot \Delta_i = 0 \tag{3}$$

where $i = 1, 2$, $\gamma_{i0} = \sqrt{k_{i2}}$, $\gamma_{i1} = \sqrt{k_{i1} + 2 \cdot \gamma_{i0}}$.

Rewrite the equations as oscillatory second-order system:

$$\tau_i^2 \ddot{\Delta}_i + 2\zeta_i \tau_i \cdot \dot{\Delta}_i + \Delta_i = 0 \tag{4}$$

$$\tau_i^2 = \frac{1}{\gamma_{0i}}, \, 2\zeta_i \tau_i = \frac{\gamma_{1i}}{\gamma_{0i}},$$

where τ_i—time constant, ζ_i—damping ratio.

The duration of transient is $t \approx \frac{3 \cdot \tau}{\xi}$ c. Then the constant γ_0, γ_1 will be equal

$$\gamma_0 = \frac{1}{\tau^2}, \gamma_1 = \frac{\sqrt{2}}{\tau}.$$

In steady motion at $t \rightarrow \infty$ are ratios for absolute coordinates:

$$x_T(t) \rightarrow x(t), y_T(t) \rightarrow y(t), z_T(t) \rightarrow z(t).$$

Thus, with some given time t, some feedback coefficients will occur the damping of transients $\Delta(t) \to 0$.

Equations of errors are :

$$\ddot{\Delta x} + \gamma_1 \dot{\Delta x} + \gamma_2 \Delta x = 0$$
$$\ddot{\Delta y} + \gamma_1 \dot{\Delta y} + \gamma_2 \Delta y = 0 \tag{5}$$
$$\ddot{\Delta z} + \gamma_1 \dot{\Delta z} + \gamma_2 \Delta z = 0$$

According to the constraint equations the actual accelerations become:

$$\ddot{x} = \ddot{x}_T + \gamma_1 \cdot (\dot{x}_T - \dot{x}) + \gamma_0 \cdot (x_T - x)$$
$$\ddot{y} = \ddot{y}_T + \gamma_1 \cdot (\dot{y}_T - \dot{y}) + \gamma_0 \cdot (y_T - y) \tag{6}$$
$$\ddot{z} = \ddot{z}_T + \gamma_1 \cdot (\dot{z}_T - \dot{z}) + \gamma_0 \cdot (z_T - z)$$

Actuator torques P_i are:

$$\mathbf{P_i} = \mathbf{A(q)} \cdot \ddot{\mathbf{q}} + \mathbf{B(q, \dot{q})} \cdot \dot{\mathbf{q}_i} + \mathbf{G(q)}, i = 1, \ldots n \tag{7}$$

where $\mathbf{A(q)}$—the $n \times n$ inertia matrix of the manipulator, $\mathbf{B(q, \dot{q})}$—$n \times 1$ vector of centrifugal and Coriolis term, $\mathbf{G(q)}$—$n \times 1$ vector of gravity terms.

Applying the principle of virtual work, we can write the following equations dynamic model:

$$m\ddot{x}\frac{\partial x}{\partial q_1}\delta q_1 + m\ddot{y}\frac{\partial y}{\partial q_1}\delta q_1 + m\ddot{z}\frac{\partial z}{\partial q_1}\delta q_1 + mg\frac{\partial z}{\partial q_1}\delta q_1 + m_1\ddot{q}_1\delta q_1 + P_1\delta q_1 = 0$$
$$m\ddot{x}\frac{\partial x}{\partial q_2}\delta q_2 + m\ddot{y}\frac{\partial y}{\partial q_2}\delta q_2 + m\ddot{z}\frac{\partial z}{\partial q_2}\delta q_2 + mg\frac{\partial z}{\partial q_2}\delta q_2 + m_2\ddot{q}_2\delta q_2 + P_2\delta q_2 = 0 \tag{8}$$
$$m\ddot{x}\frac{\partial x}{\partial q_3}\delta q_3 + m\ddot{y}\frac{\partial y}{\partial q_3}\delta q_3 + m\ddot{z}\frac{\partial z}{\partial q_3}\delta q_3 + mg\frac{\partial z}{\partial q_3}\delta q_3 + m_3\ddot{q}_3\delta q_3 + P_3\delta q_3 = 0$$

Here m is the mass of the output link, m_i are the mass of the input link, P_1, P_2, P_3 are torques in the drives, $\frac{\partial x}{\partial q_i}, \frac{\partial y}{\partial q_i}\frac{\partial z}{\partial q_i}$ are variable factors, $\delta x\ \delta x\ \delta x$ are virtual displacement of mobile platform, δq_i are virtual displacement of actuator.

Differentiating these expressions with respect to t, we obtain a system of equations of velocities of the input and output links:

$$\frac{\partial F_i}{\partial x}\dot{x} + \frac{\partial F_i}{\partial y}\dot{y} + \frac{\partial F_i}{\partial z}\dot{z} + \frac{\partial F_i}{\partial q_i}\dot{q}_i = 0, \quad i = 1, 2, 3 \tag{9}$$

The variable factors are situated by the generalized forces P:

$$\frac{\partial x}{\partial q_i} = -\frac{\partial F_i}{\partial q_i}\bigg/\frac{\partial F_i}{\partial x}, \frac{\partial y}{\partial q_i} = -\frac{\partial F_i}{\partial q_i}\bigg/\frac{\partial F_i}{\partial y}, \frac{\partial z}{\partial q_i} = -\frac{\partial F_i}{\partial q_i}\bigg/\frac{\partial F_i}{\partial z}.$$

Differentiating the equations again with respect to t, we obtain equations of accelerations. The block diagram of the computed torque control system is shown in Fig. 3.

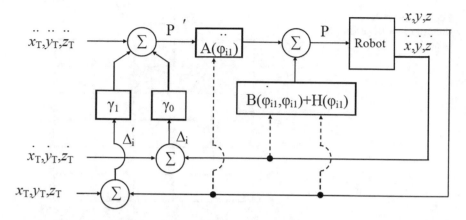

Fig. 3 Block diagram of dynamic control for a 3-PPaPa

The laws of errors are described by oscillatory second-order systems, by which control time is minimized. The most preferred mode is damping ratio $\zeta = \frac{\sqrt{2}}{2}$. The magnitude of the overshoot $\sigma \approx 5\%$. Define the transient time $t = 0.05$ s, $m = 5$ kg, $m_1 = m_2 = m_3 = 1$ kg. Thus $\tau \approx 0.011$, feedback coefficients are set as: $\gamma_0 = 7200$, $\gamma_1 = 120$.

The constraint equations the actual accelerations become:

$$\ddot{x} = \ddot{x}_T + 120 \cdot (\dot{x}_T - \dot{x}) + 7200 \cdot (x_T - x)$$

$$\ddot{y} = \ddot{y}_T + 120 \cdot (\dot{y}_T - \dot{y}) + 7200 \cdot (y_T - y)$$

$$\ddot{z} = \ddot{z}_T + 120 \cdot (\dot{z}_T - \dot{z}) + 7200 \cdot (z_T - z),$$

Let us consider an example: the desired laws of the coordinates of the moving platform are $x_T(t) = 0.1 \cdot \sin(\omega t)$, $y_T(t) = 0.12 \cdot \sin(\omega t)$, $z_T(t) = 0.15 \cdot \sin(\omega t)$.

The result of the simulation is presented on the Figs. 4 and 5. The maximal errors are about 5×10^{-3} mm.

3 Experiments with Control Algorithm

The effect of deviations of parameters of the real mechanism from its model is important [8].

Let the mass input link mechanism is not equal to the mass of the input link of the model. Let us consider the mass of input link is 30% different from the weight of the output level of the model. Error is less 5% (Fig. 6).

Fig. 4 The result of simulations of Displacement error

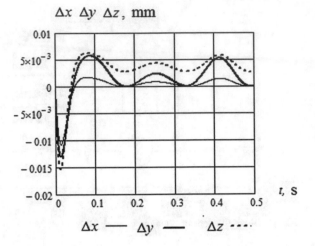

Fig. 5 The simulation result of driving torques

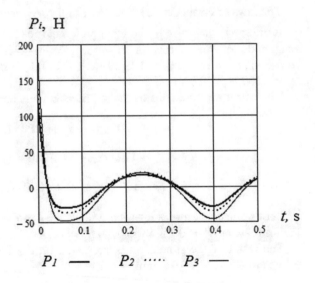

Consider the case when the mass of the output link is not equal to the mass of the output link in the model. We define the mass of the output link 30% different from the weight of the output level of the model. Error is less 3% (Fig. 7).

The deviation of the geometric parameters of the mechanism from the size of the model is important (for example, the length of the input links). How deviation is the impact on the accuracy. We define the length output link 10% different from the length output link of the model. Error is less 3% (Fig. 8).

In the presented algorithm the control error is not more than 5%.

Fig. 6 Displacement error:
1—mass of the input parts of
the model and mechanism
different, 2—mass of the
input parts of the model and
mechanism are equal

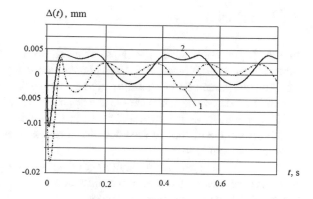

Fig. 7 Displacement errors
1—mass of the output
elements of the model and
mechanism are equal,
2—mass of the output
elements of the model and
mechanism of different

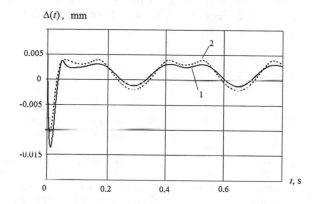

Fig. 8 Displacement error:
1—the lengths output link of
the model and mechanism are
equal, 2—the lengths output
link of the model and
mechanism of different

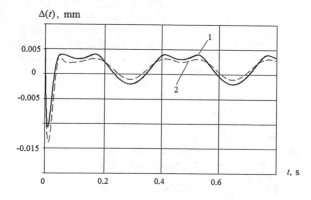

4   Conclusion

In this paper, concept of parallel root in SEMS has been proposed. A 3PPaPa translational parallel manipulator was chosen and designed to specific requirement. The kinematic analysis was performed and workspace was determined. The inverse dynamic model was established via the approach of virtual work principle.

Dynamic analysis for a parallel robot is to develop an inverse dynamic model, which enable the of the required actuator torques when a desired trajectory of the moving platform is given.

It is shown that one of the important criteria is the position error of the output link. It is presented examples of the influence of the accuracy deviation of the real mechanism from the mechanism model.

On the numerical simulations of the motion are tested and are presented efficiency of the control algorithm for the parallel structure mechanisms. The simulation results demonstrate the nice performance of the control method.

References

1. Merlet, J.-P.: Parallel Robots, 2nd edn. Springer (2006)
2. Kong, X., Gosselin, C.: Type Synthesis of Parallel Mechanisms. Springer (2017)
3. Tsai, L.: Robot Analysis: The Mechanics of Serial and Parallel Manipulators. Wiley, New York (1997)
4. Gogu, G.: Structural Synthesis of Parallel Robots, Part 1: Methodology. Springer, Dordrecht (2008)
5. Kheylo, S., Glazunov, V., Shirinkin, M., Kalendarev, A.: Possible application of mechanism parallel structure. J Mach. Manuf. Reliab. **42**(5), 359–363 (2013) (Allerton Press Inc.)
6. Kheylo, S., Glazunov, V.: Kinematics, Dynamics, Control and Accuracy of Spherical Parallel Robot. Advances on Theory and Practice of Robots and Manipulators. In: Proceedings of ROMANSY 2014 XX CISM-IFToMM Symposium on Theory and Practice of Robots and Manipulators, pp 133–140. Springer Cham Heidelberg New York Dordrecht London (2014)
7. Glazunov, V., Kheylo, S.: Dynamics and control of planar, translational, and spherical parallel manipulators. In: Dynamic Balancing of Mechanisms and Synthesizing of Parallel Robots, pp 365–403. Springer (2016)
8. Glazunov, V., Nosova, N., Kheylo, S., Tsarkov, A.: Design and analysis of the 6-DOF decoupled parallel kinematic mechanism. In: Dynamic Decoupling of Robot Manipulators, pp 125–170. Springer (2018)
9. Craig, J.J.: Introduction to Robotics: Mechanics and Control, 2nd edn. Addisson-Wesley, Reading, MA (1989)
10. Glazunov, V., Bykov, R., Esina, M.: Control of the parallel-structure mechanisms when passing through the singular positions. J. Mach. Manuf. Reliab. **2**, 73–79 (Allerton Press Inc.) (2004)
11. Kheylo, S., Glazunov, V., Sukhorukov, R.: Kinematic problem of translation manipulator parallel structure. Eng. Autom. Problem. **2**, 27–34 (2014) (in Russian)

A Coordinated Stable-Effective Compromises Based Methodology of Design and Control in Multi-object Systems

Evgeny M. Voronov and Vladimir A. Serov

Abstract In article the game hierarchical approach based optimization methodology of smart electromechanical systems group control with the principle of the coordinated stable-effective compromises application is developed. The problem of optimization of group control is decomposed into classes of problems of local, distributed and hierarchical control, taking into account structural and functional inconsistency, conflict, multicriteria, and uncertainty. Such structuring allows to consider different conditions of conflict group interaction of subsystems. The principle of coordinated stable-effective compromises, generalizing the Stackelberg hierarchical equilibrium principle, is formulated. The guaranteeing properties of stable-effective control laws are investigated.

Keywords Group control · Smart electromechanical system (SEM)
Conflict · Uncertainty · Multi-object multi criteria system (MMS)
Hierarchical system · Coordinated stable-effective compromise (COSTEC)
Structural adaptation

1 Introduction

A SEMS group control is a space distributed interconnected subsystem set control task, which implement in interaction the process of resource conversion, aimed on achieving some goals. Such a relationship inevitably creates conditions of inconsistency, conflict and uncertainty. The main reasons for these conditions are:

E. M. Voronov (✉)
Bauman Moscow State Technical University, 2nd Baumanskaya St., 5,
Moscow 105005, Russia
e-mail: emvoronov@mail.ru

V. A. Serov
Russian Technological University (MIREA), Vernadskogo Av., 78,
Moscow 119454, Russia
e-mail: ser_off@inbox.ru

© Springer Nature Switzerland AG 2019   147
A. E. Gorodetskiy and I. L. Tarasova (eds.), *Smart Electromechanical Systems*,
Studies in Systems, Decision and Control 174,
https://doi.org/10.1007/978-3-319-99759-9_12

the joint use of SEMS limited resources, multi-object architecture, multi-criteria problems, the impact of uncertain factors [1–4].

Thus, in modern representations of system analysis, the problem of SEMS group control is may be considered as the problem of MMS control under conflict and uncertainty. An important feature of SEMS group control tasks is the requirement of structural adaptation, which characterizes the possibility of changing the structural-target and functional relationships between subsystems in order to provide a specified level of efficiency and stability (balance) of the system under dynamic operating environment. The analysis of the numerous bibliography allows to draw a conclusion about the significant potential and prospects of game approaches to the development of methods of MMS control optimization, due to their universal nature [5–15]. Currently, the direction based on the aggregation of different principles of conflict equilibrium in order to form the desired properties of conflict-optimal solutions in game-theoretical models of control is actively developing. Various issues of aggregation of optimality principles are studied in [8, 11–20]. One of the promising is the direction, which is based on the formation and study of stable-effective compromises (STEC) [5, 21–24]. However, it can be concluded that many theoretical and computational aspects of the problem of synthesis of STEC in hierarchical game models of MMS control under uncertainty are insufficiently investigated, which is primarily due to their nonlinear nature and high structural, functional and information complexity order.

In this article and in [25], the following aspects of the hierarchical game approach based methodology of designing and control optimizing in MMS with the implementation of a COSTEC principle are discussed: structuring of the hierarchical two-level control system optimization problem and a COSTEC principle development, guaranteeing properties of COSTEC research; formation of the necessary conditions for the existence of COSTEC in the form of variational principles, generalizing the known Ekeland ε-variational principle on the class of hierarchical tasks of multi-criteria conflict optimization under uncertainty; the development of variational principles based adaptive evolutionary computer technology for a COSTEK formation.

2 Hierarchical Two-Level Model of MMS Control Optimization Under Uncertainty

Figure 1 shows a generalized structural-functional scheme of a two-level MMS control model, which can be considered as a basic model in the sense that [5, 26] firstly, it shows all the most significant characteristics of multilevel models; secondly, more complex multilevel group control models can be constructed from two-level, as from modules, thirdly, such a model covers classes of SEMS group control problems formed under of structural-functional inconsistency, conflict, multicriteria, uncertainty.

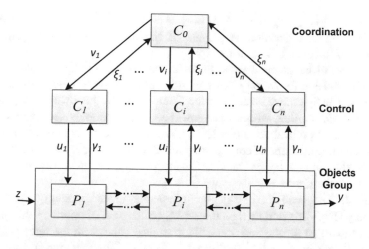

Fig. 1 A structural-functional scheme of a basic two-level MMS control model

This structure includes the levels of control and coordination.

At the control level, there are n subsystems $\{C_i, i \in \mathbf{N}\}$, $\mathbf{N} = \{\overline{1,n}\}$ that are faced with the task of developing optimal control laws $\mathbf{u}_i, i \in \mathbf{N}$ based on feedback signals γ_i and local control objectives reflecting the structuring of a multi-object system in the form of a set of objects $\{P_i, i \in \mathbf{N}\}$ and their relationships.

At the coordination level, there is a center C_0 subordinating the local contorl objectives to the global control objectives and solving the problem of optimal coordination.

Given the presence of the priority of action have subsystems top-level and inverse proportion to the results of operations of the subsystems of the upper level from the actual execution of the subsystems of the lower level of its functions, the optimization problem for two-level hierarchical control system, it is advisable to formulate as a hierarchical game under uncertainty with the right first move:

$$\Gamma = \langle \mathbf{C}_0, \mathbf{N}, \mathbf{V}_0, \{\mathbf{U}_i\}_{i \in \mathbf{N}}, \mathbf{Z}, \mathbf{F}(\mathbf{v}, \mathbf{u}, \mathbf{z})_{i \in \mathbf{N}}, \mathbf{\Omega}^0, \Gamma'(\mathbf{Z}, \mathbf{v}) \rangle. \tag{1}$$

It is assumed that in the game (1) take part the $(n+1)$ players: the coordinating Center $\mathbf{C}_0$, which controls the choice of strategy $\mathbf{v} \in \mathbf{V}_0$ and n players (control subsystems) of the lower level of the hierarchy, controlling the choice of strategies $\mathbf{u}_i \in \mathbf{U}_i, i \in \mathbf{N} = \{\overline{1,n}\}$; $\mathbf{z}$—an indefinite factor, which is only known that it takes values from a set $\mathbf{Z}$; $\mathbf{F}(\mathbf{v}, \mathbf{u}, \mathbf{z}) \in \mathbf{E}^{m_0}$—vector efficiency index of the Center $\mathbf{u} = \{\mathbf{u}_i | i \in \mathbf{N}\} \in \mathbf{U} = \prod_{i \in \mathbf{N}} \mathbf{U}_i$; $\mathbf{\Omega}^0 \subset \mathbf{E}^{m_0}$—a convex domination cone generating a binary preference relation for the Center on a set of achievable vector bounds

$\mathbf{F}(\mathbf{V}_0, \mathbf{U}, \mathbf{Z}) = \bigcup_{\mathbf{v} \in \mathbf{V}_0} \bigcup_{\mathbf{u} \in \mathbf{U}} \bigcup_{\mathbf{z} \in \mathbf{Z}} \mathbf{F}(\mathbf{v}, \mathbf{u}, \mathbf{z})$. The game $\Gamma'(\mathbf{Z}, \mathbf{v})$ describes the conditions of conflict interaction of players of the lower level of the hierarchy in conditions of uncertainty $\mathbf{Z}$ at a fixed $\mathbf{v} \in \mathbf{V}_0$.

In the case of hierarchical game under of uncertainty (1) the multistage mechanism of a decision formation can be written down in the following form.

Stage 1. The first move is made by the Center—it tells the lower level players its strategy $\mathbf{v} \in \mathbf{V}_0$. Further, with a fixed lower level players play the game $\Gamma'(\mathbf{Z}, \mathbf{v})$, forming a set of optimal solutions $\mathbf{U}^{opt}(\mathbf{v}, \mathbf{Z})$. The specific type of set $\mathbf{U}^{opt}(\mathbf{v}, \mathbf{Z})$ depends on the conditions of conflict interaction at the lower level of the hierarchy. These conditions can be set by the Center or selected by lower level players regardless of the Center.

Stage 2. The center evaluates the efficiency of the strategy $\mathbf{v} \in \mathbf{V}$ under uncertainty $\mathbf{U}^{opt}(\mathbf{v}, \mathbf{Z})$, and then on the set $\mathbf{V}_0$ builds a set of optimal solutions $\mathbf{V}_0^{opt}$.

It is known that for the decision of hierarchical game with the right of the first move the Stackelberg equilibrium principle is used. However, the formulation of problem (1) has a number of features, such as vector efficiency indicators of the participants, the presence of uncertain factors, which necessitates additional research.

The analysis of the problem statement (1) shows that in the problem of optimization of the hierarchical two-level control system it is possible to allocate the following basic having independent importance, and at the same time interconnected classes of tasks:

- local control (excluding interlevel and insidelevel interactions);
- distributed control (including intralevel interactions equivalent subsystems);
- hierarchical control (based on interlevel and insidelevel interactions of subsystems).

For certainty, we assume that all participants in group interaction seek to minimize the values of their efficiency indicators.

3 The Low Level Group Conflict Control Under Uncertainty

When solving a hierarchical game (1) in stage 1, the conditions of conflict interaction between the control subsystems at the lower level of the hierarchy with a fixed Center strategy $\mathbf{v}$ are crucial . These conditions in a fairly general form can be formalized as a coalition game under uncertainty

$$\Gamma' = \left\langle \mathbf{N}, \mathbf{P}, \left\{\mathbf{U}^{\mathbf{K}}\right\}_{\mathbf{K} \in \mathbf{P}}, \mathbf{Z}, \left\{\mathbf{J}^{\mathbf{K}}(\mathbf{u}, \mathbf{z})\right\}_{\mathbf{K} \in \mathbf{P}}, \left\{\boldsymbol{\Omega}^{\mathbf{K}}\right\}_{\mathbf{K} \in \mathbf{P}} \right\rangle. \tag{2}$$

In (2) **P** is a decomposition of the set of **N** participants in the conflict view

$$\mathbf{P} = \left\{ \mathbf{K}_1, \ldots, \mathbf{K}_l \,\middle|\, \mathbf{K}_i \bigcap_{i \neq j} \mathbf{K}_j = \varnothing; \bigcup_{j=1}^{l} \mathbf{K}_j = \mathbf{N} \right\}, \tag{3}$$

called the coalition structure of the set **N**. Coalition members are united by common interests: $\mathbf{U}^{\mathbf{K}} = \prod_{i \in \mathbf{K}} \mathbf{U}_i$—a set of admissible strategies of the coalition $\mathbf{K}$; $\mathbf{u}^{\mathbf{K}} = \{\mathbf{u}_i | i \in \mathbf{K}\} \in \mathbf{U}^{\mathbf{K}}$ the vector unites components of vectors of admissible strategies of the subsystems forming the coalition $\mathbf{K}$; $\mathbf{J}^{\mathbf{K}}(\mathbf{u}, \mathbf{z}) = \{\mathbf{J}_i(\mathbf{u}, \mathbf{z}) | i \in \mathbf{K}\} \in \mathbf{E}^{m_{\mathbf{K}}}$—a vector indicator of efficiency of the coalition $\mathbf{K}$ where $\mathbf{J}_i(\mathbf{u}, \mathbf{z})$—a vector indicator of efficiency—i control subsystem $i \in \mathbf{N}$; $\boldsymbol{\Omega}^{\mathbf{K}}$—the convex domination cone generating the coalition relation of preference for the coalition $\mathbf{K}$ on a set of achievable vector estimates $\mathbf{J}^{\mathbf{K}}(\mathbf{U}, \mathbf{Z})$.

Specific configurations of the coalition structure (3) allow to consider the following types of intralevel group interactions of subsystems: agreed (cooperative), unagreed (noncooperative), coalition, antagonistic. If in the coalition structure **P** of the species (3) $\mathbf{K}_i = \{i\}, i = \overline{1, n}$, then at the lower level of the hierarchy, the group control model has the form of a noncooperative game under uncertainty:

$$\Gamma' = \langle \mathbf{N}, \{\mathbf{U}_i\}_{i \in \mathbf{N}}, \mathbf{Z}, \{\mathbf{J}_i(\mathbf{u}, \mathbf{z})\}_{i \in \mathbf{N}}, \{\boldsymbol{\Omega}_i\}_{i \in \mathbf{N}} \rangle. \tag{4}$$

The statement of the problem (4) will be considered as the basic one in the future.

3.1 Equilibrium with Respect to the System of Domination Cones Under Uncertainty

The situation $\mathbf{u} \in \mathbf{U}$ in the game (4) we will evaluate the vector efficiency index $\mathbf{V}^{\boldsymbol{\Omega}}(\mathbf{u}) \in \mathbf{E}^m$, which is defined as follows:

$$\mathbf{J}(\mathbf{u}, \mathbf{Z}) \subset \mathbf{V}^{\boldsymbol{\Omega}}(\mathbf{u}) + \boldsymbol{\Omega}, \tag{5}$$

where $\mathbf{J}(\mathbf{u}, \mathbf{Z}) = \bigcup_{\mathbf{z} \in \mathbf{Z}} \mathbf{J}(\mathbf{u}, \mathbf{z})$, $\boldsymbol{\Omega} = \prod_{i \in \mathbf{N}} \boldsymbol{\Omega}_i \subset \mathbf{E}^m$, $\mathbf{J}(\mathbf{u}, \mathbf{z}) = \{\mathbf{J}_i(\mathbf{u}, \mathbf{z}) | i \in \mathbf{N}\}^T \in \mathbf{E}^m$, a for any other vector $\tilde{\mathbf{V}} \neq \mathbf{V}^{\boldsymbol{\Omega}}(\mathbf{u})$ satisfying inclusion

$$\mathbf{J}(\mathbf{u}, \mathbf{Z}) \subset \tilde{\mathbf{V}} + \boldsymbol{\Omega}, \tag{6}$$

take place

$$\mathbf{V}^{\boldsymbol{\Omega}}(\mathbf{u}) - \tilde{\mathbf{V}} \in \boldsymbol{\Omega} \backslash \mathbf{0}_m. \tag{7}$$

The vector efficiency index $\mathbf{V}^{\boldsymbol{\Omega}}(\mathbf{u}) = \{\mathbf{V}^{\boldsymbol{\Omega}_i}(\mathbf{u}) | i \in \mathbf{N}\}$ determines in the criterion space the "worst" point for all players (relative to the domination cone $\boldsymbol{\Omega}$) corresponding to the situation $\mathbf{u}$; $\mathbf{V}^{\boldsymbol{\Omega}_i}(\mathbf{u})$—the vector efficiency index of the player under uncertainty, determined in a similar way (5)–(7).

Definition 1 We will say that the situation $\{\mathbf{u}\|\tilde{\mathbf{u}}_i\} = \{\mathbf{u}_1, \ldots, \mathbf{u}_{i-1}, \tilde{\mathbf{u}}_i, \mathbf{u}_{i+1}, \ldots, \mathbf{u}_n\}$ $\boldsymbol{\Omega}_i$-dominates the situation $\mathbf{u}$ under uncertainty, if

$$\mathbf{V}^{\boldsymbol{\Omega}_i}(\mathbf{u}\|\tilde{\mathbf{u}}_i) - \mathbf{V}^{\boldsymbol{\Omega}_i}(\mathbf{u}) \in \boldsymbol{\Omega}_i \backslash \mathbf{0}_{m_i}. \tag{8}$$

Definition 2 The situation $\mathbf{u}^* \in \mathbf{U}$ is called $\boldsymbol{\Omega}_i$—equilibrium of a noncooperative game under uncertainty (4), if for any strategy $\mathbf{u}_i \in \mathbf{U}_i$ the situation $(\mathbf{u}^*\|\mathbf{u}_i)$ is not $\boldsymbol{\Omega}_i$-dominant in relation to the situation $\mathbf{u}^*$, i.e.

$$\mathbf{V}^{\boldsymbol{\Omega}_i}(\mathbf{u}^*\|\mathbf{u}_i) - \mathbf{V}^{\boldsymbol{\Omega}_i}(\mathbf{u}^*) \notin \boldsymbol{\Omega}_i \backslash \mathbf{0}_{m_i} \tag{9}$$

Definition 3 The situation $\mathbf{u}^e \in \mathbf{U}$ is called equilibrium with respect to the system of cones $\{\boldsymbol{\Omega}_i\}_{i \in \mathbf{N}}$ of a noncooperative game under uncertainty (4), if for any $i \in \mathbf{N}$ it is $\boldsymbol{\Omega}_i$-equilibrium.

The effectiveness of the i–player's strategy $\mathbf{u}_i \in \mathbf{U}_i$ in the game (4) in the conditions of uncertainty we will assess the vector index $\mathbf{V}_{\mathbf{u}_i}^{\boldsymbol{\Omega}_i}$, which is determined as follows:

$$\mathbf{J}_i\left(\mathbf{u}_i, \mathbf{U}^{\mathbf{N}\backslash i}, \mathbf{Z}\right) \subset \mathbf{V}_{\mathbf{u}_i}^{\boldsymbol{\Omega}_i} + \boldsymbol{\Omega}_i, \tag{10}$$

for any other vector $\tilde{\mathbf{V}} \in \mathbf{E}^{m_i}$ satisfying inclusion

$$\mathbf{J}_i\left(\mathbf{u}_i, \mathbf{U}^{\mathbf{N}\backslash i}, \mathbf{Z}\right) \subset \tilde{\mathbf{V}} + \boldsymbol{\Omega}_i, \tag{11}$$

take place

$$\mathbf{V}_{\mathbf{u}_i}^{\boldsymbol{\Omega}_i} - \tilde{\mathbf{V}} \in \boldsymbol{\Omega}_i \backslash \mathbf{0}_{m_i}. \tag{12}$$

Definition 4 The strategy $\mathbf{u}_{ig}$ is called the guaranteeing strategy of the player with respect to the domination cone $\boldsymbol{\Omega}_i$ in the noncooperative game (4) under uncertainty, if for any $\mathbf{u}_i \neq \mathbf{u}_{ig}$

$$\mathbf{V}_{\mathbf{u}_i}^{\boldsymbol{\Omega}_i} - \mathbf{V}_{\mathbf{u}_{ig}}^{\boldsymbol{\Omega}_i} \notin \boldsymbol{\Omega}_i \backslash \mathbf{0}_{m_i}. \tag{13}$$

Definition 5 The set $\mathbf{U}_{ig}$ containing all $\mathbf{u}_{ig}$ having the property (13) is called the set of guaranteeing strategies of the i—player with respect to the domination cone $\boldsymbol{\Omega}_i$ in a noncooperative game under uncertainty (4).

The guaranteeing equilibrium property with respect to the system of cones $\{\Omega_i\}_{i \in \mathbf{N}}$ of a noncooperative game under uncertainty (4) can be formulated in the form of the following statement.

Statement 1 Let it be $\mathbf{u}^e$ the equilibrium with respect to the system of cones $\{\Omega_i\}_{i \in \mathbf{N}}$ in a noncooperative game under uncertainty (4). Then for any player $i \in \mathbf{N}$ and any his guaranteeing strategy $\mathbf{u}_{ig} \in \mathbf{U}_{ig}$ regarding the domination cone Ω_i takes place

$$\mathbf{V}_{\mathbf{u}_{ig}}^{\Omega_i} - \mathbf{V}^{\Omega_i}(\mathbf{u}^e) \notin \Omega_i \backslash \mathbf{0}_{m_i}. \tag{14}$$

Comment 1 If for any $i \in \mathbf{N}$ $\Omega_i = \mathbf{E}_{\leq}^{m_i}$—is nonpositive orthant in space $\mathbf{E}^{m_i}$, then the equilibrium with respect to the system of domination cones under uncertainty is a vector Nash equilibrium under uncertainty.

3.2 Generalized Equilibrium Under Uncertainty

Definition 6 The situation $\mathbf{u}^* \in \mathbf{U}$ is called the active Ω_i-equilibrium of the non-cooperative game under uncertainty (4), if for any strategy $\mathbf{u}_i \in \mathbf{U}_i$ there is a strategy $\hat{\mathbf{u}}_{\mathbf{N} \backslash i} \in \mathbf{U}_{\mathbf{N} \backslash i}$ of counter-coalition $(\mathbf{N} \backslash i)$ such that

$$\mathbf{V}^{\Omega_i}\left(\mathbf{u}_i, \hat{\mathbf{u}}_{\mathbf{N} \backslash i}\right) \notin \mathbf{V}^{\Omega_i}(\mathbf{u}^*) + \Omega_i \backslash \mathbf{0}_{m_i}. \tag{15}$$

Definition 7 The situation $\mathbf{u}^c \in \mathbf{U}$ is called the generalized equilibrium of the noncooperative game under uncertainty (4), if for any player $i \in \mathbf{N}$ the situation $\mathbf{u}^c$ is active Ω_i-equilibrium.

The guaranteeing property of the generalized equilibrium of the game (4) can be formulated as the following statement.

Statement 2 Let it be $\mathbf{u}^c$—generalized equilibrium of a noncooperative game under uncertainty (4). Then for any player $i \in \mathbf{N}$ and any guaranteeing strategy $\mathbf{u}_{ig} \in \mathbf{U}_{ig}$ regarding the cone of domination Ω_i takes place

$$\mathbf{V}_{\mathbf{u}_{ig}}^{\Omega_i} - \mathbf{V}^{\Omega_i}(\mathbf{u}^c) \notin \Omega_i \backslash \mathbf{0}_{m_i}. \tag{16}$$

The substantive meaning of Statement 2 is that the values of vector efficiency indicators of players in the situation of generalized equilibrium under uncertainty are not worse than their guaranteed values with respect to the domination cones Ω_i, i.e. guaranteed strategy of any player $i \in \mathbf{N}$ under uncertainty is not Ω_i-dominating in relation to the situation of the generalized equilibrium of the game (4).

Comment 2 Equilibrium with respect to the system of domination cones under uncertainty is a special case of the generalized equilibrium of the game (4).

3.3 Stable-Effective Compromise Under Uncertainty

As mentioned above, SEC—a situation that combines to some extent the properties of equilibrium and efficiency. With regard to the noncooperative game model of group control under uncertainty, the following definition of SEC is possible.

Definition 8 Let be a $\mathbf{U}^c$—set of generalized equilibria of noncooperative games under uncertainty (4). The situation $\mathbf{u}^s \in \mathbf{U}^c$ is called the SEC of a noncooperative game under uncertainty (4) if for any $\mathbf{u} \in \mathbf{U}^c$, $\mathbf{u} \neq \mathbf{u}^s$ there is

$$\mathbf{V}^{\Omega}(\mathbf{u}) - \mathbf{V}^{\Omega}(\mathbf{u}^s) \notin \Omega \backslash \mathbf{0}_m. \tag{17}$$

The set $\mathbf{U}^s \subset \mathbf{U}^c$ of all generalized equilibrium with the property (17) is called SEC set under uncertainty.

The ratio (17) means that in game (4) there is no situation of generalized equilibrium more preferable than simultaneously $\mathbf{u}^s$ for all players with respect to the domination cone Ω.

The possibility of constructing a SEC on a generalized equilibrium set of the game (4) can be formulated as the following statement.

Statement 3 [15]. Let be a $\mathbf{U}^c$—a generalized equilibria set of noncooperative games under uncertainty (4).
Then:

1. A set of SEC under uncertainty $\mathbf{U}^s$ is not empty.
2. The set $\mathbf{V}^{\Omega}(\mathbf{U}^s)$ coincides with the kernel $C_{\Omega}(\mathbf{V}^{\Omega}(\mathbf{U}^c))$ of the preference relation given by the domination cone Ω on the set $\mathbf{V}^{\Omega}(\mathbf{U}^c)$.
3. Inclusion is fair

$$\mathbf{V}^{\Omega}(\mathbf{U}^s) \subseteq Opt_{\Omega}(\mathbf{V}^{\Omega}(\mathbf{U})), \tag{18}$$

where $Opt_{\Omega}(\mathbf{V}^{\Omega}(\mathbf{U}))$—a set of Ω-optimal points of the set $\mathbf{V}^{\Omega}(\mathbf{U})$.

The given statement can be used as a basis for the development of combined computational algorithms for constructing the SEC set of a no-coalition game under uncertainty (4).

The above concepts of equilibrium under uncertainty are based on the principle of vector minimax [9], develop the concept of pessimistic multivalued equilibrium [11], which allows to take into account the worst values of all performance indicators, possible on the entire set of permissible values of the uncertain factor.

4 Coordinated Stable-Effective Compromise Under Uncertainty

It seems reasonable to consider the SEC set under uncertainty $\mathbf{U}^s(\mathbf{v})$ in a fixed strategy $\mathbf{v}$ of the coordinating center as a set of optimal solutions $\mathbf{U}^{opt}(\mathbf{v})$ of the game (4).

The efficiency of the center's strategy $\mathbf{v} \in \mathbf{V}$ under uncertainty $(\mathbf{U}^s(\mathbf{v}) \times \mathbf{Z})$, we will evaluate of the vector index $\mathbf{V}^{\Omega_0}(\mathbf{v})$, which is defined as follows:

$$\mathbf{F}(\mathbf{v}, \mathbf{U}^s(\mathbf{v}), \mathbf{Z}) \subset \mathbf{V}^{\Omega_0}(\mathbf{v}) + \Omega_0, \tag{19}$$

and for any other vector $\tilde{\mathbf{V}} \neq \mathbf{V}^{\Omega_0}(\mathbf{v})$, satisfying the inclusion

$$\mathbf{F}(\mathbf{v}, \mathbf{U}^s(\mathbf{v}), \mathbf{Z}) \subset \tilde{\mathbf{V}} + \Omega_0, \tag{20}$$

take place

$$\mathbf{V}^{\Omega_0}(\mathbf{v}) - \tilde{\mathbf{V}} \in \Omega_0 \backslash \mathbf{0}_{m_0}. \tag{21}$$

The center selects a valid strategy $\mathbf{v}^s \in \mathbf{V}_0$ such that for any valid strategy $\mathbf{v} \neq \mathbf{v}^s$

$$\mathbf{V}^{\Omega_0}(\mathbf{v}) - \mathbf{V}^{\Omega_0}(\mathbf{v}^s) \not\subset \Omega_0 \backslash \mathbf{0}_{m_0}. \tag{22}$$

Definition 9 The situation $\{\mathbf{v}^s, \mathbf{u}^s\}$ called for coordinated stable-effective compromise (CSEC) of a games (1) under uncertainty if $\mathbf{u}^s$ is a SEC of noncooperative games under uncertainty (4) for a fixed $\mathbf{v}^s$ and strategy of the Center $\mathbf{v}^s$ satisfies the condition (22).

Comment 3 The hierarchical equilibrium $\{\mathbf{v}^s, \mathbf{u}^s\}$ of the game under uncertainty (1) has a stability property. CSEC $\{\mathbf{v}^s, \mathbf{u}^s\}$ stability for the Center means the implementation of the condition (22), which means the Ω_0-optimality under uncertainty. Stability for players of lower level means the implementation of the condition $\mathbf{u}^s \in \mathbf{U}_s$.

Comment 4 The CSEC principle develops the known Stackelberg equilibrium principle to a class of hierarchical game models of group control under uncertainty with vector efficiency index of the Center and subsystems.

Comment 5 Let, s consider the problem of multicriteria optimization under uncertainty

$$\langle \mathbf{V}_0, \mathbf{Y}, \mathbf{F}(\mathbf{v}, \mathbf{y}), \Omega_0 \rangle, \tag{23}$$

where $\mathbf{y} \in \mathbf{Y}$—the uncertainties factor.

Let $\mathbf{v}^*$—vector minimax decision of problems (23). This means that for anyone $\mathbf{v} \neq \mathbf{v}^*$

$$\mathbf{V}^{\boldsymbol{\Omega}_0}(\mathbf{v}) - \mathbf{V}^{\boldsymbol{\Omega}_0}(\mathbf{v}^*) \notin \boldsymbol{\Omega}_0 \backslash \mathbf{0}_{m_0}, \tag{24}$$

where $\mathbf{V}^{\boldsymbol{\Omega}_0}(\mathbf{v})$ calculated by rules (5)–(7) under the assumption that $\mathbf{Y} = \mathbf{Z}$.

If in the problem (23) $\mathbf{Y} = \mathbf{U}^s(\mathbf{v}) \times \mathbf{Z}$, where $\mathbf{U}^s(\mathbf{v})$ the SEC of the game is set (4), then CSEC acquires the properties of a vector minimax. That is, the two-level model of control optimization of a multi-object system under uncertainty can be interpreted as a model of local multicriteria control under uncertainty in the construction of more complex hierarchical control structures.

Thus, CSEC principle under uncertainty is a mathematical justification for the possibility of scaling the basic two-level model of MMS control under uncertainty in the construction of more complex hierarchical models of group control.

5 Conclusion

The task of optimization of SEMS group control is formulated as a problem of MMS control under conflict and uncertainty and formalized in the form of a hierarchical game with vector efficiency indexes of participants with the privilege of the first move under uncertainty.

CSEC principle under uncertainty, generalizing the principle of Stackelberg equilibrium, is formulated. The interrelation of the CSEC principle under uncertainty and the vector minimax principle is shown.

The guaranteeing properties of stable-effective group control laws are investigated.

It is shown that CSEC principle under uncertainty can be used as a basis of design and research methodology of hierarchical structurally complicated SEMS group control models under uncertainty providing the opportunity two-level model of MMS control optimization scaling.

References

1. Kalyaev, I.A., Gaiduk, A.R., Kapustyan, S.G: Models and Algorithms of Collective Control in Groups of Robots, 280 p. Fizmatlit (2009)
2. Jadbabaie, A., Lin, J., Morse, A.S.: Coordination of groups of mobile autonomous agents using nearest neighbor rules. In: IEEE Transaction on Automatic Control, May 2003
3. Dahl, T.S., Mataric, M.J., Sukhatme, G.S.: Emergent robot differentiation in distributed multi-robot task allocation. In: Proceedings of the 7th International Symposium on Distributed Autonomous Robotic Systems (DARS-04), Toulouse, France (2004)
4. Bruce, J., Bowling, M., Browning, B., Veloso, M.: Multi-robot team response to a multi-robot opponent team. In: Proceedings of ICRA '03, the 2003 IEEE International Conferenceon Robotics and Automation, Taiwan, May 2003

5. Voronov, E.M.: In: Egupov, D. (ed.) Methods of Optimization of Control of Multi-object Multi-criteria Systems on the Basis of Stable-Effective Game Solutions, 576 p. BMSU, Moscow (2001)
6. Vorobyiov, N.N.: Fundamentals of the Theory of Games. Noncooperative Games, 496 p. Science, Moscow (1984)
7. Vilkas, E.T.: Optimality in Games and Solutions, 256 p. Science, Moscow (1990)
8. Zhukovsky, V.I., Chikry, A.A.: Linear-Quadratic Differential Games, 320 p. Naukova Dumka, Kyiv (1994)
9. Zhukovsky, V.I., Zhukovskaya, L.V.: In: Molostvov, V.S. (ed.) Risk in Multi-criteria and Conflict Systems Under Uncertainty, 272 p. Editorial URSS, Moscow (2004)
10. Harsanyi, G., Reinhard, Z.: General Theory of Equilibrium Choice in Games, 424 p. SPb (2001)
11. Smolyakov, E.R.: Theory of Conflict Equilibria, 304 p. Editorial URSS, Moscow (2005)
12. Gorelov, M.A., Kononenko, A.F.: Dynamic models of conflict. II. Equilibrium. Autom. Telemechanics (12), 56–77 (2014)
13. Gorelov, M.A., Kononenko, A.F.: Dynamic models of conflict. III. Hierarchical games. Autom. Telemechanics (2), 89–106 (2015)
14. Gorelik, V.A., Rodyukov, A.V., Tarakanov, A.F.: Hierarchical game under uncertainty using risk players and guaranteed assessment strategies. Izv. RAS Theory Control Syst. (6), 94–101 (2009)
15. Govorov, A.N., Tarakanov, A.F.: Coalition-hierarchical game under uncertainty. Izv. RAS Theory Control Syst. (3), 75–80 (2008)
16. Moya, S.: The calculation of the Stackelberg–Nash equilibrium as a fixed point problem in static hierarchical games. Int. J. Dyn. Control 1–12 (2017)
17. Léon, E., Pape, A., Costes, M., Désidéri, J., Alfano, D.: Concurrent aerodynamic optimization of rotor blades using a Nash game method. J. Am. Helicopter Soc. 61(2), 1–13 (2016)
18. Leskinen, J., Périaux, J.: Distributed evolutionary optimization using Nash games and GPUs-applications to CFD design problems. Comput Fluids 80, 190–201 (2013)
19. Periaux, I., Greiner, D.: Efficient parallel Nash genetic algorithm for solving inverse problems in structural engineering. In: Mathematical Modeling and Optimization of Complex Structures. Computational Methods in Applied Sciences, vol. 40, pp. 205–228. Springer, New York (2016)
20. Ungureanu, V.: Pareto-Nash-Stackelberg Game and Control Theory: Intelligent Paradigms and Applications (Smart Innovation, Systems and Technologies), 579 p. Springer (2017). ISBN 10 3319751506, ISBN 13 978-3319751504
21. Serov, V.A.: Stable-equilibria control in a hierarchical game model of a structurally complex system under uncertainty. In: Popkov, Y.S. (ed.) Proceedings of the Institute for System Analysis RAS. Dynamics of Heterogeneous Systems, vol. 10(1), pp. 64–76. Komkniga, Moscow (2006)
22. Vanin, A.V., Voronov, E.M., Serov, V.A., Karpunin, A.A., Liubavskii, K.K.: Optimization of hierarchical systems "guidance-stabilization" of the aircraft with the adaptation of the stabilization system. Vestnik BMSTU Ser. Instrum. Making (4), 13–33 (2015)
23. Voronov, E.M., Karpunin, A.A., Serov, V.A.: The hierarchical equilibrium in multilevel control systems. Vestnik RUDN. Ser. Eng. Res. (4), 18–29 (2008)
24. Voronov, E.M., Karpunin, A.A., Serov, V.A.: Hierarchical equilibrium in two-level aircraft guidance-stabilization systems. In: Aerospace Guidance, Navigation and Flight Control Systems (AGNFCS'09): Proceedings of the IFAC Workshop, June 30–July 2, 2009, Samara, Russia. The International Physics and Control Society (IPACS) Electronic Library. http://lib. physcon.ru/?item=21. 6 p
25. Serov, V.A., Voronov, E.M.: Evolutionary algorithms for finding of a stable-effective compromises in multi-object control problems (in this collection)
26. Mesarovich, M., Mako, D., Takahara, I.: Theory of Hierarchical Multilevel Systems, 344 p. World, Moscow (1973)

Part III
Mathematical and Computer Modeling
Group Interaction

Estimates of the Group Intelligence of Robots in Robotic Systems

Andrey E. Gorodetskiy and Irina L. Tarasova

Abstract *Problem statement*: the questions of group interaction of intelligent Electromechanical systems (SENS) in RoboTic Systems (RTS) play an important role in the analysis of RTS capabilities to perform certain tasks. The paper proposes a solution to the problems of group intelligence assessment based on the results of RTS test modeling. This issue occurs if necessary matching the best candidates to the group of RTS from the existing set of modules SEMS, for example, when you want to perform merge/split parts of RTS, the introduction of the group new robot or withdraw the existing or producing other transformation groups SEMS associated with the implementation of technological tasks. *Purpose of research*: proposed and implemented an approach to solving the problem of test modeling of a group of robots using the apparatus of fuzzy mathematical modeling. The coefficients of vector estimation of group intelligence are developed. As the components of vector estimation, the coefficient of intellectual abilities, the creativity coefficient and the coefficient of motivational inclusion of the j-th group of robots are proposed. Results: The structure of a mathematical model for testing a group of robots is developed, including dynamic models of both interacting robots and the environment. The formulas for computing the components of the vector estimation of the group intelligence of RTS are proposed. Practical significance: the proposed solution of RTS test modeling provides high functionality of models of robot groups, taking into account the SEMS ideology, allowing to perform group intelligence estimation by computer modeling taking into account the so-called "psychological" features of the group members.

A. E. Gorodetskiy (✉) · I. L. Tarasova
Institute of Problems of Mechanical Engineering Russian Academy
of Sciences, V.O., Bolshoj pr., 61, St. Petersburg 199178, Russia
e-mail: g27764@yandex.ru

I. L. Tarasova
e-mail: g17265@yandex.ru

I. L. Tarasova
Peter the Great St. Petersburg Polytechnic University,
Polytechnicheskaya, 29, Saint-Petersburg 195251, Russia

© Springer Nature Switzerland AG 2019
A. E. Gorodetskiy and I. L. Tarasova (eds.), *Smart Electromechanical Systems*,
Studies in Systems, Decision and Control 174,
https://doi.org/10.1007/978-3-319-99759-9_13

161

Keywords Robotic complex · Robot group · SEMS · Central nervous system
Fuzzy mathematical modeling · Test computer modeling

1 Introduction

The problem of optimizing the interaction of a group of robots in the performance
of a joint task is now becoming increasingly complex, requiring the complexity and
intellectuality of technical robot control systems to be constantly expanded to
ensure the continued expansion of the scope of such robotic systems (RTS). In
complex RTS, built on the basis of intelligent electromechanical systems (SEMS)
[1], it is necessary to analyze not only the behavior of an individual robot pos-
sessing due to the presence of the central nervous system (CNS) [2] appropriate
behavior, but also the behavior of an interacting group of robots whose functions
are closely interrelated in the so-called "small group". This group can be vehicles
that carry out coordinated movements, a group of robots of collectors performing
joint operations, etc. When creating such groups, specialists face a wide range of
problems, such as assessing the ability of the group to make correct decisions under
uncertainty, determining the optimal number of team members, assessing the
compatibility of group members, taking into account the static and dynamic
characteristics of each robot, decision-making in the central nervous system of
individual robots, etc.

Procedures that are usually used to assess the performance of one robot can be
based on testing and evaluating the quality of decisions taken, according to a
particular scale of its ability to reason [3]. In assessing the quality of work performed when solving a joint task by a group of robots, testing becomes more
complicated, taking into account the so-called "psychological characteristics", just
as the work of the human operator in the group is evaluated [4].

Known is the approach to vector estimation of intelligence [5], based on the
calculation of the system of artificial intelligence estimates using test results. Such
vector estimates usually contain static probabilistic components that assess the
ability to solve fuzzy applied problems, and dynamic probabilistic components that
assess the system's ability to self-learn [6]. Obviously, this approach can be used
for an objective assessment of the suitability of certain robots to work in a particular
RTS. However, in this case, comparison and selection of the best candidates from
the set of tested robots can be based on a numerical evaluation of the results of
dynamic testing of group work, when the results of the test are recorded at different
time intervals into which the entire test run is divided and with various combinations of RTS participants. This is due to the specific differences between the group
and individual activities of the robot. Therefore, the recommended formulas for
calculating the components of the vector estimate of the intelligence of an individual robot can not be directly applied to assessing the suitability of this robot for
group interaction. In particular, as shown in [7, 8], there are factors such as the time

of the group reaction, the time necessary for making correct or important joint decisions, and other static and dynamic characteristics.

To assess the intellectual activity of groups of operators of human-machine systems (HMS), there are enough constructive approaches. For example, in [8] measurements and perceptions in the SMS are considered. This method allows you to objectively determine (using purely instrumental methods) the parameters of the human sensory system that affect the adaptability of the operator to the natural and technological environment. The approach [9] modeled the cognitive process of the operator and the behavior of the system, which is affected by the operator's actions in emergency situations of the nuclear power plant (NPP). The operator model models the cognitive behavior of the operator in random situations on the basis of the Rasmussen decision-making model and is implemented using AI-methods of distributed cooperative output with the so-called blackboard architecture. Rule-based behavior is modeled using the knowledge representation in If-then rules. Article [10] is devoted to the development of a general method for estimating the load for various tasks and workplaces. This model was developed by introducing sets of linguistic variables and applying the analytical hierarchy process (AHP) to assess the external load imposed on the human operator in the SMS. For this purpose, a five-point scale of the set of linguistic variables was constructed and hierarchical priority procedures were established. The task variables and workstation (for example, for physical, environmental, postural and mental workloads) that can include operator perception of the workload are selected as workload factors, and the AHP method is used to collect different weights. Finally, the overall work level (OWL) is calculated using a computer system to determine the work entrusted to the operator. Some other approaches in similar areas are described in the patent [11]. In the patent [11], the operator group model is connected to the system and includes a data storage device for the ideal operator model, a recognition section for inputting voice information generated by the operator, and for visual information provided to the operator. When the failure information is entered in the SMS, the data about the object and/or its operation, presented in the relevant sections on the control panel's operating panel, and when the operator responsible for the decision is selected and indicated, then this operator can quickly take corrective measures for excluding the stall in the work of the object.

By analogy with SMS, to evaluate the intelligence of the central nervous system of robots in the RTS, it is necessary to conduct its computer simulation with identification and evaluation of the decisions taken in complex dynamic systems under conditions of incomplete certainty [12]. The structure of the mathematical model should include methods for estimating the parameters characterizing the group intelligence of the RTS.

In this paper, an approach is based on fuzzy mathematical and computer simulation of the behavior of the RTS robot group and the dynamics of the selection environment [13]. Methods for estimating the parameters of the robot's intellect are considered, taking into account the work in the group. Some new approaches and results are presented. However, the methods discussed can be improved in the future, given estimates psychological characteristics group behavior, such as average scores of professional competence groups static components of intelligence candidates in a group, the components of the vector learning test candidates having a mean estimated level of intelligence and etc.

2 Test Simulation of a Robotic System

In the test simulation of RTS to obtain estimates of the group intelligence RTS first need to set in the form of fuzzy sets of the following models:

- Source environment of choice A
- End environment of choice B
- Valid i environment of choice A_i^v
- Dynamic spaces of configurations D_j of the j-th group of robots at times t_k

$$D_j(t_k) = \cup D_p(t_k). \tag{1}$$

where $D_p(t_k)$—dynamic space of configurations of the p-th robot at the time t_k,

$$T^0 < t_k < T^e, \tag{2}$$

where: T^0—start time, T^e—end time

- Set of valid configurations of the p-th robot K_p^v
- Set of valid configurations of the j-th group of robots K_j^v
- Initial j-th task environment

$$W_j\left(T_j^0\right) = A \cup D_j\left(T_j^0\right). \tag{3}$$

where: T_j^0—start time for j-th task environment

- The final j-th task environment

$$W_j(T_j^e) = B \cup D_j(T_j^e). \tag{4}$$

where: T^e_j—end time for j-th task environment

- Target set W_t

All of these fuzzy sets can be given as a set of algebraic, differential and/or difference equations, as well as logical-interval, logical-probabilistic, logical-linguistic, and sometimes purely linguistic expressions.

For each j-th group of robots (SEMS) forming RTS, the i-th task of transition from $W_{ji}(T_{ij}^0)$ to $W_{ji}(T_{ij}^e)$ is formed in the shortest possible time ΔT_{ji}

$$(W_{ji}(T_{ij}^0) \rightarrow W_{ji}(T_{ij}^e))/(DT_{ji} = T_{ij}^e - T_{ij}^0) \rightarrow \min, \tag{5}$$

where: $T^0{}_{ij}$, $T^e{}_{ij}$—start time and end time solutions of the i-th problem of the j-th group of robots

Then the binary ratio is calculated

$$W_{ji}\left(T_{ij}^e\right) q\, W_t. \tag{6}$$

where q—metric of proximity $W_{ji}(T^e{}_{ij})$ to Wt or double predicate on the analyzed sets, which can be set, for example, by specifying the formulas of the logical-mathematical language or by specifying a formalized linguistic expression [14, 15]. When designing the q formation procedure, it is necessary to obtain a quantitative estimate of proximity $W_{ji}(T^e{}_{ij})$ to Wt.

Creation of the initial base for the design q it is advisable to start with the selection in each of the compared sets of metrizable subsets, for elements of which relations and numerical measures of proximity can be specified. The next, most difficult step is ordering the elements of non-metrizable subsets. It is very likely that to solve this problem, you will need to build a new system of logical equations, the solution of which will lead either to the metrizable sets or to the ordered ones. In the first case, we immediately obtain numerical measures of proximity. In the second, these measures will have to be built anew. As possible numerical estimates can be used the power of sets, the number of matching elements, the number of groups of matching elements, etc. Any recommendations for the selection of certain assessments cannot be recommended at the present time due to the lack of knowledge of such models. In case of impossibility of ordering of non-metrizable sets the decision on the greatest proximity of any set to the standard shall be made by the person-designer performing selection of candidates for RTS group [12].

The most frequently used and easily constructed binary measures of proximity of sets include the following [13]:

– estimation on the maximum deviation of the powers of sets:

$$\sum_i x_i - \sum_i y_i = \delta_m \tag{7}$$

where: $x_i = 1$ and $y_i = 1$ for non-zero (non-empty) elements of the compared sets and respectively $x_i = 0$ and $y_i = 0$ for zero (empty) elements of the compared sets.

– evaluation of the standard deviation of the capacity sets:

$$\sqrt{\left(\sum_i x_i\right)^2 - \left(\sum_i y_i\right)^2} = \delta \tag{8}$$

- probabilistic estimation on maximum deviation of powers of sets:

$$\sum_i P(x_i = 1)x_i - \sum_i P(y_i = 1)y_i = \Delta_P \tag{9}$$

where: $P(.)$—probability

- probabilistic estimation by the standard deviation of the powers of sets:

$$\sqrt{\left(\sum_i (P(x_i = 1)x_i)^2 - \left(\sum_i P(y_i = 1)y_i\right)^2\right)} = \delta_P \tag{10}$$

The use of binary functional relations of the form (7)–(10) is most preferable, since it allows ranking models $W_{ji}(T^e_{ij})$ by their proximity to the target set W_t and at the same time entering a numerical estimate of group intelligence.

Each j-th group of robots is given n attempts (n = 1,2,3, ... N), each of which requires the following restrictions:

$$A \subseteq_i^v. \tag{11}$$

$$B \subseteq A_i^v \tag{12}$$

$$D_j(t_k) \subseteq K_j^v \tag{13}$$

$$D_p(t_k) \subseteq K_p^v \tag{14}$$

For the evaluation of learning j-th group of robots each subsequent i-th transition problem should be more complex, and to assess the sustainability (professionalism) in the decision-making of the j-th group all i-th transition problems should be about the same complexity.

Sometimes it is desirable to assess the differences in the similarity of the CNS members of the group of robots in the RTS, which may indirectly characterize the so-called "psychological compatibility" of group members. In this case, a similar test simulation is additionally performed separately for each group member.

3 Baseline Assessment of Group Intelligence in the Central Nervous System RTS

According to the results of the test simulation of the j-th group of robots in RTS solving the consistently complicated I = 1,2,3, ..., F problems can be analogous to the American psychologist Renzulli, J. S. [16–18] calculate the following estimates of the group intelligence of RTS:

– coefficient of intellectual abilities of the j-th group of robots:

$$Q_j = \frac{1}{FN} \sum_{i=1}^{F} \sum_{n=1}^{N} q_{inj} \tag{15}$$

– creative factor of the j-th group of robots:

$$V_j = \frac{1}{FN} \sum_{i=1}^{F} \sum_{n=1}^{N} \frac{1}{\Delta T_{inj}} \tag{16}$$

where: ΔT_{inj}—time of the n-th attempt to solve the i-th problem by the j-th group of robots

- the coefficient of motivational involvement M_j of the j-th group of robots is also calculated by the formula (16), however, all test problems (i = 1,2,3,...F) should be of approximately the same complexity.

Combining the entered coefficients, it is possible to calculate the intelligence coefficient of the j-th group of robots in RTS:

$$IQ_j = k_q Q_j + k_v V_j + k_M M_j. \tag{17}$$

where: k_q, k_v и k_M—weight coefficients set by the operator conducting the test simulation, depending on the purpose of the tested RTS.

Recently, additional assessments of the group intelligence of the group of operators, taking into account the psychological characteristics of the group members, are offered. Similarly, we can introduce additional factors that characterize group behaviors of robots in the RTS. For example, you can enter the coefficient of learning j-group, calculating IQj when solving each of the complex i = 1,2,3, ... F problem. Then find the maximum $\max_i\{IQ_j\}$ and the minimum $\min_i\{IQ_j\}$ value of all the calculated. Then the learning ratio will be:

$$LA = \max_i \{IQ_j\} - \min_i \{IQ_j\}. \tag{18}$$

A certain interest may be the estimation of stability in decision-making of the j-th group of robots in RTS, which can be characterized by the coefficient of stability St, calculated by the results of n = 1,2,3, ..., N attempts of the same or similar problem:

$$St = \max_n \{IQ_j\} - \min_n \{IQ_j\}. \tag{19}$$

It is also interesting to assess the similarity of participants of the j-th group of robots in RTS. This requires every i-th problem to be divided into h = 1,2,3, ..., H subtasks performed by each h-th robot of the j-th group. Then you need to calculate the IQ_h of each robot and find the maximum $\max_h\{IQ_h\}$ and the minimum $\min_h\{IQ_h\}$ value. Then the rate of unanimity will be:

$$Un = \max_h\{IQ_h\} - \min_h\{IQ_h\}. \tag{20}$$

In addition to the test results can be: average score estimates of group intelligence, average estimates of the stability of decision-making group, differences in the consensus of the Central nervous system group members of the group of robots, and the stability and variability of reasoning.

4 Conclusion

The introduced estimates of the results of the test modeling of the group of robots in the RTS allow to take into account the characteristics of the CNS of each robot group and the psychology of the group in solving the joint problem. First, these estimates have static and dynamic components that take into account both the decision-making capacity under uncertainty and the ability to learn in the process. Second, they have components that characterize the psychological stability of groups. In some cases, vector scores can be replaced by average scores, which make it easy to rank groups from best to worst. The presented version of testing of robot groups in RTS was relatively simple and did not take into account a number of design features and real operating conditions.

It seems that a more subtle modeling of RTS features and the study of the nature of variability in time of the proposed group assessments based on the results of testing of candidates will help to identify a number of other important group psychological features for work in RTS, such as indecision, nervousness, patience, scrupulousness, etc. [8]. In this case, no less important task is the generation of the most adequate test tasks for specific RTS.

It is also important to note that the assessment of group intelligence of robots needs to be improved in terms of assessing learning, professionalism, psychological stability and compatibility with the human observer or supervisor—controller of the process, which in many cases can be key in making the right decisions in conditions of uncertainty.

Acknowledgements This work was financially supported by Russian Foundation for Basic Research, Grant 16-29-04424 and Grant 18-01-00076.

References

1. Gorodetskiy, A.E.: Smart Electromechanical Systems, 277p. Springer International Publishing (2016). https://doi.org/10.1007/978-3-319-27547-5

2. Gorodetskiy, A.E., Kurbanov, V.G.: Smart Electromechanical Systems: The Central Nervous System. In: Studies in Systems, Decision and Control, vol. 95, 270p. Springer International Publishing Switzerland (2017). https://doi.org/10.1007/978-3-319-53327-8
3. Gorodetskiy, A.E., Tarasova, I.L.: Intellektual'nye programmnye sredstva dlya avtomatizirovannyh ispytanij sensornyh sistem (Intelligent software for automated testing of sensor systems). In: Gorodetskiy, A.E., Kurbanov, V.G. (eds) Physical Metrology: Theoretical and Applied Aspects, pp. 68–74. Publishing house KN (1996)
4. Gorodetskiy, A.E., Al-Rasasbeh, R.T.: Vector estimates of the group activity of operators. In: Collection of Works: Modern Problems of Socio-economic Development and Information Technology, p. 63. Baku (2004)
5. Al-Kasasbeh, R.T.: Statistical-similar model of organization work for Small Group Information System operators. In: Proceeding of International Carpathian Conference ICCC, pp. 217–224
6. Popchetelev, E.P.: Training for Studying Group, pp. 65–67. Leningrad Technical University News, Leningrad (1988)
7. Antonets, V.A., Anishkina, N.M.: Measurements and perception in man/machine systems, 12p. Preprint of IAP RAS, N 518, Nizhny Novgorod (1999)
8. Yoshida, K., Yokobayashi, M.: Development of AI-based simulation system for man-machine system behavior in accidental situations of nuclear power plant. Japan
9. Jung Hwa, S., Hyung-Shik, J.: Establishment of overall workload assessment technique for various tasks and workplaces. Int. J. Ind. Ergon. **28**, 341–353 (2001). (ISSN 0169-8141)
10. Kitaura, W., Ujita, H., Fukuda, M.: Man-machine system. United States Patent US5247433A
11. Eykhoff, P.: System Identification, Parameter and State Estimation. Wiley, London (1974)
12. Gorodetskiy, A.E, Tarasova, I.L.: Fuzzy mathematical modeling badly formalized processes and systems, 336p. SPb.: Publishing House of the Polytechnic. University Press (2010)
13. Gorodetskiy, A.E.: Foundations of the theory of intelligent control systems. LAP LAMBERT Academic Publishing GmbH @ Co. KG, 313 c (2011)
14. Gorodetskiy A.E.: Fuzzy decision making in design on the basis of the hubituality. In: Reznik, L., Dimitrov, V., Kasprzyk, J.: Fuzzy System Design, Physica Verlag (1998). ISBN 3-7908-1118-1
15. Gorodetsky A.E.: Ob ispol'zovanii situacii privychnosti dlya uskoreniya prinyatiya reshenij v intellektual'nyh informacionno-izmeritel'nyh sistemah (On the use of the habitual situation for accelerating decision-making in intelligent information and measurement systems. In: Gorodetskiy, A.E., Kurbanov, V.G. (eds) Physical Metrology: Theoretical and Applied Aspects, pp. 141–151. Publishing house KN (1996)
16. Renzulli, J.S.: What Makes Giftedness? Reexamining a Definition. Phi Delta Kappan, **60**(3), 180–184, 261 (1978)
17. Renzulli, J.S.: Schools for Talent Development: A Practical Plan for Total School Improvement. Creative Learning Press, Mansfield Center, CT (1994)
18. Renzulli, J.S., Reis, S.M.: The Schoolwide Enrichment Model: A Comprehensive Plan for Educational Excellence. Creative Learning Press, Mansfield Center, CT (1985)

Techniques to Detect and Eliminate Inconsistencies in Knowledge Bases of Interacting Robots

Boris A. Kulik and Alexander Ya. Fridman

Abstract *Problem statement*: behavior of SEMS-based robots is modeled by a knowledge base (KB) represented in different formats (semantic networks, sets of productions, etc.), and the robots interactions model can be represented as an integrated KB. During development and operation KBs, there exists a problem of their anomalies. Retarded detection of such anomalies can lead to malfunctions and breakdowns. Anomalies of inconsistency play an especially important role in faults diagnostics for robots teams. A large number of publications have been devoted to the methods of detecting these and other anomalies. At the same time, methods to eliminate anomalies lack for adequate studies. *Purpose of research*: development of new methods for detecting anomalies of inconsistency, as well as algebraic methods for eliminating some anomalies in integrated KBs of interacting robots, in particular, the contradiction anomaly during logical inference. *Results*: an overview of techniques to detect anomalies in integrated KBs of interacting robots is presented. Presupposition is shown to be one of examples of the contradiction anomaly during logical inference. Presupposition is often encountered in texts and has a property of existence of mutually exclusive conclusions from the same premise. A method for explaining and eliminating this anomaly is proposed. *Practical significance*: the proposed techniques of anomalies detection and elimination can be used to control robots teams in order to decrease failures in their functioning.

Keywords SEMS-based robot · KB anomaly · Anomaly of inconsistency
Presupposition · Anomalies detection and elimination

B. A. Kulik (✉)
Institute of Problems in Mechanical Engineering, Russian Academy
of Sciences (RAS), 61 Bol'shoi pr, V.O., 199178 St. Petersburg, Russia
e-mail: ba-kulik@yandex.ru

A. Ya. Fridman
Kola Science Centre of RAS, Institute for Informatics and Mathematical
Modelling, Fersman Str, 24A, 184209 Apatity, Russia
e-mail: fridman@iimm.ru

© Springer Nature Switzerland AG 2019
A. E. Gorodetskiy and I. L. Tarasova (eds.), *Smart Electromechanical Systems*,
Studies in Systems, Decision and Control 174,
https://doi.org/10.1007/978-3-319-99759-9_14

171

1 Introduction

Smart electromechanical systems (SEMS) provide a basis to investigate various tasks and techniques in modeling of robots' behavior, in particular, their interactions within groups [1]. Recent development of means for adaptation, telecommunications and artificial intelligence led to creation of intelligent control systems for robots, machines and equipment interconnected by communication channels in order to exchange information. Such tools with intelligent control and communication interaction are possible to consider as agents of complex robotic systems that have to collectively solve a common problem in a changing environment.

Interaction of robots (in more general case, agents) is studied in the framework of many scientific disciplines. Let list some of them.

(1) *Distributed artificial intelligence* [2]. This area of artificial intelligence deals with the most common aspects of collective behavior of agents. It is based on the results obtained in theory of distributed systems and decision-making.

(2) *Game theory* [3, 4]. This apparatus is often used to investigate interactions of agents. Cooperative games, various strategies of conducting trades (negotiations), games in placements, etc., which are analogues to a number of models of collective behavior of agents are explored.

(3) *Theory of collective behavior of automata* [5]. It studies behavior of large collectives of automata. The behavior of an automaton can be nondeterministic, which allows to constructing probabilistic models. It is possible to train an automaton by using penalties and incentives. An automaton can be endowed with a memory, which stores previous penalties and incentives in a certain form and can use this information to improve its proper behavior and collective behavior in accordance with a function of income.

(4) *Theory of multi-agent systems* (MASs) [6]. MASs can be used to solve problems that are difficult to solve with a single agent or a monolithic system. Examples of such systems are online trading, emergency response, social structure modeling, computer games, transport, logistics, network technologies, etc. Within MAS, software agents are examined usually, but robots can also be components of a MAS. A MAS can manifest self-organization and complex behavior, even if the strategy of each agent's behavior is simple enough.

Consider in more detail theory of MASs, which claims to be a basis for implementation of collective behavior of robots. As the basic formal model of an intelligent agent, the *BDI*-model (*BDI* stands for Belief-Desire-Intention) was chosen [6]. In this model, knowledge, beliefs, intentions and mechanisms of an agent's reasoning are described in terms of predicate calculus with using modal and temporal operators. This confusion of logics causes many difficulties in understanding and exploiting a logical formalization of the *BDI*-model, thus erecting a significant barrier for interpretation of MAS's basic concepts, in particular, for understanding the very term *BDI*. As noted in [6], the *BDI*-model is theoretically much more complicated than the first-order predicate calculus, and therefore it is difficult to hope for practically acceptable efficiency of the model in solving problems on industrial level.

Next, consider the possibility of representing a MAS as a knowledge base (KB) comprizing some rules. MASs consist of the following main components:

- a set of system units with a distinguished subset of active units, i.e., agents manipulating a subset of passive units (objects);
- environment, i.e. a space where agents and objects function;
- a set of tasks (functions, roles), which are assigned to agents;
- a set of relations (interactions) between agents;
- a set of actions of agents (for example, various communicative acts or operations on objects).

For psychological terms of the *BDI*-model, a fairly clear system interpretation can be formulated [6, 7].

Beliefs formalize some agent's available information about laws and the current state of an external environment and other agents. Such information can presumably be erroneous and incomplete, therefore it can be considered only as representations rather than reliable knowledge. In particular, to model the agent's representations about behavior of an external environment, we can introduce the relation $belief \subseteq S \times A \times S$, where S is a set of states of an external environment, and A is a set of actions. If a triple (s, a, s_i) enters the *belief* relation, then, according to the agent's representations, after performing an action a at a state s of the external environment, the latter can change to a state s_i. Beliefs can also be described as rules that use not only attributes A and S, but also other factors (for example, agent's own states, signals from the external environment, messages from other agents, etc.).

Desires of agents are the states to be achieved according to the agents' knowledge bases. In a number of cases, attainable states correspond to goals. Essentially, desires are a collection of tasks (or task chains) that an agent is able to perform in certain situations (for example, do a room, move from one point to another, neutralize an explosive device, etc.). It is clear that the simultaneous achievement of many goals by one agent or a group of agents is not always possible, so the choice of the set of goals to be performed in this particular situation must be logically consistent.

Intentions formalize possible actions of agents to achieve their goals. Intentions are formed depending on the specific situation (a certain team, obligations towards other agents, occurrence of an abnormal situation described in the knowledge base, etc.).

Taking into account this interpretation clarifies that it is sufficient to use rule-based KBs in order to model some versions of MASs. The difficulty lies in the fact that many MASs require not only deductive, but also defeasible (plausible) methods of reasoning analysis. The mathematical means of n-tuple algebra (NTA) [8–10] provides incorporation of these methods within the framework of classical logic, particularly, for using in robotics.

One of the problems in modeling interactions of robots is the possibility of conflicts. Local beliefs of one agent can, for example, contradict beliefs of other agents. An agent can form a goal that conflicts with goals of other agents. Main *types of conflicts* in MASs are as follows [6]:

(1) conflicts within the *agents' belief system* that may arise when an agent receives false information;
(2) conflicts caused by incompleteness of the agent's knowledge regarding the model of the surrounding world and models of other agents;
(3) conflicts associated with *competition for shared resources* or with *conflicting goals*.

One of possible ways to prevent conflicts is formalization of various kinds of constraints, whose fulfillment allows either to completely avoid, or at least to significantly reduce the number of conflict situations. As a tool for solving such problems in the framework of artificial intelligence, the methodology of constraint programming is widely used [11]. Application of NTA to constraint satisfaction problems allows to substantially reduce required computational resources [12].

Usage of rule-based KBs in groups of interacting agents implies studying anomalies in such KBs; a brief analysis of these anomalies is provided in the next section.

2 KBs' Anomalies

KBs' anomalies are considered in many publications, for example, in [13–15]. The first results on methods for recognizing and eliminating KBs' anomalies were related to the works on KBs' verification [16]. There exist anomalies associated with *integrity violations* (errors in description of types and values of attributes) and with *consistency violations* (errors in rules themselves). Let us briefly consider some kinds of anomalies of the second type.

The *duplication anomaly* is recognized when some two rules are identical and differ only in the order of their variables. A *partial inclusion anomaly* is detected when some two rules have the same consequent, but the antecedent of one rule is a subset of the antecedent of another rule, or, on the contrary, antecedents of the rules are the same, and the consequents satisfy an inclusion relation.

Researchers distinguish several types of *inconsistency anomalies*. Consider two of them, namely incompatibility of an antecedent and contradiction anomaly [15].

A rule r_p usually has the structure $B_1 \wedge B_2 \wedge \ldots \wedge B_n \to A$. It contains an *incompatible antecedent* if there is at least one pair of inconsistent atoms B_i and B_j in this antecedent. This means that $B_i = C_i(x)$, $B_j = C_j(x)$ and $C_i(x) \subseteq \overline{C_j(x)}$, i.e., these atoms contain the same variable and do not intersect.

Contradiction anomaly. Suppose there are two rules:

$$r_1 : B_1 \wedge B_2 \wedge \ldots \wedge B_n \rightarrow D;$$

$$r_2 : C_1 \wedge C_2 \wedge \ldots \wedge C_n \rightarrow F.$$

Moreover, $C_i \subseteq B_i$ for each i ($i = 1, 2, \ldots, n$) and $D \cap F = \emptyset$. Then the rules r_1 and r_2 initiate an anomaly of contradiction.

A particular case of contradictory rules are the rules with contradictory consequents and coinciding antecedents. For example, the following rules can exist in a robot's KB:

$$r_p : B_1 \wedge B_2 \rightarrow D;$$

$$r_q : B_1 \wedge B_2 \rightarrow F;$$

where B_1 means "there is an obstacle ahead," B_2 says "the target is behind the obstacle," D advises to get around the obstacle on the right, F requires to get around the obstacle on the left.

These rules clearly contradict each other, although both are feasible. We can consider them a contradiction anomaly and make corresponding corrections in the KB (for instance, delete one of the rules). However, there are situations when removal of one of the rules can lead to undesirable consequences, since the contradiction anomaly can not only be a signal of a fault in a KBs' system. It turns out that reasoning can lead to situations when conflicting rules become completely correct after adding some perhaps missed data. An example of this kind is an unusual situation, discovered in linguistics relatively recently and called presupposition. Let us consider it in more detail.

3 Presupposition

The concept of presupposition (English equivalent is assumption) can be very often found in literature on logic and philosophy [17–19], linguistics [20], cognitive science [21], neuro linguistic programming (NLP) [22], etc. There are many different definitions of this concept. Let us accept the following one.

Presupposition is a statement that is implied (or perceived as true) when the main statement or question is actualized, and the negation (or falsity) of the basic statement does not violate the trueness of the presupposition.

For instance, the sentence "John has returned back in his family" stipulates that John left the family once, and this is a presupposition. Clearly, negation of the basic statement ("John has not returned to his family") does not question the truth of the presupposition. At the same time, if the sentence "Peter bought a smartphone in a store" is considered a presupposition for the statement "Peter had enough money to buy," another situation arises. If we deny the main statement, we can assume that one of the possible reasons for refusing to buy was that Peter did not have enough money to pay at that time. Therefore, it is incorrect to use the sentence "Peter had enough money to buy" as a presupposition. It is not uncommon for a presupposition to be expressed in a sentence explicitly. For example, the statement "Richard did not know that wolves were found in this forest" has an evident presupposition "There are wolves in this forest."

Often, implicitly expressed presuppositions yield to perceiving some statements at the subconscious level. This can be used by specialists to form hidden attitudes in humans, in other words, to manipulate consciousness [22]. Presuppositions in advertising, as well as in polemics for formulation of "tricky" questions, in which the opponent's fault is implicitly asserted, are used very effectively. For instance, "Do you continue to beat your father?" or "Are you going to return the stolen goods?".

In artificial intelligence, the notion of presupposition was studied in detail by Popov [23]. One of his important conclusions in the research of communication with a computer in natural language was that neglecting presuppositions can distort meaning of texts during automatic translation. Pospelov [24] drew attention to importance of analysis of presuppositions in reasoning models. It should be noted that modern literature on artificial intelligence (AI) do not practically mention presuppositions. In particular, this word is not present in such representative AI monographs as [25, 26]. Further on we follow [19] in naming a main statement as *assertion*.

4   Logical Analysis of Presuppositions

Pospelov [24] noted that problems of linking expressed and unspoken facts excited as early as medieval logicians. However, linguists believe that one of the first who attracted attention of many scientists to unspoken statements in logical analysis was G. Frege. In particular, he proposed to distinguish between what is asserted in a statement and what is supposed [27]. His understanding of presuppositions was some simplistic yet, he regarded as what is now called presupposition, only statements about existence of a subject of reference. For example, the phrase "Mozart died in poverty" assumes that the name "Mozart" means existence of a man with the surname Mozart.

A detailed logical analysis of presuppositions is contained in works of Strawson [17] and van Fraassen [18]. Strawson defines presupposition as follows: a sentence P is a presupposition of S, if trueness of P is a necessary condition for S to have a truth value (that is, it can be either true or false). If P is false, S is neither true nor false.

Van Fraassen examines the relationships that arise between presupposition and implication. As one of the possible options, he proposes such a definition:

P is a presupposition of S, if and only if:

(a) if S is true, then P is true,
(b) if (non-S) is true, then P is true.

This relation can be expressed in the form of formulas in propositional calculus: P is a presupposition of S if $(S \supset P)$ and $(\neg S \supset P)$.

In some publications, presupposition is determined not by means of implication, but by using the notion of "sequence": "Proposition P is a presupposition of S, if it is a sequence both from S and from negation of S." However, such interpretations of presupposition are questionable. First, as noted in [19], the formula $(S \supset P) \wedge (\neg S \supset P)$ is equivalent to the formula P, i.e., the assertion S turns out to become a fictitious logical variable under this interpretation. Secondly, numerous examples of presuppositions show that, in the sense of the sentences analyzed, presupposition usually serves a precondition for an assertion, but not vice versa. As a rule, events, expressed in an assertion, are a continuation of the events described in a presupposition, and therefore their consideration as a precondition (or antecedent) is incorrect.

Often a search for a presupposition P from a given assertion S is considered a result of a logical analysis reminiscent of derivation a corollary. For instance, the following reasoning seems correct: "From the fact that John continues to beat his father (S), it follows that John used to beat his father (P)." However, the stricter approach shows out that this case is not a question of deductive derivation of P from S, rather, it is restoring a preceding event.

In view of what has been said, there are reasons to suggest a logical definition of presupposition different from the cited works. Consider the sentence (S) "Anthony was late for school." Its presupposition is clearly the sentence (P) "Anthony was going to school." This sentence is true also in the case when Anthony was not late. If we assume that P is false, then S does not make any sense. It cannot be called false, because its denial ("Anthony was not late for school"), is also false in fact.

If presupposition is considered a precondition, the formal approach leads to a paradox. Indeed, the formula $(P \supset S) \wedge (P \supset \neg S)$ is equipotent to the formula $\neg P$ that confirms the unconditional falsity of the presupposition. Conversely, informal analysis of all examples of presupposition shows that there is no paradox. To understand the problem, let us look more closely at examples of presuppositions. Consider the example "Anthony was late for school." The event "Anthony was going to school" evidently preceded Anthony's being late (or not late) to school.

To explain presupposition as a precondition within the framework of classical logic, a new factor needs to be added to reasoning. It is clear that Anthony, when going to school, could be late for various reasons (overslept, met friends and talked with them, helped an old woman to cross the road, etc.). At the same time, if there were no such interfering factors, Anthony would not be late for school. Hence, correctness of a presupposition as a prerequisite for a basic sentence can be ensured by introducing at least one new factor (attribute, variable). In our example, such a factor may be a logical variable R denoting presence or absence of reasons for being late for school. Obviously, it is impossible to substantiate some "oddities" of presupposition without such a factor.

Let us give a new logical definition of presupposition. Assume S is a sentence, and P is its presupposition. We add a new attribute R, which we call the *relay of an assertion*. Then, in the propositional calculus we get:

Definition 1 P is a *presupposition* of S if there exists a parameter (attribute) R such that $(P \wedge R) \supset S$ and $(P \wedge \neg R) \supset \neg S$.

Analysis shows that there are no dummy variables in the formula $((P \wedge R) \supset S) \wedge ((P \wedge \neg R) \supset \neg S)$, i.e., all variables including the main sentence (S), presupposition (P) and relay (R) are meaningful. In addition, this formula initiates no paradox, since values of P in it can be true as well as false.

Let us consider applicability of this definition by another example. A presupposition for the sentence "Alex did not pass the contest to an institute" can be formulated as the sentence "Alex entered the Institute." If we deny the original sentence, the truth of this presupposition does not change. In order to explain what contributed to the fact that Alex passed (or did not pass) through the contest, at least one more attribute is needed to characterize a reason for passing (or not passing) through the competition (for instance, the level of training, Alex's abilities).

At the same time, the proposed approach does not explain yet another "strangeness" of presupposition: if a presupposition is false or denied, the sense of both assertion and its negation loses its meaning. For instance, presupposition for the assertion "Jones has recovered" is the sentence "Jones has hitherto been sick." If we deny the presupposition, then neither assertion, nor its denial make sense (Jones did not recover). The same is revealed in the analysis of many other examples of presupposition.

Linguists believe that it is necessary to go beyond the two-valued logic in order to explain this phenomenon. In other words, we have to use nonclassical logic as a tool for analysis of this case. However, it is possible to solve this problem in the framework of classical logic, it is enough to assume that the number of values of some logical variables can be more than two. Hence, explanation is achievable by means of predicate calculus rather than propositional calculus.

Let the values of attributes $P = \{0, 1\}$, $R = \{0, 1\}$, $S = \{a, b, c\}$ be given. Then we can define presupposition as follows.

Definition 2 P is a presupposition of S if the below-listed rules are satisfied:

(1) if $P = 1$ and $R = 1$, then $S = a$;
(2) if $P = 1$ and $R = 0$, then $S = b$;
(3) if $P = 0$, then $S = c$.

Rule (3) simulates the "loss of meaning" of an assertion under falsity of the presupposition. So, if Jones *was not sick* ($P = 0$), then at the moment he is *neither recovered* ($S \neq a$), nor *continues to be sick* ($S \neq b$). Conversely, he simply *continues to be healthy* ($S = c$) or maybe he *fell ill* ($S = c_1$), which we do not wish him, of course. Almost all examples of presuppositions are easily analyzed using this introduced scheme.

5 Considering Contradiction Anomaly as Presupposition

Let us consider an example of contradictory rules from Sect. 1:

r_p if an obstacle is ahead, and the goal is behind the obstacle, go around the obstacle on the right;

r_q if an obstacle is ahead, and the goal is behind the obstacle, go around the obstacle on the left.

We can say that this case arises no anomaly, rather, this situation is similar to presupposition. Then, correction of the KB will not require for removing one of the rules, we will need to find an additional attribute (the assertion relay, see Definitions 1 and 2). In particular, values of such an attribute can describe a list and locations of other obstacles disposed to the right or left of the main obstacle.

It is easy to see that the logical definition of contradictory rules and presupposition is the same. The only difference is that the precondition of a presupposition contains only one attribute, while matching (or nested) preconditions in contradictory rules can contain more than one attribute. However, this does not prevent us from considering the contradiction anomaly not only as a source of a paradox, but also as a signal to try finding some additional attributes of contradictory rules that remove the paradox, i.e., finding an assertion relay.

Consider some approaches to develop methods for finding a relay of an assertion. Let a knowledge base contain conflicting rules:

$$r_D : \quad : B_1 \wedge B_2 \wedge \ldots \wedge B_n \rightarrow D;$$

$$r_F : \quad : B_1 \wedge B_2 \wedge \ldots \wedge B_n \rightarrow F,$$

and $D \cap F = \varnothing$.

In these rules, each atom B_i is represented by a set of values of an attribute X_i. For example, the expression "If $X_i = a$ or $X_i = b$, then ..." means that B_i in this rule is represented by the set of values $\{a, b\}$. Similarly, the consequent of every rule can be modeled as a set of values of a specific attribute (for instance, Y). In our example, the value of the attribute Y in the rule r_D is the set D, and for the rule r_F this value equals the set F.

Each rule is specified in a specific relation diagram, and each relation diagram is characterized by a set of attribute names. For a rule r_m, we denote the relation diagram of its antecedent as $Ant(r_m)$; $Cons(r_m)$ will stand for the relation diagram of its consequent, and $Val(X_i, r_m)$ means the value of the attribute X_i in this rule.

Let X_{Contr} be the set of attributes in antecedents of contradictory rules r_D and r_F, i.e., $Ant(r_D) = Ant(r_F) = X_{Contr}$.

Then the search algorithm of an assertion relay for the above contradictory rules will be as follows.

(1) Among the set of all rules in the KB, find the set S of rules r_m for which $Cons$ $(r_m) = \{Y\}$;

(2) In the set S, form the subsets of rules S_D and S_F, for which values of the attribute Y are equal to the sets D and F correspondingly;

(3) Using the sets S_D and S_F, find the set P of pairs (r_m, r_n) such that $r_m \in S_D$, $r_n \in S_F$ and $(Ant(r_m) \cap Ant(r_n)) \setminus X_{Contr} \neq \emptyset$;

(4) In the set P, test the correlation $Val(X_i, r_m) \cap Val(X_i, r_n) = \emptyset$ for each pair (r_m, r_n) and each attribute X_i;

(5) If in step 4 yields any "yes," the attribute X_i is an assertion relay for conflicting rules r_D and r_F.

For negative result of the search, a similar algorithm can be used for the case when pairs, triples, etc. of attributes are considered as assertion relays rather than single attributes.

6 Conclusion

A connection between presupposition and the contradiction anomaly in knowledge bases of interacting agents is shown. To explain presupposition as a precondition within the framework of classical logic, it is suggested to supply reasoning with a new factor(s) formalizing presence or absence of reasons for trueness or falsity of an assertion. To simulate the "loss of meaning" of an assertion with falsity of its presupposition, it is proposed to use the model of predicate calculus instead of propositional calculus in order to define the assertion as a logical variable with more than two values. For contradictory rules, an algorithm for finding assertion relays is developed.

Acknowledgements The authors would like to thank the Russian Foundation for Basic Researches (grants 16-29-04424, 16-29-12901, 18-07-00132, 18-01-00076) for partial funding of this research.

References

1. Gorodetskiy, A.E. (ed.): Smart Electromechanical Systems. Springer International Publishing (2016)
2. Bond, A., Gasser, L.: Readings in Distributed Artificial Intelligence. Morgan Kaufmann, San Mateo, CA (1988)
3. Luce, R.D., Raiffa, H.: Games and Decisions. Wiley, New York (1957)

4. Myerson, R.: Game Theory: Analysis of Conflict. Harward University Press, Cambrige, Massachusetts (1991)
5. Varshavsky, V.I.: Collective Behavior of Automata. Nauka, Moscow (1973). (in Russian)
6. Wooldridge, M.: An Introduction to Multi-agent Systems. Wiley (2009)
7. Gorodetsky, V.I., Skobelev, P.O.: Industrial applications of multi-agent technology: reality and perspectives. SPIIRAS Proc. **6**(55) (2017) (in Russian)
8. Kulik, B.A., Zuenko, A.A., Friedman, A.Ya.: Deductive and defeasible reasoning on the basis of a unified algebraic approach. Sci. Tech. Inf. Process. **42**(6), 402–410 (2015)
9. Kulik, B.A., Fridman, A.Ya.: Logical analysis of data and knowledge with uncertainties in SEMS. In: Gorodetskiy, A.E. (eds.) Studies in Systems, Decision and Control. Smart Electromechanical Systems, vol. 49, pp. 45–59. Springer International Publishing Switzerland (2016)
10. Kulik, B.A., Fridman, A.Ya.: Unified logical analysis in robots' CNS based on N-Tuple algebra. In: A.E. Gorodetskiy, V.G. Kurbanov (eds.) Smart Electromechanical Systems: The Central Nervous System. Series: Studies in Systems, Decision and Control, vol. 95, Chapter 2, pp. 17–30. 1st edn. Springer (2017)
11. Ruttkay, Z.: Constraint satisfaction—a survey. CWI Q. **11**, 163–214 (1998)
12. Zuenko, A.A.: Constraint inference upon matrix representations of finite predicates. Artif. Intell. Decis. Making **3**, 21–31 (2014). (in Russian)
13. Preece, A.D.: Validation of knowledge-based systems: the state-of-the-art in North America. J. Comm. Cogn. Artif. Intell., **1**(№ 4) (1994)
14. Felfernig, A., Friedrich, G., Jannach, D., Stumptner, M.: Consistency-based diagnosis of configuration knowledge bases. AI J. **152**(2) (2004)
15. Baumeister, J., Seipel, D.: Anomalies in ontologies with rules. Web Semant. Sci. Serv. Agents World Wide Web **8**(1) (2010)
16. Nguyen, T.A., Perkins, W.A., Laffey, T.J., Pecora, D.: Knowledge base verification. AI Mag. **8**(№ 2) (1987)
17. Strawson, P.: Introduction to Logical Theory. London (1952)
18. van Fraassen, B.: Presupposition, implication and self-reference. J. Philos. **65**(№ 5) (1968)
19. Beaver, D.: Presupposition and Assertion in Dynamic Semantics. Ph.D. Dissertation, University of Edinburgh (1995)
20. Karttunen, L., Peters, S.: Requiem for presupposition [microform]. Lauri Karttunen and Stanley Peters, Distributed by ERIC Clearinghouse (1977)
21. Lakoff, G.: Women, Fire, and Dangerous Things. The University of Chicago Press, Chicago, IL (1987)
22. Bandler, R., Grinder, J.: The Structure of Magic I: A Book About Language and Therapy. Science & Behavior Books, Palo Alto, CA (1975)
23. Popov, E.V.: Communication with a Computer in a Natural Language. Nauka, Moscow (1982). (in Russian)
24. Pospelov, D.A.: Modeling of Reasoning. Experience in Analysis of Mental Acts. Radio i Svyaz', Moscow (1989). (in Russian)
25. Russel, S., Norvig, P.: Artificial Intelligence: A Modern Approach, 2nd edn. Prentice Hall (2003)
26. Thayse, A., Gribomont, P., Hulin, G., et al.: Approche logique de l'intelligence artificielle, vol. 1. De la logique classique a la programmation logique, Paris (1988)
27. Frege, G.: Sinn und Bedeutung. In: Frege G. Funktion, Begriff, Bedeutung. Fünf logische Studien, pp. 38–63. Vandenhoeck & Ruprecht, Göttingen (1962)

Group Interaction of SEMS Modules at Control of an Adaptive Surface of the Main Dish

Andrey Yu. Kuchmin

Abstract *Problem statement*: Further improvement of quality of dish systems (DS) of large radiotelescope (RT) is bound to development of more perfect systems of the active correction of shapes of dishes and their relative position as it is impossible to create the metal construction having the required rigidity. The active correction assumes existence of the composite control system of DS using the modern measurement system with an accuracy $\lambda/10$ capable to predict deformations of DS and to develop cooperative control of drives of the DS elements for compensation of these deformations. Therefore SEMS modules can be used for these purposes. There are operating antennas with adaptive reflectors which facets move by means of adaptive platforms of Gough-Stewart, for example, as it is realized in the 500th meter radio telescope of FAST. Also similar systems are used for positioning of dishes of Naismith and counterreflectors (GTM, RT-13, etc.). In addition there is realization of basic fulcrum arrangements on Gough-Stewart's platform, for example, the massif of anisotropy of relict radiation of Lie Yuanzhe, a radio interferometer from 13 elements (with a possibility of expansion to 19) on the rotary 6-m platform, the MMA project, a space radio telescope of Millimetron, etc. *Purpose of research*: development of algorithms for control of an adaptive surface of the main dish of the radio telescope consisting of adaptive facets which executive mechanisms are SEMS modules. *Results*: The algorithm of finding of the approximating paraboloid of the main dish taking into account features of executive mechanisms of SEMS modules, such as restriction for movements of rods of actuators and the line of their action. The algorithm of calculation of lengthenings of actuators as the setting influences in a control system of an adaptive surface is offered. *Practical significance*: the proposed algorithms for control of adaptive surface can be used in creating modern antennas for compensation of various external influences, such as weight and uneven heating.

A. Yu. Kuchmin (✉)
Institute of Problems of Mechanical Engineering,
Russian Academy of Sciences, St. Petersburg, Russia
e-mail: radiotelescope@yandex.ru

© Springer Nature Switzerland AG 2019
A. E. Gorodetskiy and I. L. Tarasova (eds.), *Smart Electromechanical Systems*,
Studies in Systems, Decision and Control 174,
https://doi.org/10.1007/978-3-319-99759-9_15

Keywords Radio telescope · SEMS · Adaptive surface · Approximating paraboloid

1 Introduction

SEMS modules were widely adopted in control systems of antenna installations in quality of elements of the positioning mount (PM) of the dish systems of radio telescopes [1–7]. For all history of a radio astronomy various designs of radio telescopes were developed, but practice showed that optical-type antennas with full-rotary reflectors—analogs of optical telescopes reflectors were the most convenient in work. Problems of creation and control of large radio telescopes are reflected in works of scientists and designers: Arkhipov [8], Bakhrakh [9], Vinogradov [10], Voskresensky [11], Gurbanyazova [12], Kalachev [13], Kozlov [12], Sokolovsky [14, 15], Belyansky [16], Polyak [17], Tarasov [12], Dubarenko [18–21]. Basic element of antennas of this type is the dish which collects the radiation falling on it in a focal point or on the focal line. In focus the irradiator in the form of a loud-hailer or a matrix of loud-hailers is installed. The directional diagram of the irradiator is formed so that to irradiate all dish, but to exclude radiation of space out of it. It reaches maximal use of a surface of a dish and the minimum level of noise.

Depending on the scheme of DS the radio telescope consists of the following basic elements:

1. DS RT elements:
   1.1. Main Dish (MD), or reflector;
   1.2. A counterreflector (CR) (for Kassegren, Gregory and Naismith's schemes);
   1.3. The Periscopic Mirror (PM), or Naismith's mirror (only for Naismith's scheme);
   1.4. Receiver, or irradiator;
2. PM RT elements:
   2.1. The platform providing the turn DS of rather azimuthal axis;
   2.2. The basis providing turn of rather elevation angle axis;
   2.3. The farm framework supporting the DS elements;
   2.4. A crossbar to which to fasten CR (for Kassegren, Gregory and Naismith's schemes);
   2.5. The counterbalances compensate weight a farm framework, the basis, a crossbar and DS.

Full-rotary antennas have larger advantages before the fixed: they can be sent to any point of a palate and also to carry out with their help keeping track of by a radiation source—to accumulate a signal that allows to exceed limit of sensibility of the receiver and to allocate space radio signals of extremely low power against the background of various noise.

Further improvement of quality of dish systems of large radiotelescope is bound to development of more perfect systems of the active correction of shapes of dishes and their relative position as it is impossible to create the metal construction having the required rigidity. The active correction assumes existence of the composite control system of DS using the modern measurement system with an accuracy λ/10 capable to predict deformations of DS and to develop cooperative control of drives of the DS elements for compensation of these deformations. Therefore SEMS modules can be used for these purposes. There are operating antennas with adaptive reflectors which facets move by means of adaptive platforms of Gough-Stewart, for example, as it is realized in the 500th meter radio telescope of FAST. Also similar systems are used for positioning of dishes of Naismith and counterreflectors (GTM, RT-13, etc.). In addition there is realization of basic fulcrum arrangements on Gough-Stewart's platform, for example, the massif of anisotropy of relict radiation of Lie Yuanzhe, a radio interferometer from 13 elements (with a possibility of expansion to 19) on the rotary 6-m platform, the MMA project, a space radio telescope of Millimetron, etc.

2 Principles of Creation of Adaptive Dish Systems

As a prototype we will consider the dish system of Naismith consisting of 3 mirrors: parabolic reflector, elliptic counterreflector and plane periscopic mirror. The similar system is supposed to be used on a radio telescope a 70-m radio telescope of RT-70. The radio telescope of RT-70 is under construction on the mountain Suffa plateau in spurs of the Turkestan Range in the Republic of Uzbekistan. A working range of frequencies of the accepted radiation of 5–300 GHz (6 cm–1 mm). Taking into account unique radio astroclimate of the region of the RT plateau will mainly work in a short-wave part of millimetric wave band.

The radio telescope of RT-70 has adaptive MD—a paraboloid of rotation with a diameter of 70 m. Targeting of MD is carried out on a azimuth angle and a elevation angle by means of two electric drives.

Realization of the principle of homologous deformations which is been the basis for calculation and projection of MD imposed a number of distinctiveness on a design of dish system: compensation of change of a focal distance and the provision of a focal axis when targeting on a elevation angle is made automatically by the linear and angular movement of CR so that the phase distortions caused by this movement completely compensated the phase distortions caused by deformation of MD.

The counterreflector with a diameter of 3 m has the form of an ellipsoid of revolution it will be collected from single sheets on a framework. Control of the provision of CR is exercised of a control system. The counterreflector is equipped with the six-degree drive allowing CR to move on three linear and three angular coordinates.

As it was already noted, the main source of decrease in effectiveness of larger RT are static and dynamic deformations of the DS elements.

The modern requirements to accuracy of reflective surfaces of DS of larger radio telescopes lead to creation of means of their adaptation to an external loading, such as weight, wind load capacities, thermo deformations etc. Global trends in projection of similar DS are directed to introduction of the SEMS universal modules that is bound to design and economic advantages of their use. For antennas of small and effective diameter (up to 25 m) the PM can be executed on the adaptive platform which provides both angular, and linear movements of the antenna on six degree of freedoms that can be used, for example, for rolling compensation if the antenna is installed by the ship, such antenna has no dead sectors and can provide the continuous maintenance of space objects at larger elevation angles. At addition of the antenna access to the send-receive equipment without servicing tower is provided.

Similar mounts in the basis use the adaptive platforms of Gough-Stewart (hexapod) moved with electromechanical actuators, position of the platform is defined by the firmware laser measuring system.

The next level of adaptation is use of the methods of active correction of reflective surfaces. Many modern antennas in the basis use the principle of a homology and there is a need of correction of a form and the provision of secondary dishes: counterreflector and periscopic mirror. For this purpose of CR and PM can be also installed on adaptive platforms or suspensions with the parallel kinematic scheme, for example, the hexapod. The purpose of movement of CR is combination of focuses of MD and CR on three linear and two angular coordinates and also scanning exercise, for the purpose of capture of sources on maintenance, measurement of sky noise temperature etc. By means of PM switching from one receiver to another is implemented and fast correction of errors of pointing of MD and CR, can be also carried out quickly scanning, with speeds larger, that it can be done by CR. For compensation of deformations of MD the surface of CR can be made from separate boards, everyone them which is also established on the tripod or the hexapod that provides on the one hand a high level of adaptation, with another the required rigidity of a design and a universalization of its components.

For work in the millimetric range of radiowaves the extreme accuracies of a surface of MD which cannot be received only for the account of effects of a homology are required, and it is required to use adaptive surfaces of MD. For this purpose the surface of MD is carried out made of separate boards, by analogy with the surface of the CR. These boards the systems of electromechanical jacks are also installed. So in adaptive surfaces of MD and CR of a first generation (Green Bank, the USA) the group of four boards is fastened with corners on a support, and each support are moved by the actuator. Thus each board has four points of fixing that leads to its deformation when moving, and the design allows only the linear movements and has low rigidity. Adaptive surfaces of a second generation are made on the basis of tripods and hexapods by means of which each board moves. In this case boards are mechanically untied and both angular, and linear movements of boards without their deformation are provided. The third generation of adaptive mirrors provides use of the actuators allowing to correct surfaces of boards MD and CR. The main problem in control of similar DS is calculation of approximating surfaces taking into account properties of an electrodynamics and lines of action of operating mechanism.

The aspiration is correct to formulate the purpose of control, resulted in need of the analysis of standard duties of RT and a research of the DS RT electrodynamic properties of mm range for receiving estimates of influence of deformations of its elements on distribution of an electromagnetic field in the region of secondary focus where the irradiator is installed. The electrodynamic model of DS (EDM) was for this purpose developed. This model is used for calculation of position of focuses of dish system as function of deformations that allows to make corrections to positions of elements of dish system. It reaches upgrading of receiving of radio signals. Usually EDM represents the functional converter which entrances are coordinates of the reference points of dish system, and exits of coordinates of focuses.

3   Control of an Adaptive Surface of the Main Dish

For calculation of lengthenings of rods of actuators of an adaptive surface of the main dish it is necessary to calculate the approximating paraboloid (AP) of MD, use of the method of least squares (MLS). In MD surface observed data, the point set which are characterized by coordinates $x_{0,i}$, $y_{0,i}$, $z_{0,i}$ in basic coordinate system (BCS) through which it is necessary to carry out a paraboloid is received and to choose its numerical parameters, so that the sum of squares of deviations of the calculated points was minimum, taking into account the finite size of the lengthenings of rods of actuators and the line of their action.

Let's formulate criteria of finding of AP MD. A primal problem of an adaptive surface is ensuring the required root mean square deviation of a surface of MD from a surface created due to the principle of homology. On the other hand it is necessary to provide physical feasibility of this AP as the size of movement of boards is limited and it is necessary to provide that movement happened near justified values of displacements of a rod of actuators. Therefore it is expedient to choose the following as criteria: a minimum of average deviation of errors of AP from the measured field of points on a normal to a surface AP MD and a minimum of the sum of the displacements of rods of actuators:

$$J_1 = \min_{\mathbf{p}} \xi_n^2(\mathbf{p}), \quad J_2 = \min_{\mathbf{p}} \sum_{j=1}^{m} l_{a,j}^2(\mathbf{p}),$$

where ξ_n^2—the sum of squares of discrepancies on a normal to AP MD, $l_{a,j}^2$—square of length of j-th actuator, $\mathbf{p}$—vector of parameters AP MD: $\mathbf{p} = [x_v \; y_v \; z_v \; \beta \; \theta \; a_p]^{\mathrm{T}}$, $|p_i| \leq p_i^{\max}$, $v_0 = [x_v, y_v, z_v]$—coordinates of apex of AP in BCS, β, θ—angles of orientation of focal axis in BCS, a_p—focal distance.

As restrictions the model of AP which is set in AP coordinate system (CS AP) by the initial equation of a paraboloid is chosen:

$$\mathbf{x}_{p,i}^{\mathsf{T}}\mathbf{A}_p\mathbf{x}_{p,i} + 2a_p\mathbf{B}_p^{\mathsf{T}}\mathbf{x}_{p,i} = 0,$$

$$\mathbf{x}_{p,i} = \begin{bmatrix} x_{p,i} \\ y_{p,i} \\ z_{p,i} \end{bmatrix}, \; \mathbf{A}_p = \begin{bmatrix} 1 & 0 & 0 \\ 0 & 1 & 0 \\ 0 & 0 & 0 \end{bmatrix}, \; \mathbf{B}_p = \begin{bmatrix} 0 \\ 0 \\ -1 \end{bmatrix},$$

where $\mathbf{x}_{p,i}$—paraboloid point coordinates in CS AP. As the measured coordinates of points of a surface are given in BSK it is more conveniently to represent of AP in BCS:

$$\mathbf{x}_{p,i} = \mathbf{c}(\beta, \theta)^{\mathsf{T}}\left[\mathbf{x}_{p,0,i} - \mathbf{v}_0\right],$$

$$\left[\mathbf{x}_{p,0,i} - \mathbf{v}_0\right]^{\mathsf{T}}\mathbf{A}_0\left[\mathbf{x}_{p,0,i} - \mathbf{v}_0\right] + 2a_p\mathbf{B}_0^{\mathsf{T}}\left[\mathbf{x}_{p,0,i} - \mathbf{v}_0\right] = 0, \tag{1}$$

$$\mathbf{A}_0 = \mathbf{c}\mathbf{A}_p\mathbf{c}^{\mathsf{T}}, \mathbf{B}_0^{\mathsf{T}} = \mathbf{B}_p^{\mathsf{T}}\mathbf{c}^{\mathsf{T}},$$

where $\mathbf{x}_{p,0,i}$—paraboloid point coordinates in BCS.

To calculate discrepancies between points of a surface of MD and AP it is necessary to enter a normal to AP which is defined by the equation in BCS:

$$\mathbf{B}_p^{\mathsf{T}}\mathbf{x}_{p,i} - \mathbf{B}_p^{\mathsf{T}}\mathbf{c}^{\mathsf{T}}\left[\mathbf{x}_{0,i} - \mathbf{v}_0\right] + a_p\sqrt{\frac{\left[\mathbf{x}_{0,i} - \mathbf{v}_0\right]^{\mathsf{T}}\mathbf{c}\mathbf{A}_p\mathbf{c}^{\mathsf{T}}\left[\mathbf{x}_{0,i} - \mathbf{v}_0\right]}{\mathbf{x}_{p,i}^{\mathsf{T}}\mathbf{c}\mathbf{A}_p\mathbf{c}^{\mathsf{T}}\mathbf{x}_{p,i}}} = 0,$$

$\mathbf{x}_{0,i}$—coordinates of a point of a surface of the main dish. Then the discrepancy equation on a normal has an appearance:

$$\xi_n^2 = \sum_i \left[\mathbf{c}^{\mathsf{T}}\mathbf{x}_{0,i} - \mathbf{c}^{\mathsf{T}}\mathbf{v}_0 - \mathbf{x}_{p,i}\right]^{\mathsf{T}}\left[\mathbf{c}^{\mathsf{T}}\mathbf{x}_{0,i} - \mathbf{c}^{\mathsf{T}}\mathbf{v}_0 - \mathbf{x}_{p,i}\right].$$

It is necessary to define restrictions for the displacements of rods of actuators. The line of operation of actuators is set in BCS in parametrical form as follows:

$$\mathbf{x}_{p,b0,j} = \mathbf{x}_{b0,j} + l_{a,j}\mathbf{e}_{a0,j}, \tag{2}$$

where $\mathbf{x}_{p,b0,j}$—cross point of the line of operation of the actuator with AP MD, $\mathbf{x}_{b0,j}$—coordinates of justified position of base of the actuator, $l_{a,j}$—actuator rod length, $\mathbf{e}_{a0,j}$—basis vector of the line of operation of the actuator. Having substituted (2) in (1) we will receive expression for lengths of rods of actuators:

$$\left[\mathbf{e}_{a0,j}^{\mathsf{T}}\mathbf{A}_0\mathbf{e}_{a0,j}\right]l_{a,j}^2 + \left[2\mathbf{e}_{a0,j}^{\mathsf{T}}\mathbf{A}_0\mathbf{x}_{b0,j} - 2\mathbf{e}_{a0,j}^{\mathsf{T}}\mathbf{A}_0\mathbf{v}_0 + 2a_p\mathbf{B}_0^{\mathsf{T}}\mathbf{e}_{a0,j}\right]l_{a,j}$$
$$+ \left[\mathbf{x}_{b0,j}^{\mathsf{T}}\mathbf{A}_0\mathbf{x}_{b0,j} - 2\mathbf{x}_{b0,j}^{\mathsf{T}}\mathbf{A}_0\mathbf{v}_0 + \mathbf{v}_0^{\mathsf{T}}\mathbf{A}_0\mathbf{v}_0 + 2a_p\mathbf{B}_0^{\mathsf{T}}\mathbf{x}_{b0,j} - 2a_p\mathbf{B}_0^{\mathsf{T}}\mathbf{v}_0\right] = 0. \tag{3}$$

The length of a rod of the actuator has to be in range $0 \le l_{a,j} \le l_{a,j}^{\max}$, where $l_{a,j}^{\max}$—maximal value of the length of a rod of the actuator.

Usually in practice errors of a deviation of an actual surface from AP MD are set that allows to pass to one goal function, having introduced additional restrictions, then we will receive the following problem definition of creation of AP MD:

$$J_2 = \min_{\mathbf{p}} \sum_{j=1}^{m} l_{a,j}^2(\mathbf{p})$$

at restrictions:

$$\mathbf{x}_{p,i} = \mathbf{c}(\beta, \theta)^{\mathrm{T}} \left[\mathbf{x}_{p,0,i} - \mathbf{v}_0 \right],$$

$$\left[\mathbf{x}_{p,0,i} - \mathbf{v}_0 \right]^{\mathrm{T}} \mathbf{A}_0 \left[\mathbf{x}_{p,0,i} - \mathbf{v}_0 \right] + 2a_p \mathbf{B}_0^{\mathrm{T}} \left[\mathbf{x}_{p,0,i} - \mathbf{v}_0 \right] = 0,$$

$$\mathbf{A}_0 = \mathbf{c} \mathbf{A}_p \mathbf{c}^{\mathrm{T}}, \mathbf{B}_0^{\mathrm{T}} = \mathbf{B}_p^{\mathrm{T}} \mathbf{c}^{\mathrm{T}},$$

$$\mathbf{B}_p^{\mathrm{T}} \mathbf{x}_{p,i} - \mathbf{B}_p^{\mathrm{T}} \mathbf{c}^{\mathrm{T}} \left[\mathbf{x}_{0,i} - \mathbf{v}_0 \right] + a_p \sqrt{\frac{\left[\mathbf{x}_{0,i} - \mathbf{v}_0 \right]^{\mathrm{T}} \mathbf{c} \mathbf{A}_p \mathbf{c}^{\mathrm{T}} \left[\mathbf{x}_{0,i} - \mathbf{v}_0 \right]}{\mathbf{x}_{p,i}^{\mathrm{T}} \mathbf{c} \mathbf{A}_p \mathbf{c}^{\mathrm{T}} \mathbf{x}_{p,i}}} = 0,$$

$$\xi_n^2 = \sum_i \left[\mathbf{c}^{\mathrm{T}} \mathbf{x}_{0,i} - \mathbf{c}^{\mathrm{T}} \mathbf{v}_0 - \mathbf{x}_{p,i} - \mathbf{x}_{s,i} \right]^{\mathrm{T}} \left[\mathbf{c}^{\mathrm{T}} \mathbf{x}_{0,i} - \mathbf{c}^{\mathrm{T}} \mathbf{v}_0 - \mathbf{x}_{p,i} - \mathbf{x}_{s,i} \right] \le \varepsilon^2,$$

$$\left[\mathbf{e}_{a0,j}^{\mathrm{T}} \mathbf{A}_0 \mathbf{e}_{a0,j} \right] l_{a,j}^2 + \left[2\mathbf{e}_{a0,j}^{\mathrm{T}} \mathbf{A}_0 \mathbf{x}_{b0,j} - 2\mathbf{e}_{a0,j}^{\mathrm{T}} \mathbf{A}_0 \mathbf{v}_0 + 2a_p \mathbf{B}_0^{\mathrm{T}} \mathbf{e}_{a0,j} \right] l_{a,j}$$
$$+ \left[\mathbf{x}_{b0,j}^{\mathrm{T}} \mathbf{A}_0 \mathbf{x}_{b0,j} - 2\mathbf{x}_{b0,j}^{\mathrm{T}} \mathbf{A}_0 \mathbf{v}_0 + \mathbf{v}_0^{\mathrm{T}} \mathbf{A}_0 \mathbf{v}_0 + 2a_p \mathbf{B}_0^{\mathrm{T}} \mathbf{x}_{b0,j} - 2a_p \mathbf{B}_0^{\mathrm{T}} \mathbf{v}_0 \right] = 0,$$

$\mathbf{x}_{s,i}$—coordinates of points of the moved boards, ε—mean squared deviation of a surface. Displacements of rods $l_{a,j}$ calculate as functions of the AP MD parameters it agrees (3).

4 Group Interaction of SEMS Modules

The adaptive surface of MD RT consists of huge number of effectors therefore there is a question of their group interaction for the purpose of implementation of the laws of control calculated proceeding from the solution of the optimizing task described above. Depending on type of an adaptive surface various types of SEMS modules, such as packages of parallel actuators, tripods, hexapods can be used. The existing designs of adaptive surfaces give necessary information on advantages and shortcomings of each type of an operating mechanism. The greatest simplicity of a design is reached in case of use of packages of parallel actuators. In this case on one actuator 4 boards meet, and there is a problem of calculation of deformations of boards when moving one actuator as driving of one actuator involves four boards at once. In such design each board has four points of fastening and can be deformed during its movement. These deformations have to be controlled and be in an elastic range, otherwise it can lead to damages of a board. This type of an operating mechanism has considerably smaller rigidity, than, for example, at the tripod or the hexapod.

Operating mechanism in the form of adaptive platforms on the basis of tripods and hexapods are more convenient in use and control, though considerably raise the price of a design. In this case the separate board can be established on the similar adaptive platform and move independently that does not lead to a hyperstatic task, as for the previous case. Tripods and hexapods besides actuators still have the systems of hinges therefore monitoring of their situation has to be provided with the composite measuring system. Group control of massifs of tripods or hexapods comes down to searching of provisions of boards, most close corresponding AP MD. However it is necessary to control gaps between boards. Similar operating mechanism allow not only the linear, but also angular movements of boards therefore are perspective as operating mechanism of adaptive surfaces of the reflecting elements.

5   Conclusion

In article questions of creation and control of an adaptive surface of the main dish of a full-rotary radio telescope of millimetric range of radiowaves are considered. The method of calculation of an approximating paraboloid and a technique of definition of optimum laws of displacements of rods of actuators taking into account the line of their action and restrictions for the range of lengthenings is presented. As criterion the minimum of movements of separate boards of rather average position of displacements of actuators is chosen. It is specified that depending on type of operating mechanism of an adaptive surface (tripods, hexapods, the systems of parallel actuators) various techniques of interaction between SEMS modules are required. It is shown, it is the most preferable to use hexapods which represent the adaptive platforms moved with six actuators as operating mechanism. As this design has the considerable rigidity and can hold a separate board, this case the mechanical sheaf with the next boards is not required, and problems with their collateral deformation at control are as a result fixed.

References

1. Artemenko, Y.N., Agapov, V.A., Dubarenko, V.V., Kuchmin, A.Y.: Co-operative control of subdish actuators of radio telescope. Informatsionno-upravliaiushchie sistemy **4**, 2–9 (2012) (In Russian)
2. Gorodetsky, A.E., Kurbanov, V.G., Tarasova, I.L., Kuchmin, A.Y.: Electric drives of system of logical control of the position of a counterreflector of the space radio telescope. Antenna **4**, 52–55 (2011) (In Russian)

3. Gorodetsky, A.E., Kurbanov, V.G., Tarasova, I.L., Kuchmin, A.Y.: Structure of the system of logical control of the position of a counterreflector of the space radio telescope. Antenna **4**, 56–59 (2011) (In Russian)
4. Artemenko, Y.N., Gorodetsky, A.E., Dubarenko, V.V., Kuchmin, A.Y., Tarasova, I.L.: Problems of development of space radio-telescope adaptation systems. Informatsionno-upravliaiushchie sistemy **3**, 2–8 (2010) (In Russian)
5. Kingsley, J.S., Martin, R.N., Gasho, V.L.: A Hexapod 12 m antenna design concept for the MMA. MMA Memo 263, 7 May 1999
6. Koch, P.M., et al.: The AMiBA Hexapod Telescope Mount. arXiv.org>astro-ph>arXiv:0902.2335v1, 13 Feb 2009. https://doi.org/10.1088/0004-637x/694/2/1670
7. Artemenko, Y.N., Gorodetsky, A.E., Dubarenko, V.V., Kuchmin, A.Y., Agapov, V.A.: Analysis of dynamics of automatic control system of space radio-telescope subdish actuators. Informatsionno-upravliaiushchie sistemy **6**, 2–6 (2011) (In Russian)
8. Arkhipov, M.Y., Telepnev, P.P., Kuznetsov, D.A.: To a question of numerical modeling of dynamics of a design of the SPEKTR-R spacecraft. In: Lavochkina, S.A. (ed.) The Bulletin of NPO IM, vol. 3, issue 24, pp. 96–99 (2014) (In Russian)
9. Bakhrakh, L.D., Galimov, G.K.: The mirror scanning antennas. In: Theory and Methods of Calculation, 300 pages. Nauka (1981) (In Russian)
10. Vinogradov, I.S.: Radiation cooling of a mirror of the large-size space telescope. In: Radio-astronomical Equipment and Methods. FIAN (Works FIAN; T.228) (2000) (In Russian)
11. Voskresensky, D.I., Kanashchenkov, A.I.: The active phased antenna lattices. Radiotechnika **488** (2004) (In Russian)
12. Gurbanyazov, M.A.: Modern problems of creation of mirror antennas. In: Gurbanyazov, M. A., Kozlov, A.N., Tarasov, V.B., Bakhrakh, L.D. (eds.) AN of Turkmenistan, NPO Solntse, Institute of Mathematics and Mechanics, 416 pages. Ylym, Ashgabat (1992) (In Russian)
13. Kalachev, P.D.: Research of elastic properties of the full-rotary parabolic antenna of the text radio telescope. In: Kalachev, P.D., Dyachkov, V.E. (eds.) Radiotelescopes. Submillimetric and x-ray Relescopes. Works FIAN. T.77. Nauka (1974) (In Russian)
14. Bortsov, Y.A.: The automatic electric drive with elastic links. In: Bortsov, Y.A., Sokolovsky, G.G. (eds.) Energoatomizdat, 288 pages (1992) (In Russian)
15. Sokolovsky, G.G., et al.: Control of the radio telescope electric drive with use of the simplified observer. Izv. LETI, L.: LETI publishing house (344), 23–33 (1984) (In Russian)
16. Belyansky, P.V.: Control of land antennas and radio telescopes. In: Belyansky, P.V., Sergeyev, B.G. (eds.) Soviet Radio, 279 pages (1980) (In Russian)
17. Polyak, V.S.: Evolution of development of precision designs of radio telescopes for radio astronomy, long-distance and satellite space communication. PGS: Magazine. PGS **5**, 14 (2005) (In Russian)
18. Dubarenko, V.V., Kuchmin, A.Y.: Identification of complex mechanical systems as objects of control. In: Regional Bulletin of Young Scientists: Collection of Articles of Young Scientists and Graduate Students, vol. 2, pp. 7–9 (2006) (In Russian)
19. Dubarenko, V.V., Kuchmin, A.Y.: An approach to improve the quality of pointing a millimeter wave range large radiotelescope with an adaptive dish system. Informatsionno-upravliaiushchie sistemy **5**, 14–19 (2007) (In Russian)
20. Dubarenko, V.V., Kuchmin, A.Y.: Modeling and identification of complex mechanical systems as objects of control. In: Bulletin of Young Scientists: Collection of Articles of Young Scientists and Graduate Students, pp. 23—26 (2007) (In Russian)
21. Dubarenko, V.V., Kuchmin, A.Y., Artemenko, Y.N.: Radiotelescopes; IPMASH RAS, 546 pages. Polytechnical University Publishing House (2014) (In Russian)

Identification of Dynamics
of Modules SEMS

Andrey Yu. Kuchmin

Abstract *Problem statement*: SEMS modules are the composite non-linear non-stationary objects. The central problem of creation of models adequate to an actual object is identification not only object parameters, but also its structure and a state that represents a non-linear problem of mathematical programming of big dimension in the common statement. Therefore for simplification of a problem of identification it can be divide into a number of the independent stages: estimation of structure of model, parameters and state. This work is focused on modification of methods of least squares. *Purpose of research*: development of algorithms of identification of dynamics of the basic SEMS module with use of the identifier on the basis of a method of least squares. The primal problem of identification is formed as finding of a transfer matrix of the basic SEMS module. *Results*: The algorithm of identification of parameters and structure of the model of the basic SEMS module consisting in application of the modified method of least squares is developed and its effectiveness on the example of the hexapod of PI M-810 at its driving in small deviations concerning the given trajectory is proved. *Practical significance*: the offered algorithm of identification can be used at synthesis of adaptive controllers in the composite systems on the basis of SEMS modules, for example, during creation of the modern antennas where similar modules are used for compensation of the weight deformations and deformations caused by heating.

Keywords Identification · SEMS · A transfer matrix

A. Yu. Kuchmin (✉)
Institute of Problems of Mechanical Engineering,
Russian Academy of Sciences, St. Petersburg, Russia
e-mail: radiotelescope@yandex.ru

© Springer Nature Switzerland AG 2019
A. E. Gorodetskiy and I. L. Tarasova (eds.), *Smart Electromechanical Systems*,
Studies in Systems, Decision and Control 174,
https://doi.org/10.1007/978-3-319-99759-9_16

1 Introduction

Recently there was interest in use of electromechanical systems of parallel architecture, for example n-pods [1–9], in high-precision instrument making, robotics, adaptive antennas etc. Advantages, ranges of application and the description of these mechanisms are reflected in publications [3–9].

The perspective direction in creation of robotic, high-precision and distributed systems is use of a paradigm of cyberphysical systems [1, 2]. The main difference from classical configurations is use of the self-organized modules with parallel kinematic structure and parallel distributed by the control system which is developed on the principles of optimal control, self-organization and agent approach. Such configurations are demanded in various areas of technical applications such as: antennas, mobile robots, trunk robots manipulators, high-precision machines, simulators etc. Advantages of such systems are:

- small metal consumption,
- high robustness,
- ability to self-organization and adaptation to change of an external loading,
- the movements with a large number of degree of freedoms,
- high carrying capacity,
- simplicity of an adjustment,
- simplicity of manufacture (restricted number of standard sizes).

As shortcomings it should be noted:

- the complex control algorithms,
- existence of special positions (jammings and loss of stability).

As one rapidly developing direction. It should be noted intellectual electromechanical systems. Which basic component is the mainframe representing the adaptive platform moved with electromechanical jacks (actuators). Creation of linearized models of the adaptive platforms on the relative frame basis moved with packages of actuators taking into account change of the line of operation of these actuators working in the mode of tracking and identification of their parameters is a relevant task which plays an important role during the calculating of parameters of controllers and estimation of not measurement components of state vector of the platform [10, 11].

2 Problems of Identification of Dynamic Objects

Creation of models adequate to an actual object is the central problem identification not only object parameters, but also its structure and a state that represents in the common statement a non-linear problem of mathematical programming of big dimension even for a case of the linear nonstationary dynamic object [12–14].

Therefore for simplification of a problem of identification it can be broken into a number of the interdependent stages: determination of structure of model, parameters and state. This work is focused on modification of methods on the basis of least-squares [12, 15], and the set-functional methods [16, 17]. Methods of identification can be carried to the category of optimizing tasks which demand the difficult calculations and specialized hardware for their realization. Until recently application of similar approaches in actual tasks restrained for this reason. As a turning point it is possible to note emergence of a algorithm of phase training [15] that gave the chance to describe the complex non-linear systems in the form of multilayer network structure. Later this approach was realized in neural network technologies.

All existing approaches to identification can be broken into two groups: the set-functional and statistical [12–16]. Such division is caused by differences in accounting of the disturbance operating on system.

Many algorithms of parametrical identification are constructed taking into account the generalized function chart reflecting the main informational streams used in the course of a parameter estimation of an object [18]. This scheme is shown in the Fig. 1.

Apparently from the figure, on an input of the studied object some test influence $g(t)$, for example, step, pulse, harmonic, polynomial, noise etc. moves. The output signal $y(t)$ represents reaction to test influence $g(t)$. The signal, the more information is more dynamic it comprises. Means, in the course of identification it is necessary to provoke artificially the informational saturation of an output signal $y(t)$ by essential evolutions of the test setting action $g(t)$.

Let's notice that process of parametrical estimation conflicts to a task of control of an object. Stabilization of driving of an object, for example, complicates parametrical identification because of "impoverishment" of informational streams. On the contrary, saturation of informational streams leads to destabilization of driving of an object.

Very often as test step influence is used. It is known that reaction to such influence is a transfer characteristic of an object. The informational streams created thus are easy in realization, but are less informative. The largest informational saturation takes place at the beginning of processes, and by the time of achievement of the setting state the informational content of processes decreases to zero. For this reason for padding "swing" of the studied object stochastic process (noise) $f_1(t)$

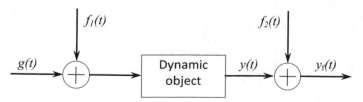

Fig. 1 The generalized function scheme of model of parametrical estimation: $g(t)$—the setting action; $y(t)$—the output signal; $f_1(t)$—the input Gaussian noise; $f_2(t)$—the Gaussian noise of measurements of output; $y_f(t)$—the noisy output signal

promoting increase in informational content of input and output signals is added to the setting action. Unlike input noise $f_1(t)$ playing a positive role in the course of identification, output noise $f_2(t)$ is a hindrance or is caused by inaccuracy of measurements of an output signal $y(t)$. Output noise $f_2(t)$ reduces accuracy of the solution of a problem of identification therefore its role is negative, and it is necessary to seek for decrease in the size $f_2(t)$.

As we see, the procedure of identification demands giving on an input of an object of test influence, receiving reaction at the output of an object, filing of processes on an input and an output with the subsequent analysis of results. There are methods of the analysis of parameters of an object on his continuous an input-output characteristics. However these methods are badly formalized, and, therefore, their algorithmic realization on the computer is difficult. The most prime represent collecting and processing of discrete information. Dynamic models which parameters are calculated in the course of such identification are, therefore, discrete.

Too big period of quantization leads to loss of information, and, so to deterioration of the received model. Decrease of the period of quantization complicates processing of information obtained in an observed data as the majority of machine time is spent for poll of sensors.

Thus, to start process of identification it is necessary to estimate dynamic properties of an object beforehand roughly. If it is difficult, then the period of quantization is selected iterations.

As we see, for identification not the continuous processes, but selections of their measurements which are carried out with the quantization period are used. Respectively, and the used algorithms of parametrical identification are focused on processing of discrete selections.

For the analysis of thread-specific data in control systems with statistical identifiers, there is a problem of the choice of time frames of "a floating window" that is usually badly formalized for an analytical research and has algorithmic character. Showed practical researches of data from dynamic objects that there is an opportunity for the discrete description of similar dynamic objects to enter a matrix shift operator in a time domain:

$$\mathbf{A}^\tau = \begin{cases} \mathbf{A}^{\text{fix}\left(\frac{\tau\omega_0}{2\pi}\right)}\left[1 - \frac{\tau\omega_0}{2\pi} + \text{fix}\left(\frac{\tau\omega_0}{2\pi}\right)\right] + \left(\frac{\tau\omega_0}{2\pi} - \text{fix}\left(\frac{\tau\omega_0}{2\pi}\right)\right)\mathbf{A}^{\text{fix}(\tau)+1}, \tau \geq 0 \\ \left(\mathbf{A}^\mathrm{T}\right)^{\text{fix}(-\tau)}\left[1 + \frac{\tau\omega_0}{2\pi} + \text{fix}\left(\frac{-\tau\omega_0}{2\pi}\right)\right] + \left(\frac{\tau\omega_0}{2\pi} + \text{fix}\left(\frac{-\tau\omega_0}{2\pi}\right)\right)\left(\mathbf{A}^\mathrm{T}\right)^{\text{fix}(-\tau)+1}, \tau < 0 \end{cases},$$

$$\mathbf{A} = \begin{bmatrix} 0 & 0 & & 0 \\ 1 & 0 & & 0 \\ & & \ddots & \\ 0 & 0 & 1 & 0 \end{bmatrix}_{n \times n}$$

where τ—the shift in a time domain, ω_0—the frequency of sampling, $\text{fix}(\ldots)$—the function of rounding with a deletion of a fractional part, n—the number of samples. For example, for a problem of a parameter estimation of the discrete filter:

$$\sum_{i_1=0}^{m_1} a_{i_1} x[t_k - \tau_{i_1}] + \sum_{i_2=1}^{m_2} \sum_{i_3=0}^{m_{3,i_2}} b_{i_2,i_3} u_{i_2} \left[t_k - \tau_{i_2,i_3}\right] = 0,$$

where a_{i_1}, b_{i_2,i_3}—the filter coefficients, $i_1, i_2,$ i_3—the corresponding indexes, m_1, m_2, m_{3,i_2}—the maximal values of indexes, t_k—the current time, τ_{i_1}, τ_{i_2,i_3}—the time shifts, x—output coordinate, u_{i_2}—the control inputs, it is possible to use compact analytical form:

$$\sum_{i_1=0}^{m_1} \mathbf{A}^{-\tau_{i_1}} \mathbf{x} a_{i_1} + \sum_{i_2=1}^{m_2} \sum_{i_3=0}^{m_{3,i_2}} \mathbf{A}^{-\tau_{i_2,i_3}} \mathbf{u}_{i_2} b_{i_2,i_3} = \varepsilon, \mathbf{Fp} = \varepsilon,$$

$$[a_0, a_1, \ldots, a_{m_1}, b_{1,0}, b_{1,1}, \ldots, b_{m_2,m_3,m_2}]^{\mathrm{T}} = \mathbf{p},$$

$$[\mathbf{A}^{-\tau_0}\mathbf{x}, \mathbf{A}^{-\tau_1}\mathbf{x}, \ldots, \mathbf{A}^{-\tau_{m_1}}\mathbf{x}, \mathbf{A}^{-\tau_{1,0}}\mathbf{u}_1, \mathbf{A}^{-\tau_{1,1}}\mathbf{u}_1, \ldots, \mathbf{A}^{-\tau_{m_2,m_3,m_2}}\mathbf{u}_{m_2}] = \mathbf{F},$$

$$(1)$$

where $\mathbf{x}, \mathbf{u}_{i_2}, \varepsilon$—corresponding a vector of samples of output coordinate, the input influences and discrepancies, $\mathbf{p}$—vector of the estimated parameters.

The algorithms of parametrical identification based on a method of least squares (LS-identifier) find broad practical application. The LS-identifier favourably differs from the determined identifier in a possibility of flexible change of volume of regression selection and noncriticality to existence of noise in measurements.

Transition to a form (1) allows to use methods of LS-identifier and to find estimates of parameters of the filter through minimization of target function $J(\mathbf{p}) = \mathbf{p}^{\mathrm{T}}\mathbf{Hp}$. The analysis of eigenvalues of a matrix and vectors of $\mathbf{H} = \mathbf{F}^{\mathrm{T}}\mathbf{F}$ allows to execute a reduction of initial model taking into account the prior restrictions for the frequencies range and informational saturation of data.

In case of a low ratio signal/noise and the low informational saturation of data it is expedient to pass from the description (1) to the system of the linear inequalities taking into account interval assessment β of mismatches between the measured values of output coordinate dynamic object and received from model $\mathbf{e}$:

$$\begin{cases} \mathbf{F}(\beta_1)\mathbf{p} \geq 0 \\ \mathbf{F}(\beta_2)\mathbf{p} \leq 0 \end{cases}, \text{where } \sum_{i_1=0}^{m_1} a_{i_1} \geq 0, \quad \text{or} \quad \begin{cases} \mathbf{F}(\beta_1)\mathbf{p} \leq 0 \\ \mathbf{F}(\beta_2)\mathbf{p} \geq 0 \end{cases}, \quad \text{where } \sum_{i_1=0}^{m_1} a_{i_1} \leq 0,$$

$$\sum_{i_1=0}^{m_1} [\mathbf{A}^{-\tau_{i_1}} \mathbf{x} + \mathbf{Ie}] a_{i_1} + \sum_{i_2=1}^{m_2} \sum_{i_3=0}^{m_{3,i_2}} \mathbf{A}^{-\tau_{i_2,i_3}} \mathbf{u}_{i_2} b_{i_2,i_3} = 0, e_i \in \beta = [\beta_1 \quad \beta_2],$$

$$\mathbf{F}(\beta) = [\mathbf{A}^{-\tau_0}\mathbf{x} + \xi\beta, \mathbf{A}^{-\tau_1}\mathbf{x} + \xi\beta, \ldots, \mathbf{A}^{-\tau_1}\mathbf{x} + \xi\beta, \mathbf{A}^{-\tau_{1,0}}\mathbf{u}_1, \mathbf{A}^{-\tau_{1,1}}\mathbf{u}_1, \ldots, \mathbf{A}^{-\tau_{m_2,m_3,m_2}}\mathbf{u}_{m_2}],$$

$$(2)$$

where $\mathbf{I}$—identity matrix, ξ—vector of units.

3   Identification of Parameters of the Basic Module of SEMS

Object of a research is the module SEMS on the basis of the hexapod of PI M-810.

The M-810 hexapod of the Physik Instrumente (PI) company (Fig. 2) with the platform with a diameter of 10 cm is one of the most compact six-coordinate systems of micropositioning working by the principle of a parallel kinematics. The systems of a parallel kinematics in comparison with the systems of a serial kinematics have larger a rigidity, smaller dimensions, do not possess moving cables that reduces a sliding friction and increases accuracy of positioning.

The relative frame platform of the hexapod is set in motion by means of six independent high-precision engines thanks to what it is possible to carry out positioning on three linear coordinates (X, Y, Z) and to three coordinates of rotation around the corresponding axes (θ_X, θ_Y, θ_Z). In the M-810 hexapod the principle of vector control of engines is used, at the same time, the maximal speed of positioning is 2.5 mm/sec.

The M-810 hexapod is delivered complete with the controller. The M-810.D12 model has the C-887.21 controller

The user software of PIMikroMove with a broad set of opportunities and intuitively the clear interface, LabView drivers and libraries of functions for development of proprietary applications is included in the package of the software. There is an opportunity by program methods to set a rotation point that is especially important for problems of an adjustment.

The controller can be connected to the computer on means of TCP/IP or RS-232 interfaces.

Fig. 2 General view of the hexapod of PI M-810

Let's consider a problem of identification of parameters of the hexapod at its driving in small deviations concerning the given trajectory. The hexapod can move on six degree of freedoms to three linear movements (x, y, z) and to three corners $(\theta_X, \theta_Y, \theta_Z)$ therefore it makes sense to enter division of movements on six independent channels proceeding from superposition of small movements. Let's consider identification of parameters on the example of the channel of movement on the linear coordinate x. Let's define structure of model for identification. Output variable will be the coordinate of the linear movement x. Six lengthenings of actuators u_1, ..., u_6 will be the input variables. The model for problems of identification of the hexapod of coordinate x will have a form:

$$\sum_{i_1=0}^{m_1} a_{i_1} x[t_k - \tau_{i_1}] + \sum_{i_2=1}^{6} \sum_{i_3=0}^{m_2} b_{i_2,i_3} u_{i_2}[t_k - \tau_{i_2,i_3}] = 0.$$

Using the LS-identifier described in the previous section and at small deviations of movements of the hexapod, we will construct a transfer matrix of system.

Algorithm of operation of the identifier:

1. Start
2. Collecting measuring information and formation of selections on each of coordinates of movement of the hexapod (in an example x) and six lengthenings of actuators u_1, ..., u_6.
3. Formation of a matrix $\mathbf{F}$.
4. Formation of a matrix $\mathbf{H}$.
5. Finding of eigenvalues $\mathbf{H}_{eig}$ and eigen vectors $\mathbf{V}_{eig}$ of a matrix $\mathbf{H}$.
6. System reduction by means of the analysis of eigen frequencies of a matrix $\mathbf{H}$. Only informative part of a spectrum of system remains. The matrix $\mathbf{H}_{cut} = \mathbf{H}_{eig}*\mathbf{V}_{eig}^T$ is for this purpose formed, from which the rows and columns corresponding to the chosen frequencies are cut out and matrixes $\mathbf{A}_{cut}$ and $\mathbf{B}_{cut}$ are formed.
7. Parameters of the filter are on a formula: $\mathbf{P} = -\text{inv}(\mathbf{A}_{cut})*\mathbf{B}_{cut}$.
8. End.

In case of a low ratio signal/noise it is more efficient to use the identifier set by expression (2).

Algorithm of operation of the identifier:

1. Start
2. Collecting measuring information and formation of selections on each of coordinates of movement of the hexapod (in an example x) and six lengthenings of actuators u_1, ..., u_6.
3. The choice of interval assessment of a discrepancy for the identifier.
4. Creation of system of the linear inequalities.
5. Application of methods of the linear programming for the solution of the received system of inequalities.

6. Model operation of the difference filter for the purpose of definition of a discrepancy between a response of an object and the filter.
7. If quality of identification accepted to pass into point 8, differently into point 3.
8. End.

Transfer function for an input x and an output u_1:

$$\frac{-3.022z^3 + 6.558z^2 - 4.734z + 1.136}{z^4 - 2.183z^3 + 1.577z^2 - 0.3763z + 0.0001721}$$

Transfer function for an input x and an output u_2:

$$\frac{2.501z^3 - 5.407z^2 + 3.885z - 0.9274}{z^4 - 2.154z^3 + 1.547z^2 - 0.3706z + 0.0001328}$$

Transfer function for an input x and an output u_3:

$$\frac{0.7856z^3 - 1.694z^2 + 1.213z - 0.2883}{z^4 - 2.155z^3 + 1.543z^2 - 0.367z + 2.749e - 005}$$

Transfer function for an input x and an output u_4:

$$\frac{1.587z^3 - 3.43z^2 + 2.464z - 0.5877}{z^4 - 2.156z^3 + 1.548z^2 - 0.3702z + 9.034e - 005}$$

Transfer function for an input x and an output u_5:

$$\frac{1.231z^3 - 2.659z^2 + 1.908z - 0.4547}{z^4 - 2.157z^3 + 1.548z^2 - 0.3693z + 5.396e - 005}$$

Transfer function for an input x and an output u_6:

$$\frac{-1.782z^3 + 3.868z^2 - 2.793z + 0.6708}{z^4 - 2.177z^3 + 1.573z^2 - 0.3767z + 7.558e - 005}$$

Results of identification are given in the Fig. 3.

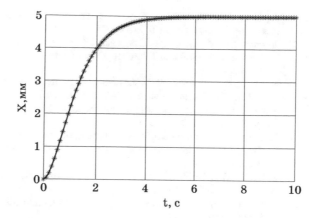

Fig. 3 Result of identification of the hexapod. (- an input of the identified model, * an output of an object)

4 Conclusion

In article questions of identification of SEMS modules on the example of the hexapod of PI M-810 are considered. Two types of identifiers are offered and algorithms of their functioning are described. It is shown that the quick action of the first identifier is higher, but it is more sensitive to measuring noise. The second type of the identifier requires the solution of the difficult optimizing task of the linear programming. Therefore it is expedient to use the first at the tactical level for fast tracking of changes of parameters of an object, the second for strategic level taking into account noise of a measuring system and for correction of operation of the tactical identifier.

Acknowledgements This work was financially supported by Russian Foundation for Basic Research, Grant 16-29-04424, Grant 18-51-06003 and Grant 18-01-00076, the Russian National Fund (grant 18-19-00005).

References

1. Gorodetskiy, A.E., Kurbanov, V.G. (eds.): Smart Electromechanical Systems: The Central Nervous System, 266 p. Springer International Publishing AG (2017)
2. Gorodetskiy A.E. (ed.): Smart Electromechanical Systems, 277 p. Springer International Publishing Switzerland (2016)
3. Volkomorov, S.V., Kaganov, Y.T., Karpenko, A.P.: Modelling and Optimization of Some Parallel Mechanisms, 32 pages. New Technologies, Moscow (2010) (In Russian)
4. Glazunov, V.A., Koliskor, A.S., Kraynev, A.F.: Spatial mechanisms of parallel structure. Science, 96 pages (1991) (In Russian)
5. Zenkevich, S.L., Yushchenko, A.S.: Bases of control of handling robots. MSTU, 480 pages (2004) (In Russian)
6. Merlet, J.P.: Parallel Robots (Solid Mechanics and Its Applications). Springer, Berlin (2004)
7. Heylo, S.V., Glazunov, V.A., Palochkin, S.V.: Handling mechanisms of parallel structure. The dynamic analysis and management. MGUDT, 86 pages (2014) (In Russian)

8. Agapov, V.A., Gorodetsky, A.E., Kuchmin, A.Y., Selivanova, E.N.: Medical Microrobot. Patent for the Invention of RUS 2469752, 20 May 2011 (In Russian)
9. Artemenko, Y.N., Badu, E.I., Gorodetsky, A.E., Dubarenko, V.V., Kuchmin, A.Y., Tarasova, I.L.: Antenna of the Radio Telescope. Patent for the Invention of RUS 2421765, 9 Feb 2010 (In Russian)
10. Kuchmin, A.Y., Dubarenko, V.V.: Linearized model of the mechanism with parallel structure. In: Gorodetskiy, A.E., Kurbanov, V.G. (eds.) Smart Electromechanical Systems: The Central Nervous System, 266 p. Springer International Publishing AG (2017). https://doi.org/10.1007/978-3-319-53327-8_13
11. Kuchmin, A.Y., Dubarenko, V.V.: Definition of a Rigidity of a hexapod. In: Gorodetskiy, A.E. (ed.) Smart Electromechanical Systems, 277 p. Springer International Publishing Switzerland (2016)
12. Isermann, R., Münchhof, M.: Identification of dynamic systems. In: An Introduction with Applications, 705 p. Springer (2011)
13. Mzyk, G.: Combined Parametric-nonparametric Identification of Block-oriented Systems, 238 p. Springer (2014)
14. Boutalis, Y., Theodoridis, D., Kottas, T., Christodoulou, M.A.: System identification and adaptive control. In: Theory and Applications of the Neurofuzzy and Fuzzy Cognitive Network Models, 313 p. Springer (2014)
15. Grop, D.: Methods of Identification Systems, 302 p. Springer (1979)
16. Karabutov, N.N.: Strukturnaia identifikatsiia sistem: Analiz dinamicheskikh struktur [Structural identification of systems: analysis of dynamic structures]. MGIU, 160 p (2008) (In Russian)
17. Pervushin, V.F.: On nonparametric models of linear dynamic objects. Bull Tomsk State Univ. Manage Comput Facil Inform. 4(25), 95–104 (2013) (In Russian)
18. Gusev, S.A.: Parametrical Identification of Dynamic Objects. Methodical Instructions to Performance of Laboratory Works, 24 p. GUAP (1997)

Implementation of a Joint Transport Task by a Group of Robots

Valery G. Gradetsky, Ivan L. Ermolov, Maxim M. Knyazkov,
Eugeniy A. Semenov, Sergey A. Sobolnikov and Artem N. Sukhanov

Abstract *Problem statement*: in a lot applications it is emerging to perform a joint transportation task of a load by a group of robots. In order to implement target trajectory of a load it is necessary to ensure coordination of motion of robots—members of a group. Moving an object by a large number of mobile robots requires information about the geometrical center of the object. However there may occur a situation that various robots—members of a group will be in contact with different types of soil. This will affect resulting transportation trajectory of a load and must be considered during motion planning and control. *Purpose of research*: moving through cross-country a mobile robot meets different types of the ground. That changes different ground-wheel's interaction parameters such as width and height of a wheel's tire, pressure in the contact area, the resistance coefficient, the track depth and magnitude of wheel's bending. In order to move a load according to desired trajectory corresponding changes should be done to motion control signal to SEMS-based group of robots considering with which type of soil robot—member of a group is in contact each specific moment. *Results*: Paper presents the calculated nomograms for determining the influence of soil and external forces on the robot's parameters and the passability depending on the soil. The simulation of the given parameters is performed. Simulation results allow creating control algorithms for mobile robots to provide joint transportation task. Also in order to solve problems arising during the movement of robots in the group there was proposed a new type of propulsion type wheel with a variable geometry. Each wheel is capable of changing its configuration at the critical changes of the wheel's attachment with the

V. G. Gradetsky · I. L. Ermolov (✉) · M. M. Knyazkov · E. A. Semenov
A. N. Sukhanov
Ishlinsky Institute for Problems in Mechanics, Russian Academy
of Sciences, Moscow, Russia
e-mail: ermolov@ipmnet.ru

A. N. Sukhanov
e-mail: sukhanov-artyom@yandex.ru

S. A. Sobolnikov
MSTU STANKIN, Moscow, Russia

© Springer Nature Switzerland AG 2019
A. E. Gorodetskiy and I. L. Tarasova (eds.), *Smart Electromechanical Systems*,
Studies in Systems, Decision and Control 174,
https://doi.org/10.1007/978-3-319-99759-9_17

ground. *Practical significance*: the robot control technique and the control for a group of robots with a wheels moving on the terrain with different traction properties of the soil are discussed. These methods use correction by applying feedback force (at rigid hitch) or feedback on the deviation of the point of attachment of the load from the nominal (at non-rigid hitch) within SEMS-based system. In this paper we present some simulation results that can be used to control a group of robots, considering target trajectory, type of load, load distribution among agents, type of terrain and specifics of soil.

Keywords Group mobile robotics · Transportation task · Wheel-ground interaction · Cross-country moving · Types of the ground · Forces in soil contact area SEMS-systems modelling

1   Introduction

A number of robotics research centers investigate transportation task performed by a group of robots. This includes a number of scientific problems to be solved [1–3].

Among a lot of tasks for single robot-transporter the following tasks were solved: consideration of geometrical and inertia properties of the object to be transported, object's slippage effect referring to UGV, consideration of undulations of ground while keeping desired orientation of transported object, interaction of soil with wheels. These tasks were discussed throughout scientific community during last 15 years [4]. However we have not met such solutions for transportation task performed by a group of robots.

In a lot of multi-robot application a difficult task is to provide reliable communication. Hence it is desirable to ensure coordination of robots in group without communication. A new approach was proposed in [4]. Group of robots includes a leader which directs a transported load into goal position. Other robots can define the direction of load's transportation. Once they "understand" the direction they also start moving the load into same direction by their own propulsion. However orientation of load is not discussed in this work.

In practice there are a lot of cases of transporting deformable loads. This is a case of [5] which discusses transportation of deformable load by a group of wheeled UGVs equipped with manipulators. There they study centralized and decentralized control approaches. An algorithm presented there secures obstacle avoidance for both static and dynamic obstacles. A remarkable point is that during whole transportation task a desired shape of deformable load is secured. However all studies were done only for statics, not for dynamics.

Concluding this analysis what was done in this area we can state the following:

- Load transportation by a group of robots on uneven terrain has not been studied completely. Additional research in this area is required.

- Within our analysis we have not detected studies on control of group of transportation robots considering their interaction with soil and uneven terrain.
- It was approved that among main tasks to solve this problem are the following: effective "wheel-soil" model compilation, identification of parameters of this model and consequent generation of control signals.
- In order to move a load according to desired trajectory corresponding changes should be done to motion control signal to SEMS-based group of robots considering with which type of soil robot—member of a group is in contact each specific moment.

2 Robots' Interaction Within Group

In order to move an object by a large number of UGVs it is necessary to consider the geometrical center of the object. This will define desired rotational velocities of wheels, however their slippage along soil should be also considered. For this goal one should be aware of the uncertainties of the environment and variety of soil. Starting 2000 this task was investigated in Omnimate project [6]. Within this project a special system correcting wheels' position error was developed (IPEC). However this system was realized only for flat surfaces.

In paper [7] which was devoted to transportation of a car by a group of robots an autonomous search and further transportation of a car was performed by a group of UGVs lifting a car by its wheels and then transporting it to desired position. Their prototype called AVERT was able to move cars with various velocities into various directions. AVERT was equipped with omniwheels, this gave a capacity to move into any direction. UGV's group had common control system which took decisions regarding trajectories. Each transporting agent had built-in collision avoidance sensors, allowing rescheduling their trajectory automatically. However omniwheels can be regarded as a shortage of this system as it doesn't allow using it on uneven terrain.

Automata theory was used to multi-agent robotic systems' synthesis, including environment modeling and man-machine interaction [8]. New control methods and autonomous UGVs' navigation were discussed in [9]. Aggregating behavior based algorithms based on criteria of autonomous agent goal approach and their mutual utility was studied in [10]. Application of social behavior models for multi-agent self-organization and control was suggested in [11].

New approaches in fast and reliable leader's selection in cases of local communication among agents were presented in [12]. Paper [13] studies various strategies for centralized control to get together an UGVs group in a specific point of flat surface basing on measured distances among robots. Paper [14] deals with selection of methods and algorithms for jam-resistant communication between robots in group. Paper [15] studies capability of UGVs' group to transport an object in case when goal position changes sporadically.

Analysis we have done has shown that for motion planning and control for group-based transportation usually assumptions are done which are far from reality in cross-country conditions. E.g. wheel-soil interaction is not considered or soil considered being non-deformable, only slippage factors are considered, some works include only 2 robots into transportation. A lot of studies do not include practical experimentation.

In this paper we perform some simulation results that can be used to control a group of SEMS-based robots, considering target trajectory, type of load, load distribution among agents, type of terrain and specifics of soil (Fig. 1).

Figure 1 shows the group of three mobile robotic platforms caring an object. The group contains Master-robot (MR-Master) that plans trajectory and gives tasks to other Slave-robots (MR-Slave). Each mobile robot has rotary platform that provides one degree of freedom between each robot and the object. That allows turning for each robot to change direction.

Within this model following assumptions were accepted:

- during transportation we consider that all robots have stable mechanical contact with load;
- during transportation distance among robots does not change;
- each robot has information regarding its position regarding other agents;
- declination away from desired coordinates will affect in slippage of load regarding transporting agent (in case of flexible hitch) or generation of torques and forces in contact points (in case of rigid hitch);
- also it is assumed that the load is attached to robot in point located on vertical axis passing through geometrical center of robot.

In terms of these assumptions there has been developed a method to control each of the robots which transports a load in group (Fig. 2). This method is based on decrease of positions error for contact zones coordinates. Here the operator downloads the local map to the environment model. The environment model uses the ground database to form map grid with areas which have their own parametrical vector. That vector consists of metric parameters, soil parameters and passability. This environment model is used by MR-Master for trajectory planning and initial

Fig. 1 Schematic view of a group of robots

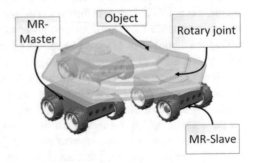

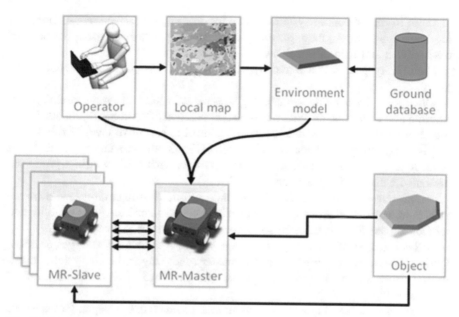

Fig. 2 Functional model of elements' interaction

optimization. It sends trajectory data to each MR-Slave. Each MR-Slave forms its own trajectory based on its position according to the object and MR-Master.

A special motion algorithm for system of transporting SEMS-based agents which secure lack of slippage was developed. It uses determination of instantaneous center of velocities for whole system and current wheels' position. This algorithms should be used by slave robots.

Once having information regarding motion trajectory and target velocity of load master robot should also have information regarding system's angular velocity on current path segment. If it's equal to zero, straightforward motion is performed. Hence target velocity of slave robots will be equal to velocity of load and of the master robot.

During coordinated functioning of robots a multi-agent control system should solve following tasks: decomposition of common task into subtasks, subtasks distribution between agents within group, motion planning for each robot and coordinated motion control within group.

Basing on these functions a generalized multi-agent control model was formed. This model uses combined approach to group control. Group control system consists of control post (master robot) with central computational device and local control systems of slave robots.

Task distribution is done at strategic control level. There local sub-tasks are generated for each SEMS-based robot of the group. Once each robot receives its subtask robot performs local control task of tactical level also interacting with neighboring robots of the group.

Input to central control system is done by task of load transportation. Once it's received it performs group planning considering position of group, model of environment and models of robots.

Preliminary plan is transferred by communication system to each robot of the group.

Once robot receives its subtask it processes it in local planning module, considering also data from navigation system and position of neighboring robots. Using this data local planning module generates control to executive level of robot.

The designed mathematical model of robot (Fig. 3) allows analyzing mobility of robot in terms of state space. Joint coordinates for each robot within a group are presented with position coordinates (X_i, Y_i, Z_i).

Parameters as input to each robot include information regarding load's trajectory, its orientation in space (R—distance between geometrical center of robot and rotation center, R_{oi}—distance between i-th wheel of robot to rotation center).

MR-Master initially considers this trajectory as a shortcome to target position. During motion in uneven environment master robot may make corrections to trajectory depending on surface's properties and existence of obstacles in front of robot or load.

A task of remote soil type identification is not studied in this paper. However we assume that this done basing on maps or sensors including sophisticated data fusion algorithms [16].

During its motion robot identifies soil type and consequently calculates necessary parameters of the surface, i.e. rolling friction coefficients, slippage and soil elasticity coefficients. These are taken from on-board data base. Also they can be obtained from tire's deflection of each wheel equipped with integrated deflection sensors.

Friction coefficients data are processed in order to obtain minimal slippage friction force, which may cause slippage, and rolling friction force. Soil elasticity parameter is used to determine contact surface between soil and wheel.

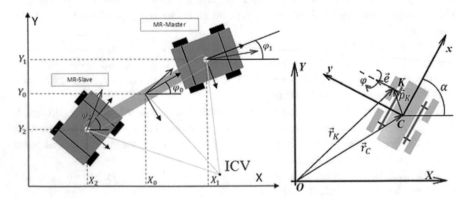

Fig. 3 Mathematical models for wheels rotation calculus

Basing on soil parameters from database SEMS-based robot determines possible deflection of wheels. In case changing of orientation of the load robot receives information regarding instantaneous velocity center. Consequently it receives necessary changes for wheels' velocities V_{oi}.

Rotation calculus presented in Fig. 3 allows determining rotation speeds of each wheel in group. Comparing desired and actual speeds V_p, measured by sensor this algorithm generates control response to motors in order to zero this difference.

Once slave robot has a motion command from master robot, it generates its own velocity vector. During motion of robot position errors due to external and internal factors may occur.

Motion control for slave robots should consider velocity vector, current coordinates of instantaneous velocity center and data coming from displacement sensor or strain gauge in load contact area.

For UGVs' motion on uneven terrain usually wheels and crawlers are used. Crawlers compared to wheels have better cross-country ability however have serious restrictions in velocity and lower efficiency.

In terms of this work also a new propulsion system was developed [17]. It presents a wheel with varying geometry (Fig. 4). While moving on various soils this propulsion system may function in 2 modes: rotating and stepping. This allows keeping high robot velocity with improved cross-country ability.

Figure 4 shows the proposed technical solution to increase mobile robot's passability. The mover consists of the outer disk moving around the shaft. The outer disk and the shaft are equipped with grousers. They are actuated and can turn on desired angle to perform better propulsion.

The degree of actuation of these grousers depends on the type of the ground. Thus control of the actuation should be considered due to differences between various types of the ground.

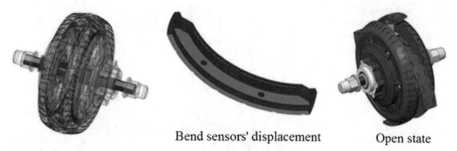

Bend sensors' displacement   Open state

Fig. 4 Wheel's configuration change and blade with deflection sensor

3 Interaction Between the Wheel and the Ground

Moving through cross-country SEMS-based mobile robot meets different types of the ground. That specifies the degree of the grousers' rotation. That changes different ground-wheel's interaction parameters such as width and height of a wheel's tire, pressure in the contact area, the resistance coefficient, the track depth and magnitude of wheel's bending. Figures 5, 6 and 7 present the calculated nomograms for determining the influence of soil and external forces on the robot's parameters and the passability depending on the soil.

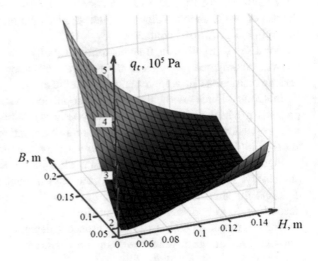

Fig. 5 The dependence of the pressure in the contact area from the geometric parameters of the tire

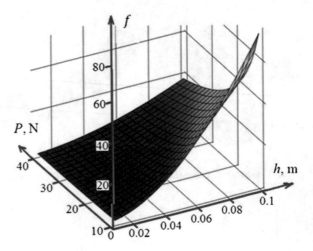

Fig. 6 The resistance coefficient change depending on the magnitude of wheel's bending and on the load

Fig. 7 The dependence of
the axial load from
deformation of the tire and the
soil type

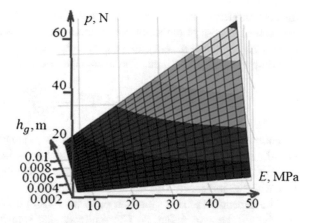

Balanced action of the forces is controlled in the process of moving. Suppose that the wheels of the robot equipped with spikes. The change of direction of any object can only be achieved by the application of additional external forces. The robot is subjected on many forces when driving. Thus tires perform important functions: every change of speed or movement direction of the robot causes the reactive forces. The tire is the communication element between the robot and the ground. Through tires all forces and torques influence on the system during acceleration and deceleration of the robot. The tire takes the action of lateral forces, keeping the robot on the operator selected trajectory. Therefore, the physical conditions of the tire friction with the road surface define the boundaries of the dynamic loads acting on the mobile robot.

The dependence of the pressure in the contact area q_t on the geometric parameters of the tire is shown in Fig. 5. This dependence is given in accordance with variation of the width and height of a wheel's tire of the robot.

The pressure in the contact area q_t depends on a number of parameters. It is expressed as follows:

$$q_t = \pi h_g \frac{(p_0 + p_w)}{2b} \left(\frac{B}{H} + \frac{3H}{2B}\right)\left(1 - \frac{h}{B}\right) \tag{1}$$

Here p_0 is the pressure caused by the stiffness of the tire carcass on the ground, 105 Pa; p_w is the internal air pressure in the tire, 105 Pa; b is width of the tire tread, m; B—width of the profile of tire, m; H is the height profile of the tire, m, h_g is the track depth, h is magnitude of wheel's bending.

Here one can see that the slight increase of the height of the tire profile with increase of the width of the tire profile leads to increase of the internal air pressure in the tire. Further increasing of the height of the tire profile with increase of the width of the tire profile decreases the internal air pressure in the tire.

Figure 6 shows the change of resistance coefficient f depending on the magnitude of wheel's bending h and on the load. The soil type on which the robot moves through influences on the resistance coefficient f. After determining the soil type, the control system of the robot can generate the necessary control for drive system of the robot to overcome the possible obstacles, or to plan a new trajectory, avoiding the dangerous area. The resistance coefficient is determined as follows:

$$f = 3.5p_0 \psi h^2 \left(B^2 + 1.5H^2\right) \frac{B - 0.3h}{PHB^2} + 0.6qz \frac{b + b_{Ki}}{P} \tag{2}$$

Here ψ is the coefficient of free traction; P is payload, H; q is roughness vertical coordinate, m; z is RMS displacement value; b_{Ki} is the width of the tire tread, m.

This figure presents increasing of the wheel's bending according to the quadratic law with the increasing of the resistance coefficient and payload.

The dependence of the load on the wheel from deformation of the tire and the soil type is shown in Fig. 7. The payload also influences the parameters of the wheel and, as a consequence, its interaction with the soil.

$$p = \frac{Eh_g}{\frac{2Eh_g}{\pi p_{so}} \arctg\left(\frac{\pi(H_g - h_g)}{2b_t}\right) + ab_t \arctg \frac{H_g - h_g}{ab_t}} \tag{3}$$

Here E is deformation modulus; h_g is the track depth; H_g is the depth of soft soil layer; p_{so} is inner pressure; b_t is tread width; $a = 0,64 \left(1 + b_t/H_g\right)$ is passability coefficient.

When the load increases than the area of contact with the surface increases too, but in loose soil it can lead to slippage and penetration of the wheel into the soil with the formation of ruts.

Table 1 shows parameters of different types of soils, which a mobile robot can move through.

Movement of loaded robot should ensure the implementation of the given tasks, which is possible when the robot meets certain requirements [18, 19]. These requirements and characteristics of movements depend on a combination of interacting forces. These requirements include maneuverability, handling, stability, passability, braking.

Table 1 Parameters of different types of soils

Types of soils	Deformation modulus E (MPa)
Loam, smooth	6
Loam, medium density	19
Middle-sized sand, loose	20
Middle-sized sand, medium density	28
Sand, medium, loose	15
Fine sand, medium density	25
Fine sand, dense	35

Moving through cross-country mobile robot meets different types of the ground. That specifies the degree of the grousers' rotation. That changes different ground-wheel's interaction parameters such as width and height of a wheel's tire, pressure in the contact area, the resistance coefficient, the track depth and magnitude of wheel's bending.

Thus the decisive evaluation criteria are:

- to sustain rectilinear movement under the action of the robot's lateral forces;
- to ensure stable turning;
- to provide reliable grip on various surfaces;
- to provide reliable grip under different weather conditions;
- to ensure reliable control of a mobile robot;
- to provide damping and smoothness.

Wheel slip occurs from the difference between the theoretical speed of a mobile robot resulting from the rotation of the wheel and the actual speed provided by the force between wheel and the road.

4 Conclusion

The robot control technique and the control for a group of robots with a wheeled chassis moving on the terrain with different traction properties of the soil are discussed. These methods use correction by applying feedback force (at rigid hitch) or feedback on the deviation of the point of attachment of the load from the nominal (at non-rigid hitch).

To solve problems arising during the movement of robots in the group there was proposed a new type of propulsion type wheel with a variable geometry. Each wheel is capable of changing its configuration at the critical changes of the wheel's attachment with the ground.

The simulation of the given parameters is performed. Simulation results may help to create control algorithms for mobile robots with such type of movers to provide joint transportation task. This will improve passability for mobile robots on different types of the ground.

Acknowledgements This research was supported by a grant №16-29-04199 from Russian Fund of Fundamental Research.

References

1. Sreenath, K., Kumar, V.: Dynamics, control and planning for cooperative manipulation of payloads suspended by cables from multiple quadrotor robots. In: Robotics: Science and Systems (RSS) (2013)

2. Brumitt, B.L., Stentz, A.: GRAMMAPS: a generalized mission planner for multiple robots in unstructured environments. In: IEEE International Conference on Robotics and Automation, vol. 2, pp. 1564–1571. Leuven, Belgium, 16–20 May 1998
3. Yamaguchi, H.: A cooperative hunting behavior by mobile robot troops. In: IEEE International Conference on Robotics and Automation, vol. 4, pp. 3204–3209. Leuven, Belgium, 16–20 May 1998
4. Wang, Z., Schwager, M.: Kinematic multi-robot manipulation with no communication using force feedback. In: International Conference on Robotics and Automation (ICRA). Stockholm, Sweden (2016)
5. Alonso-Mora, J., Knepper, R., Siegwart, R., Daniela, R.: Local motion planning for collaborative multi-robot manipulation of deformable objects. In: International Conference on Robotics and Automation (ICRA). Washington State Convention Center Seattle, Washington (2015)
6. Borenstein, J.: The omnimate: a guidewire- and beacon-free AGV for highly reconfigurable applications. Int. J. Prod. Res. **38**(9), 1993–2010 (2000)
7. Amanatiadis, A., Henschel, C., Birkicht, B., Andel, B., Charalampous, K., Kostavelis, I., May, R., Gasteratos, A.: Avert: an autonomous multi-robot system for vehicle extraction and transportation. In: 2015 IEEE International Conference on Robotics and Automation (ICRA), pp. 1662–1669, 26–30 May 2015
8. Lokhin, V.M., Man'ko, S.V., Romanov, M.P.: Development of technologies of application of the theory of automatic control of multi-agent robotic systems. Rob. Tech. Cybern. **2**(11), 3–7 (2016)
9. Mikhailov, B.B., Nazarova, A.V., Yushchenko, A.S.: Autonomous mobile robot navigation. Proc. South. Fed. Univ. Tech. Sci. **2**(175), 48–67 (2016)
10. Kulinich, A.A.: Swarm algorithms of formation and functioning for groups of robots. In: Congress on Intelligent and Information Technologies IS & IT' 16. Proceedings of the Congress on Intellectual Systems and Information Technologies AIS-IT' 16, vol. 1, pp. 301–310. SFEDU, Taganrog (2016)
11. Karpov, V.E.: Patterns of social behavior in group robotics. Large Syst Control (59), 165–232. IPU RAS, Moscow (2016)
12. Vorob'ev, V.V., Moscowsky, A.D.: The algorithm for choosing the leader in the systems with changing topology. In: The Fifteenth National Conference on Artificial Intelligence with International Participation (CAI-2016), pp. 149–157 (2016)
13. Karpov, V., Migalev, A., Moscowsky, A., Rovbo, M., Vorobiev, V.: Multirobot exploration and mapping based on the subdefinite models. In: The 1st International Conference on Interactive Collaborative Robotics, pp. 143–152 (2016)
14. Shchegoleva, L.V., Zhukov, A.V.: The problem of gathering a group of mobile robots. Mod. Sci. Res. Innov. № 8 (2016). http://web.snauka.ru/issues/2016/08/70249
15. Arkhipkin, A.V., Kopchenkov, V.I., Korolkov, D.N., Petrov, V.F., Simonov, S.B., Terent'ev, A.I.: The problem of group control of robots in a robotic fire complex. In: Proceedings of SPIIRAS, issue 45, pp. 116–129 (2016)
16. Ermolov, I.L.: Hierarchical data fusion architecture for unmanned vehicles. In: Smart Electromechanical Systems: The Central Nervous System. Springer
17. Gradetsky, V., Ermolov, I., Knyazkov, M., Lapin, B., Semenov, E., Sobolnikov, S., Sukhanov, A., Ronzhin, A., Shishlakov, V.: Highly passable propulsive device for UGVs on rugged terrain. MATEC Web of Conferences vol. 161, p. 03013 (2018)
18. Parker, L.E.: Current state of the art in distributed autonomous mobile robotics. In: Distributed Autonomous Robotic System, vol. 4, pp. 3–12. Springer, Tokyo, Japan (2000)
19. Groß, R., Dorigo, M.: Group transport of an object to a target that only some group members may sense. In: 8th International Conference, Birmingham, UK, 18–22 September 2004. Proceedings, Parallel Problem Solving from Nature—PPSN VIII Volume 3242 of the Series Lecture Notes in Computer Science, pp 852–861

Part IV
Safety, Flexibility, and Adaptability of SEMS Group

Synthesis of the Program Motion of a Robotic Space Module Acting as the Element of an Assembly and Servicing System for Emerging Orbital Facilities

Y. N. Artemenko, Anatoliy P. Karpenko and P. P. Belonozhko

Abstract *Problem statement*: the group use of emerging SEMS devices, i.e., robotic assembly and servicing space modules (RASSM), involves solving the task of synthesizing a program motion for each RASSM, taking into account its own inertial motions in terms of internal mobility degrees. It was previously shown for the model case of the plane motion of an RASSM equipped with an one degree of freedom manipulator with a load in the gripper that in the absence of the forces and moments external to the "base-manipulator-load" system in the study of a controlled relative motion, it is appropriate to introduce a given system into consideration. For the initial system's nonzero kinetic moment determined by the initial conditions and remaining constant due to the momentum conservation principle, the given system is a nonlinear oscillatory system qualitatively similar to a mathematical pendulum. In this case, the kinetic energy of the initial system can be interpreted as the total energy of the given system, which is the sum of the kinetic and potential components. In absence of a control moment in the hinge, there is an energy integral of the equation for the given system, i.e., the equation of a phase trajectory family. The problems of the synthesis of program motions for an RASSM-associated nonlinear oscillating given system have been investigated. In particular, the pulse control option, which provides a transition to the required point of a phase plane by transferring the image point to the corresponding phase trajectory over a negligible time range, has been studied. *Practical significance*: the results obtained are of interest from the viewpoint of implementing the important principle of organizing the movement of robots—coordination of free and forced

Y. N. Artemenko (✉)
The Astrospace Center of P. N. Lebedev Physical Institute,
Russian Academy of Sciences, Moscow, Russia
e-mail: artemenko.akc@yandex.ru

A. P. Karpenko · P. P. Belonozhko
Bauman Moscow State Technical University, Moscow, Russia
e-mail: apkarpenko@mail.ru

P. P. Belonozhko
e-mail: byelonozhko@mail.ru

© Springer Nature Switzerland AG 2019
A. E. Gorodetskiy and I. L. Tarasova (eds.), *Smart Electromechanical Systems*,
Studies in Systems, Decision and Control 174,
https://doi.org/10.1007/978-3-319-99759-9_18

manipulator movements—in the synthesis of control for a robotic space module as an element of an assembly and servicing system for emerging orbital facilities.

Keywords Robotic assembly and servicing space modules · The given system Optimum control in terms of energy consumption

1 Introduction

Following the outcomes of the analysis of various space robotics devices and the problems addressed thereby [1–8], a emerging class of assembly and servicing robotic space modules (RASSMs) that are equipped with a manipulator (manipulators) capable of independently traveling in space and adapted for a contact interaction with mounted (serviced) facilities [1–3, 6, 9, 10] has been singled out. A characteristic feature of such modules is the availability of dynamic modes wherein the manipulator-driven load movement relative to the base is combined with the base movement. RASSM's own inertial motions along the internal degrees of freedom are possible, too. These are enabled in the absence of both RASSM-related external forces and moments and control actions in the manipulator hinges. The study of these motions is of interest from the viewpoint of implementing the important principle of organizing the movement of robots—coordination of free and forced manipulator movements. In this case, introducing for consideration of a given system, whose dynamics equations can be obtained in the form of Routh equations [4–6, 9, 10] under certain conditions, has turned out to be effective. The paper deals with the synthesis of program motions for a particular type of an RASSM-associated nonlinear oscillating given system.

2 Initial and Given Systems

Adhering to [4–6, 9, 10], let us consider a system of two rigid bodies (a base and a load), which is free in the inertial space, connected by an ideal one degree of freedom rotational hinge, and making a plane motion, as an initial system. The bodies' masses are m_1 and m_2, J_1 and J_2 are the bodies' inertia moments relative to their mass centers, l_1 and l_2 are distances from the bodies' mass centers to the hinge. We consider a motion with respect to a non-rotating coordinate system XCY originating from the mass center of the system C, which will be inertial unless external forces and moments are acting on the system. The system position relative to XCY is determined by the angle φ_1 characterizing the absolute motion of the base and by the hinge angle q characterizing the motion of the load relative to the base. The control moment M is applied in the hinge. This design scheme can be assigned to an RASSM with one controlled degree of mobility of the manipulator (e.g., in a mode where the rest of the hinges are motionless for a specific RASSM configuration).

The expression for the kinetic energy of the initial system as a quadratic form of the independent generalized coordinates φ_1 and q is given as

$$T = \frac{1}{2}\dot{\varphi}_1^2(\gamma_1 + \gamma_2 + 2\gamma_3 \cos q) + \dot{\varphi}_1\dot{q}(\gamma_2 + \gamma_3 \cos q) + \frac{1}{2}\dot{q}^2\gamma_2, \tag{1}$$

where

$$\gamma_1 = J_1 + \tilde{m}l_1^2, \gamma_2 = J_2 + \tilde{m}l_2^2, \gamma_3 = \tilde{m}l_1l_2, \tilde{m} = \frac{m_1 m_2}{m_1 + m_2}. \tag{2}$$

In the absence of external forces and moments, the coordinate q is positional, while the coordinate φ_1 is cyclic. We have the cyclic integral

$$\frac{\partial T}{\partial \dot{\varphi}_1} = \dot{\varphi}_1(\gamma_1 + \gamma_2 + 2\gamma_3 \cos q) + \dot{q}(\gamma_2 + \gamma_3 \cos q) = L = const, \tag{3}$$

where L is the angular momentum of the system.

It is shown in [6, 9, 10] that the initial system unaffected by external forces and moments can be associated with a given nonlinear oscillatory system, whose controlled motion dynamics is described by an independent differential equation. Thus, the problem of load motion control with respect to the base can be confined to investigating the problem of the motion control of the given system.

The kinetic energy of the given system is thus

$$E_k^* = \frac{1}{2}\frac{\gamma_1\gamma_2 - \gamma_3^2 \cos^2 q}{\gamma_1 + \gamma_2 + 2\gamma_3 \cos q}\dot{q}^2. \tag{4}$$

The potential energy of the given system is thus

$$E_p^* = \frac{1}{2}\frac{1}{\gamma_1 + \gamma_2 + 2\gamma_3 \cos q}L^2 - \frac{1}{2}\frac{1}{\gamma_1 + \gamma_2 + 2\gamma_3}L^2. \tag{5}$$

The dynamics of the given system is described in Lagrange's equations of the second kind

$$\frac{d}{dt}\left(\frac{\partial(E_k^* - E_p^*)}{\partial \dot{q}}\right) - \frac{\partial(E_k^* - E_p^*)}{\partial q} = M. \tag{6}$$

Inserting (4) and (5) into (6), we obtain

$$\ddot{q}\frac{\gamma_1\gamma_2 - \gamma_3^2 \cos^2 q}{(\gamma_1 + \gamma_2 + 2\gamma_3 \cos q)} + \dot{q}^2 \frac{\gamma_3 \sin q(\gamma_1 + \gamma_3 \cos q)(\gamma_2 + \gamma_3 \cos q)}{(\gamma_1 + \gamma_2 + 2\gamma_3 \cos q)^2}$$
$$+ L^2 \frac{\gamma_3 \sin q}{(\gamma_1 + \gamma_2 + 2\gamma_3 \cos q)^2} = M. \tag{7}$$

The total energy of the given system is thus

$$E^* = E_k^* + E_p^*. \tag{8}$$

The kinetic energy of the initial system can be represented in the next form

$$T = E^* + T_L = E_k^* + E_p^* + T_L. \tag{9}$$

The last term in (9) is a constant value caused by the presence of the system's nonzero angular momentum L, which cannot be changed due to the control moment M:

$$T_L = \frac{1}{2}\frac{1}{\gamma_1 + \gamma_2 + 2\gamma_3}L^2. \tag{10}$$

3   Requirements for Control. Pulse Control Features

It is clear from (4), (5), and (7) that the given system is a nonlinear oscillatory system. The hinge ($M = 0$) with no control action has its own ballistic motions. The phase portrait is shown in Fig. 1. The potential energy E_p^* of the given system equals to zero with $q = 0$, while the kinetic energy E_k^* is maximum. The potential energy E_p^* is maximum with $q = \pm\pi$, while the kinetic energy E_k^* is zero.

Two types of own ballistic motions can be singled out: oscillations and rotations. The corresponding groups of phase trajectories are separated by a separatrix. The limiting motion along the separatrix corresponds to the total energy value of the given system

$$Es = E_{p\,\text{max}}^* = \frac{1}{2}\frac{1}{\gamma_1 + \gamma_2 - 2\gamma_3}L^2 - \frac{1}{2}\frac{1}{\gamma_1 + \gamma_2 + 2\gamma_3}L^2. \tag{11}$$

Taking into account (7), the time derivative of the total energy (8) of the given system is

$$\dot{E}^* = \frac{d}{dt}E^* = \frac{d}{dt}(E_k^* + E_p^*) = M\dot{q}. \tag{12}$$

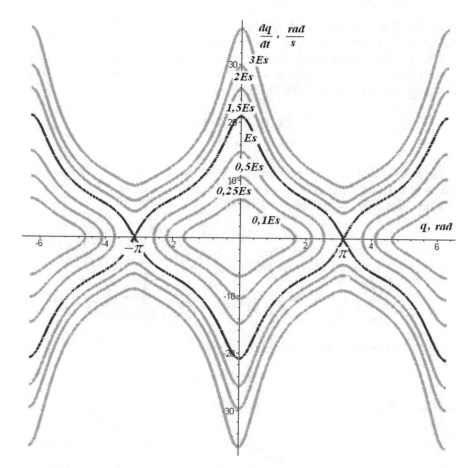

Fig. 1 Phase trajectory family for the case of the given system's own ballistic motions

We will assume that the hinge mobility drill irreversibly consumes energy, and the energy consumption for control purposes is determined by the sum of the mechanical work absolute values of the control moment for the sections, on each of which the total energy of the given system (8) against time is a monotonic function. Then the control quality in the time range $[t_0, t_k]$ can be naturally assessed [11] by a functional of the form

$$J(t_0, t_k) = \int_{t_0}^{t_k} |M\dot{q}| dt. \qquad (13)$$

The control $M(t)$, which provides a minimum of the functional (13), will be optimum in terms of energy consumption. In this case, the minimum possible value of the functional (13) is obviously equal to the modulus of the difference in the energy levels of the phase trajectories containing start and end image points.

The rate of the given system experiencing the pulse control at a certain position $q*$ will jump (over a negligible time range $[t^-, t^+]$) accompanied by the vertical displacement of the image point on the phase plane.

$$
\begin{aligned}
q(t^-) &= q(t^+) = q*, \\
\dot{q}(t^-) &\neq \dot{q}(t^+).
\end{aligned}
\tag{14}
$$

Here, the value of the functional (13) is determined by the difference in the values of the kinetic energy of the given system at the instants of time t^- and t^+.

If the sign of the speed does not change

$$
\dot{q}(t^-)\dot{q}(t^+) \geq 0,
\tag{15}
$$

the value of the functional (13) is defined by

$$
J(t^-, t^+) = \left| \frac{1}{2} \frac{\gamma_1\gamma_2 - \gamma_3^2 \cos^2 q*}{\gamma_1 + \gamma_2 + 2\gamma_3 \cos q*} \dot{q}(t^+)^2 - \frac{1}{2} \frac{\gamma_1\gamma_2 - \gamma_3^2 \cos^2 q*}{\gamma_1 + \gamma_2 + 2\gamma_3 \cos q*} \dot{q}(t^-)^2 \right|.
\tag{16}
$$

If the sign of the speed changes

$$
\dot{q}(t^-)\dot{q}(t^+) < 0,
\tag{17}
$$

the value of the functional (13) is defined by

$$
J(t^-, t^+) = \left| \frac{1}{2} \frac{\gamma_1\gamma_2 - \gamma_3^2 \cos^2 q*}{\gamma_1 + \gamma_2 + 2\gamma_3 \cos q*} \dot{q}(t^+)^2 + \frac{1}{2} \frac{\gamma_1\gamma_2 - \gamma_3^2 \cos^2 q*}{\gamma_1 + \gamma_2 + 2\gamma_3 \cos q*} \dot{q}(t^-)^2 \right|.
\tag{18}
$$

Obviously, in the case (16), the pulse control will be optimum in terms of minimizing the functional (13), and in the case (18), the pulse control will not be optimum in this sense.

4 Examples of Optimum Control

Considering the control problem as a problem of transferring the image point from the position $(q_0, \dot{q}_0)$ to the position $(q_k, \dot{q}_k)$ on the phase plane, we can assert that an optimum control in terms of minimizing the functional (13) will be any smooth function $M(t)$, whose sign coincides with the sign of the speed $\dot{q}(t)$ at each instant of time or is opposite to the sign of the speed $\dot{q}(t)$ at each instant of time. Let us consider controls of the form

$$M_1(t) = \frac{\partial E^*}{\partial \dot{q}} = \dot{q}(t) \frac{\gamma_1 \gamma_2 - \gamma_3^2 \cos^2 q(t)}{(\gamma_1 + \gamma_2 + 2\gamma_3 \cos q(t))} \tag{19}$$

and

$$M_2(t) = -\frac{\partial E^*}{\partial \dot{q}} = -\dot{q}(t) \frac{\gamma_1 \gamma_2 - \gamma_3^2 \cos^2 q(t)}{(\gamma_1 + \gamma_2 + 2\gamma_3 \cos q(t))}, \tag{20}$$

transferring the image point to the position

$$q_k = q(t_k) = \frac{\pi}{2}, \tag{21}$$
$$\dot{q}_k = \dot{q}_k(t_k) = 0.$$

in a fixed time

$$t_u = t_k - t_0. \tag{22}$$

The predetermined control laws (19) and (20), the fixed control time (22), and the image point end position (21) are made use of to expressly find the image point start positions $(q_{10}, \dot{q}_{10})$ and $(q_{20}, \dot{q}_{20})$. The control of the form (19) and (20) in accordance with [12] will be called collinear. Figure 2 shows the phase trajectories corresponding to the controlled motion of the given system, which moves the image point from the position $(q_{10}, \dot{q}_{10})$ to the position (21) over the time $t_u = 0.9$ c by the control (19) and from the position $(q_{20}, \dot{q}_{20})$ to the position (21) by the control (20).

The values of the parameters (2) are

$$m_1 = 10 \text{ kg}, m_2 = 10 \text{ kg}, J_1 = 1 \text{ kg m}^2, J_2 = 1 \text{ kg m}^2,$$
$$l_1 = 1 \text{ m}, l_1 = 1 \text{ m}, L = 22 \frac{\text{kg m}^2}{\text{s}} \tag{23}$$

Figure 2 illustrates: E_{01}^*—the energy level of the phase trajectory containing the start point $(q_{10}, \dot{q}_{10})$, E_{02}^*—the energy level of the phase trajectory containing the start point $(q_{20}, \dot{q}_{20})$, E_{kk}^*—the energy level of the phase trajectory containing the end point (21).

Considering that

$$\gamma_1 > 0, \gamma_2 > 0, \gamma_3 > 0,$$
$$\gamma_1 + \gamma_2 - 2\gamma_3 > 0, \tag{24}$$
$$\gamma_1 \gamma_2 - \gamma_3^2 > 0,$$

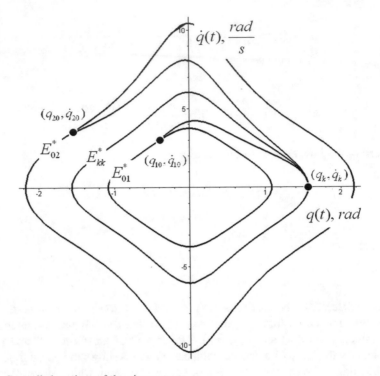

Fig. 2 Controlled motions of the given system

we can write down

$$M_1(t)\dot{q}(t) = \dot{q}^2(t)\frac{\gamma_1\gamma_2 - \gamma_3^2\cos^2 q(t)}{(\gamma_1 + \gamma_2 + 2\gamma_3\cos q(t))} > 0, t < t < t_k, \tag{25}$$

$$M_2(t)\dot{q}(t) = -\dot{q}^2(t)\frac{\gamma_1\gamma_2 - \gamma_3^2\cos^2 q(t)}{(\gamma_1 + \gamma_2 + 2\gamma_3\cos q(t))} < 0, \ t < t < t_k. \tag{26}$$

In case of the control (19), the time derivative of the total energy of the given system (12) over the time interval $[t_0, t_k]$ is positive, and the total energy of the given system monotonically increases. In case of the control (20), the time derivative of the total energy of the given system (12) over the time interval $[t_0, t_k]$ is negative, and the total energy of the given system monotonically decreases. In both cases, the control is optimum in terms of minimizing the functional (13).

In case of the control (19), the control energy consumption is

$$J(t_0, t_k) = \int\limits_{t_0}^{t_k} |M_1(t)\dot{q}(t)|dt = E_{kk}^* - E_{01}^*. \tag{27}$$

In case of the control (20), the control energy consumption is

$$J(t_0, t_k) = \int\limits_{t_0}^{t_k} |M_2(t)\dot{q}(t)| dt = E_{02}^* - E_{kk}^*. \tag{28}$$

Figure 3 includes the selected sections of the two phase trajectories shown in Fig. 2. The first section corresponds to the movement of the image point from the position $(q_{10}, \dot{q}_{10})$ to the position $(q_k, \dot{q}_k)$ using the control (19) during the time $t_u = 0.9$ s. The second section corresponds to the free movement of the image point along a ballistic trajectory from the position $(q_{b0}, \dot{q}_{b0})$ to the position $(q_k, \dot{q}_k)$ during the same time $t_u = 0.9$ s. It is obvious that the combination of the pulse control transferring the image point from the position $(q_{10}, \dot{q}_{10})$ to the position $(q_{b1}, \dot{q}_{b1})$, $q_{b1} = q_{10}$ over a negligible time range with the subsequent free movement along a ballistic trajectory will also be optimum in terms of minimizing the criterion (13) because the energy consumed for control will still be equal to the value (27). Besides, it is obvious that in the second case the controlled motion of the image point from the position $(q_{10}, \dot{q}_{10})$ to the position $(q_k, \dot{q}_k)$ will take less time.

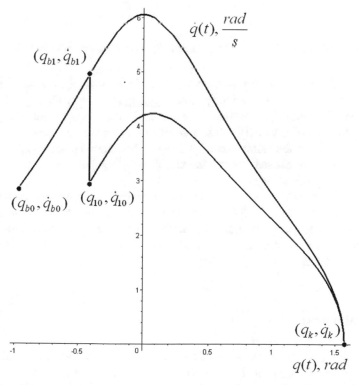

Fig. 3 Comparison of the time consumptions optimum in terms of energy consumption for continuous and pulse control

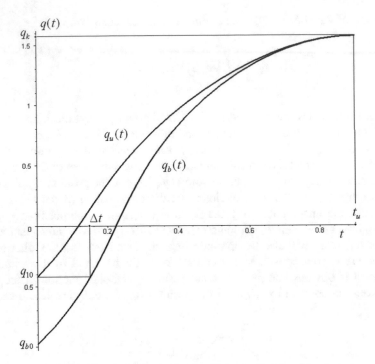

Fig. 4 The controlled $q_u(t)$ and the free ballistic $q_b(t)$ motions of the given system

Figure 4 illustrates the laws of the variation of the hinge coordinate $q(t)$ for the controlled movement $q_u(t)$ from the position $(q_{10}, \dot{q}_{10})$ to the position $(q_k, \dot{q}_k)$ using the control (19) and for the free ballistic movement $q_b(t)$ from the position $(q_{b0}, \dot{q}_{b0})$ to the position $(q_k, \dot{q}_k)$. It is assumed that $t_0 = 0$ s and $t_k = t_u = 0.9$ s. In case of combining the pulse control with the free movement along a ballistic trajectory, the time needed for the transferal from the position $(q_{10}, \dot{q}_{10})$ to the position $(q_k, \dot{q}_k)$ will be equal to $t_{u1} = t_u - \Delta t$.

The given example shows that the considered criterion does not impose strict constraints on the control $M(t)$, which is optimum in terms of minimizing the functional (13). Thus, the problem of searching for some optimum controls, which satisfy the additional criteria, among those that are optimum within a given sense can be formulated.

5 Conclusion

We considered a task of the synthesis of the program motion of a robotic space module acting as the element of an assembly and servicing system for emerging orbital facilities. We used an approach based on introducing for consideration of a given nonlinear oscillating system and on the analysis of its ballistic phase trajectories. We performed the visual comparative assessments of different types of control optimum in terms of energy consumption.

The requirements to the control, which would be optimum in terms of minimizing the functional in question, do not impose stringent restrictions on the class of the functions used, which allows to take into account additional considerations determined by the technical nature of the problem, such as restrictions on the relative motion of the manipulator links, on the value of the control moment, the speed, or the response time.

References

1. Belonozhko, P.P.: Emerging assembly and servicing robotic space modules. Rob. Tech. Cybern. **2**(7), 18–23 (2015)
2. Belonozhko, P.P.: Space robotics. Modern state, emerging tasks, development trends. Analytical review. Science and education. Bauman Moscow State Technical University. Electron. J. (12), 110–153 (2016). https://doi.org/10.7463/1216.0853919
3. Belonozhko, P.P.: Space robotics. Experience and development prospects. Aerosp. Sphere **1** (94), 84–93 (2018)
4. Artemenko, Y.N., Karpenko, A.P., Belonozhko, P.P.: Features of manipulator dynamics modeling into account a movable platform. In: Gorodetskiy, A.E. (ed.) Smart Electromechanical Systems. Studies in Systems, Decision and Control, vol. 49, pp. 177–190. Springer International Publishing Switzerland (2016). https://doi.org/10.1007/978-3-319-27547-5_17
5. Artemenko, Y.N., Karpenko, A.P., Belonozhko, P.P.: Synthesis of control of hinged bodies relative motion ensuring move of orientable body to necessary absolute position. In: Gorodetskiy, A.E., Kurbanov, V.G. (ed.) Smart Electromechanical Systems: The Central Nervous System. Studies in Systems, Decision and Control, vol. 95, pp. 231–239. Springer International Publishing Switzerland (2017). https://doi.org/10.1007/978-3-319-53327-8_16
6. Belonozhko, P.P.: Methodical features of acquisition of independent dynamic equation of relative movement of one-degree of freedom manipulator on movable foundation as control object. In: Gorodetskiy, A.E., Kurbanov, V.G. (ed.) Smart Electromechanical Systems: The Central Nervous System. Studies in Systems, Decision and Control, vol. 95, pp. 261–270. Springer International Publishing Switzerland (2017). https://doi.org/10.1007/978-3-319-53327-8_19
7. Karpenko, A.P., Sayapin, S.A., Hiep, D.X.: Dodekapod as universal intelligent structure for adaptive parallel spatial self-moving modular robots. Nature-inspired mobile robotics. In: Proceedings of the 16th International Conference on Climbing and Walking Robots and the Support Technologies for Mobile Machines, pp. 163–167, University of Technology, Sydney, Australia, 14–17 July 2013
8. Karpenko, A.P., Sayapin, S.A., Hiep, D.A.: Universal Adaptive Spatial Parallel Robots of Module Type Based on the Platonic Solids. Lecture Notes in Engineering and Computer Science, vol. 2, pp. 1365–1370 (2014)
9. Belonozhko, P.P.: Study of flat inertia movements of space manipulator on movable base as nonlinear oscillatory system. Rob. Tech. Cybern. **4**(13), 52–58 (2016)
10. Belonozhko, P.P.: Analysis of given nonlinear oscillatory system corresponding to one degree of freedom manipulators on movable and gimble-mounted base. Rob. Tech. Cybern. **3**(16), 44–52 (2017)
11. Formalskiy, A.M.: Motion control of unstable objects. Physmatlit, 232 (2014). ISBN 978-5-92221-1460-8
12. Smolnikov, B.A.: Robots mechanics and optimization problems. Nauka. Phys. Math, 232. Literature Main Office (1991)

Automatic Control Panel Module SEMS

Vladimir V. Dubarenko, Andrey M. Kornuyshin
and Vugar G. Kurbanov

Abstract *Problem statement*: the results of the control system simulation SEMS module, which has more than three parallel kinematic constraints. Described control objects form the basis manipulator robots, hexapods, Stewart platforms, simulators and others. *Purpose of research* A distinctive feature of the control system is the use of the reduced status of the device identifier in conjunction with sensors of linear elongated stem, which at the relative simplicity of the measuring system, ensure the required quality of dynamic processes. To improve the quality of automatic dynamic control of multilink mechanism requires the use of position sensors and gyroscopic solid velocity. Also to set sensors relative angles of rotation hinge elements. *Results*: By the results of the work include the creation of a computer model hexapod in the Simulink computing environment, taking into account the many nonlinear features multilink mechanisms and allows the study and synthesis of control systems with a minimum amount of time to change the parameters of the initial conditions and the structure of the model. *Practical significance* Multilink mechanisms can be applied in the management of the mutual

V. V. Dubarenko (✉) · A. M. Kornuyshin · V. G. Kurbanov
Institute of Problems of Mechanical Engineering, Russian Academy
of Sciences, Bolshoy pr, 199178 St. Petersburg, Russia
e-mail: vladimir.dubarenko@gmail.ru

A. M. Kornuyshin
e-mail: mrloled@hotmail.com

V. G. Kurbanov
e-mail: vugar_borchali@yahoo.com

A. M. Kornuyshin
Peter the Great St. Petersburg Polytechnic University, Polytechnicheskaya,
29, 195251 St. Petersburg, Russia

V. V. Dubarenko · V. G. Kurbanov
Saint-Petersburg State University of Aerospace Instrumentation,
67, B. Morskaia St, 190000 St. Petersburg, Russia

© Springer Nature Switzerland AG 2019
A. E. Gorodetskiy and I. L. Tarasova (eds.), *Smart Electromechanical Systems*,
Studies in Systems, Decision and Control 174,
https://doi.org/10.1007/978-3-319-99759-9_19

229

position of the elements of the system mirror telescopes. On the basis of the Hexapod can be constructed control system positions the main mirror second reflector, a receiving device and shields adaptive reflecting surface.

Keywords SEMS · Automatic control · Dynamic objects · A system with parallel

1   Introduction

Structural identification methods and computational optimization methods are widely used in the analysis and synthesis of high-precision information-measuring and control systems (IMCS). A special feature of modern dynamic robotic systems, such as SEMS, is the use of such systems in the In this work group under the automatic control module SEMS (dynamic object) is understood a physical task movable in three-dimensional Euclidean space of a solid state (SS), under the action on it of several (more than three) of concentrated forces with respect to the center of mass spaced points of application. At the points of the application there is no torque, so these points will be called joints, and very solid—a mobile platform (MP). The forces acting on the PCB is transmitted via rods called in feet (Legs) literature or actuators (AC). AC MP bind to the fixed base (FB) by means of hinges of Hooke [1]. AK can change their length by extension rods. When nominating rods in the joints, forces directed along these rods. The forces act on the MP as a SS, forcing it to move in space. Obviously, the extension rods should be coordinated to avoid jamming. The group agreed automatic extension rods is considered one of the main object control tasks.

When the action of forces from the PM by AK rods, it will perform the translational motion, in addition, its center of mass relative torque occurs under the action of which the PM will perform a rotational movement. At relatively low speeds of translational and rotational motions of the objects, we assume that when evaluating overall results of their dynamics, superposition principle is valid, which allows to investigate separately the translational and rotational movement of the SS.

SS movement will explore the state space in [2] for the hexapod (Fig. 1).

To study the dynamics hexapod usually administered two Euclidean coordinate system (SC): SC fixed—*xyz*, and the movable SC—*uvw*, constant unit vectors with length equal to 1. Any arbitrary vector of length n, emanating from the center of mass P, called the directional vector. Its projections on the coordinate axes do not exceed 1, and are called the direction cosines.

If PP turn around the direction vector n at the angle Q (in the positive direction in the counterclockwise direction when viewed from the end of the vector), then along with it will turn and SC—uvw, individual wherein the unit vectors will be the projection on the original SC—uvw, in the form of direction cosines. Direction unit vectors starting SC—uvw coincide with the directions of respective axes fixed SK —xyz. The values of the direction cosines rotated around the direction vector n at the angle of Q MP, can be determined using a matrix $H(n, Q)$ [3]. Multiply matrix

Fig. 1 Block diagram
hexapod. 1

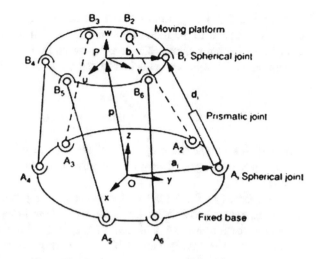

$H(\mathbf{n}, Q)$ for every vector of *i-th* point $\mathbf{x}_i0$, belonging to MP in the rest position, resulting in a vector of the same point xi in the new rotated position MP.

$$\mathbf{x}_i = II(\mathbf{n}, Q) * \mathbf{x}_i 0.$$

$H(\mathbf{n}, Q)$ matrix is called the rotation matrix and is a convenient means of identifying new coordinates hinges MP, after its rotation.

SS state vector is the vector $\mathbf{x}p$, whose components are.

In translation:

- the linear position of the center of mass movable SC coordinate system SS relative to the fixed coordinate system,
- linear velocity center of mass SS relative to the fixed coordinate system.

In rotary motion:

- the angular position of the movable SC coordinate system SS relative to the fixed coordinate system,
- the angular velocity changes of the angular position of the movable coordinate system SS relative to the fixed coordinate system.

Changing the position of the center of mass of SS is provided by a force applied at the same point, and the angular position—relative to the torque inertia center whose coordinates for the model coincide with the coordinates of the center of mass.

Dynamic SS in translational motion dynamics is considered as a point centered therein and the mass m a parallel translation (no rotation), the movable coordinate system SC—uvw.

Dynamic CT during rotational movement is regarded as its rotation around the direction vector n to the moment of inertia J and the corresponding rotation of the coordinate system moving SC—uvw by an angle Q.

Force f and torque mp, attached to the SS is in the form of:

$$f = mf * nf,$$
$$mp = \text{Mmp} * \text{n},$$

where f—force, mf—the magnitude of the force, nf—directional force vector; mp—vector torque, Mmp—the magnitude of the torque, n—directional vector torque.

Presentation of the force f and torque mp as a product module on the steering vector makes it possible to integrate with their integration modules only, and direction vectors are not included in the integration process, as their direction cosines are taken into account when calculating the vectors modules. The Euler vectors model integration is performed separately for each coordinate axis, and then is summed in a certain sequence using Euler angles. This algorithm leads to ambiguity TT position determination depending on the sequence of its rotation around the coordinate axes. Therefore, the representation of force f and torque mp shaped product reduces the number of integrators 3 times and thereby increase the computational efficiency.

2   Aims Management

The purpose of the control MP is to agree with the state vector xp software state vector $xp*$ and tracking it under the action of external disturbances.

In most practical cases the exact requirement translation system from an initial state $xp(0)$ to the desired $xp*(t)$ may well be replaced with the requirement of the system asymptotic stability relative $xp*(t)$.

This circumstance makes it possible to synthesize a control system, as a system of c the linear control law, in a stationary state feedback

$$u(t) = K[xp^*(t) - xp(t)]$$

where K—constant matrix.

Selection feedback matrix K, which provides the desired dynamics aspiration vector $xp(t)$ to a state of $xp*(t)$ is usually carried out either at the specified location the characteristic numbers of the closed system using known algebraic techniques or by solving the matrix Riccati equation. Using these methods encounters difficulties associated with the problem of optimal allocation of the characteristic numbers of the closed system on a complex plane or selecting weighting coefficients in the quadratic quality criteria.

The main problem is the MP control communication equations finding MP angular displacement with a linear displacement hinges consequent angular movement. That is, for a given rotation of the steering angle and rotation vector MP around this vector, it is necessary to find a new position of the hinges, to calculate their new position. The solution of this problem allows to unambiguously determined by measuring the MP hinges coordinates, the coordinates of its center of mass, the guide rotation vector and angle of the vector relative to [4].

3   Driving Control Hexapod

Control hexapod system (GP), as a dynamic object, includes:

- mathematical model of GP (ID state or in other words—"Observer"), consisting of a body—a mathematical model of the mobile platform (MP) and a block of 6 mathematical models of linear actuators (legs) (Fig. 2);
- measuring system for measuring the coordinates of the vector $\mathbf{X_u}$
- the feedback loop of the measured coordinates of the state vector $\mathbf{X_u}$ through the matrix L to obtain estimates of unmeasured coordinates $\overline{\mathbf{X}}_\mathbf{H}$;
- a feedback loop from the reconstructed state vector $\overline{\mathbf{X}}$ through the matrix K_{oc} for vector control actions u on the actuator drives;
- «Computer» for setting the initial state vector values and parameters;
- Generator desired state MP $\mathbf{X}^*$ as a vector function of spatial position coordinates of software tracking of the object.

4   Mathematical Model Hexapod

Mathematical model hexapod as dynamic management object is represented in Figs. 3, 4, 5, 6, 7, 8, 9, 10, 11, 12, 13, 14, 15 and 16. Notation in Simulink MatLab. Essentially in these figures are given algorithm and program control hexapod system simulation that adding initial data allows to study the dynamics of iterative mechanisms and structure and principles of the universal controller to control them in real time.

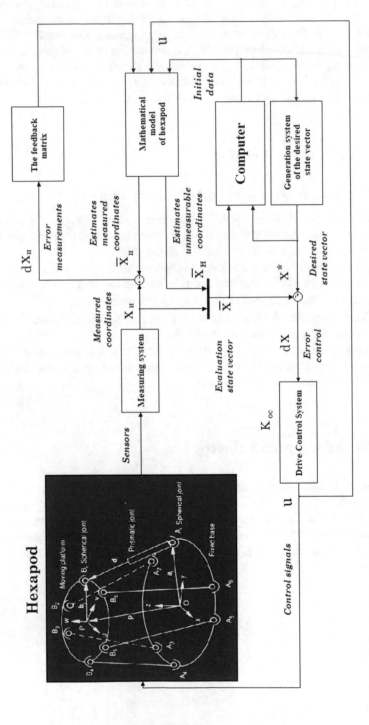

Fig. 2 Block diagram of the control by hexapod

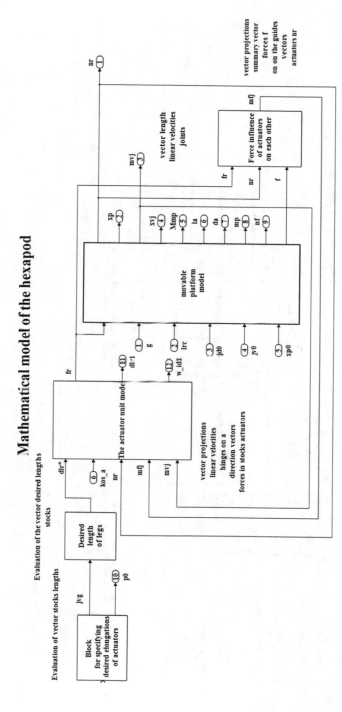

Fig. 3 Mathematical model hexapod

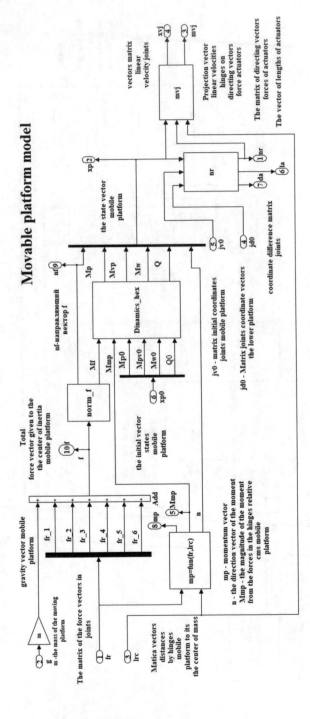

Fig. 4 Model movable platform

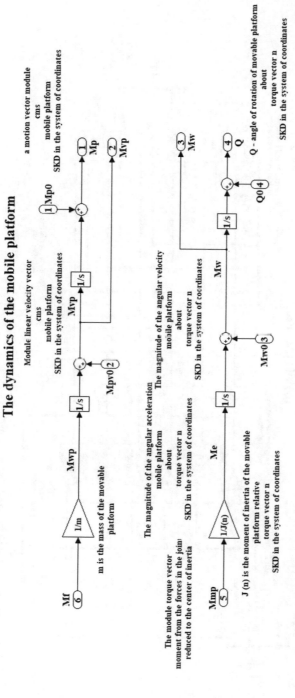

Fig. 5 The dynamics of the mobile platform

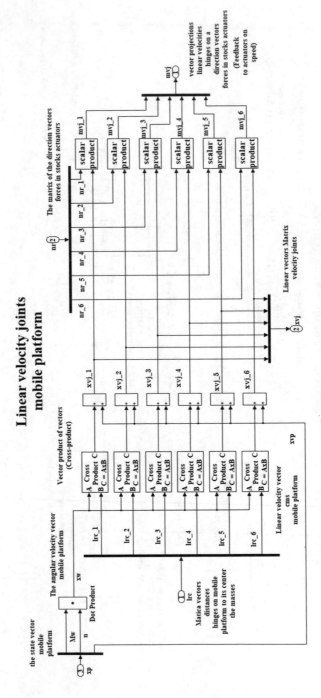

Fig. 6 Calculating the linear velocity scheme movable platform hinges

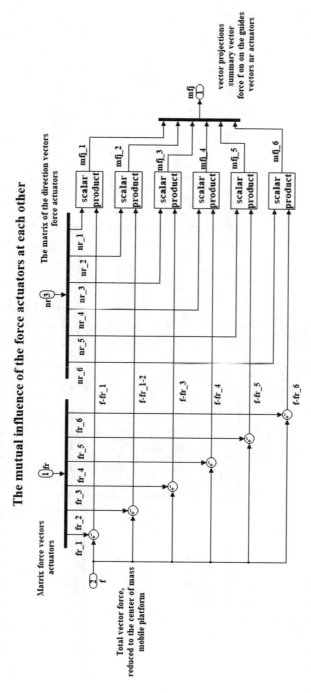

The mutual influence of the force actuators at each other

Fig. 7 Mutual influence of actuators

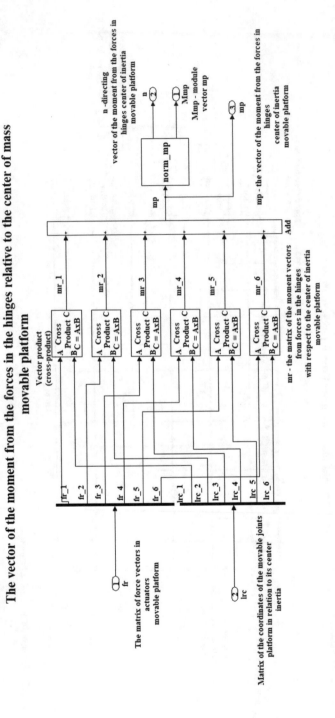

Fig. 8 Calculating the time scheme of the vector forces in the hinges

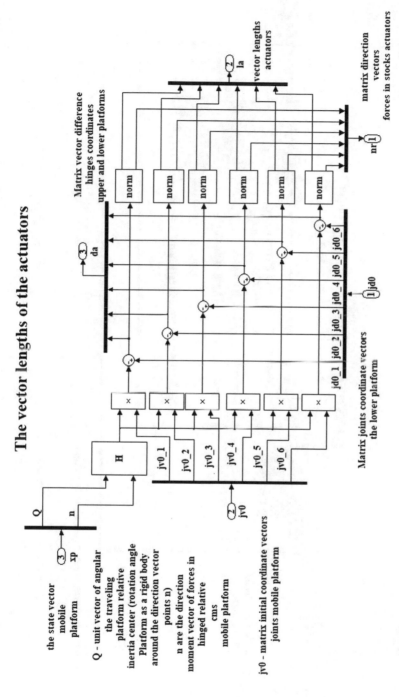

Fig. 9 Scheme vector calculating lengths actuators

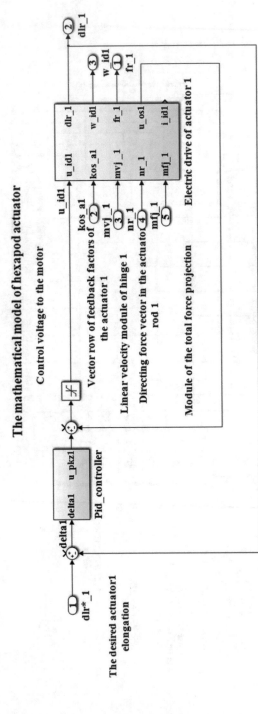

Fig. 10 A mathematical model of the actuator hexapod

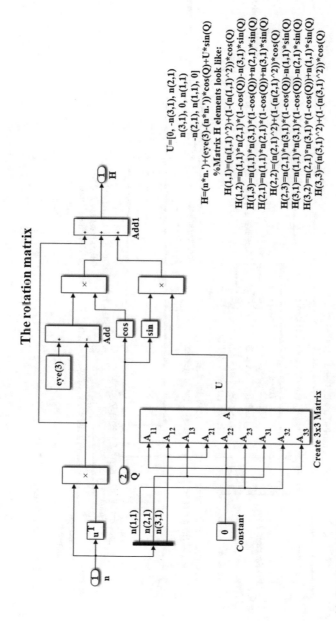

Fig. 11 Rotation matrix

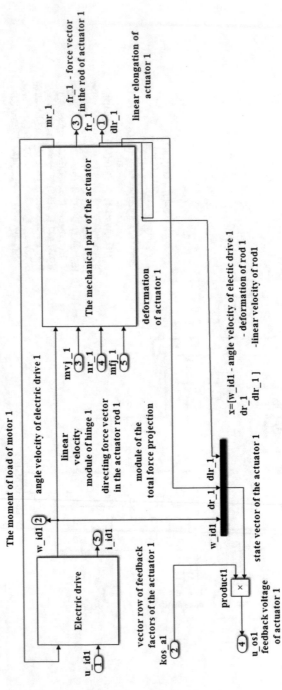

Fig. 12 Model motor and mechanical actuator

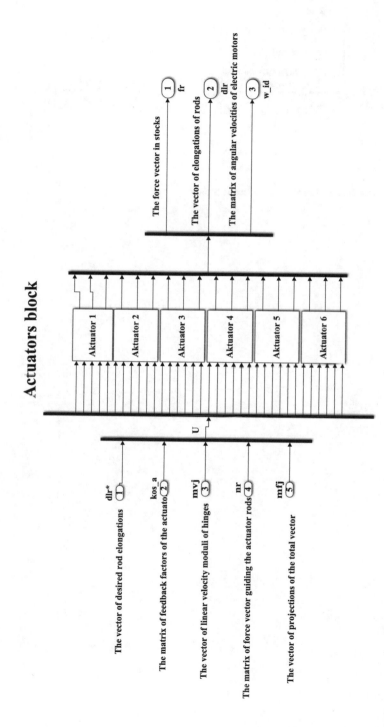

Fig. 13 Block actuators hexapod

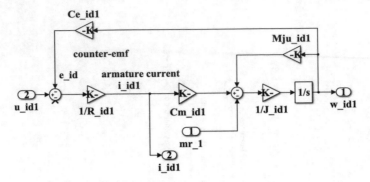

Fig. 14 A mathematical model of the electric motor

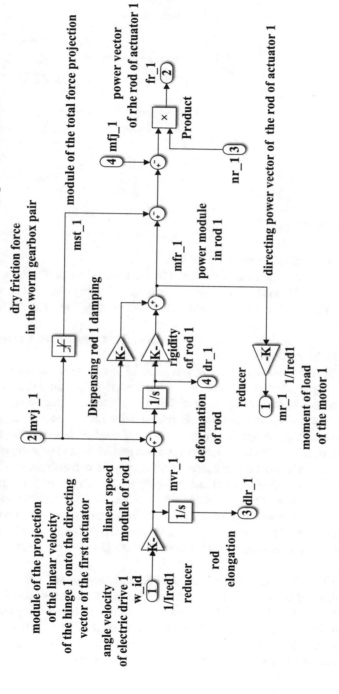

Fig. 15 Model mechanical actuator

Fig. 16 PID—regulator

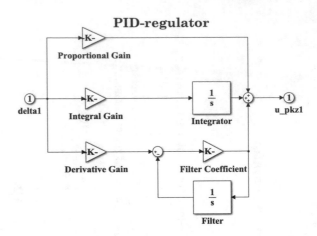

5 Conclusion

We obtain the following results:

Hexapod motion equations formulated in the form of block diagrams of system simulation models Matlab Simulink, allowing to study the dynamics of the multi-link mechanism as nonlinear automatic control system with variable parameters.

It was found that the study of the dynamics of translational and rotational motions can be carried out independently.

The stationary points at the "frozen" model parameters, hexapod system of differential equations can linerization, find the roots of the characteristic polynomial, and reduce by decreasing the model order by discarding high-frequency roots.

The output scalar coordinates determined with respect to which the quality control system, are, for the rotary movement—the angle between the target (desired) position of the axis of symmetry hexapod and its current position obtained by reversing this axis around the center of gravity. For translational motion of the output coordinates of the selected distance between the lengths of the actuator in the desired and actual positions determined by the magnitude stem elongation.

Hexapod remarkable property is the ability to define new hinges coordinates by multiplying the rotation matrix initial vector coordinates hinges. A rotation matrix is defined as a predetermined function of the rotation axis direction and rotation angle [5].

The main node hexapod kinematic couple is a "screw-nut" designed for converting rotational motion into linear. This device has a substantial disadvantage—a tendency to jam, which makes it impossible to apply it in the tracking mode, so the realization is considered hexapod linear motor precluding this adverse effect.

Further work is to conduct a test of the simulation model and hardware-based hexapod management models for their industrial application.

References

1. Dubarenco, V.V., Artemenko, Y.N., Gorodetsky, A.E., Kuchmin, A.Y., Kurbanov, V.G.: Radio telescopes based on a parallel kinematic mechanisms bonds. In: Transactions 7th Russian Multiconference Management Problems (RMKPU 2014) (XXIX conference in memory of a prominent designer of gyroscopic instruments N. N. Ostryakova), 7–9 October 2014
2. Gorodetsky, A.E., Dubarenco, V.V., Kuchmin, A.Y., Agapov, V.A.: Control systems PT 70 kontrreflektorom (Suffa) using parallel computing and measuring mechanical structures. In: Proceedings of the Russian National Radio Astronomy Conference (WRC—2014) "Radio Telescopes, Instruments and Techniques of Radio Astronomy", 22–26 September 2014. Pushchino, PRAO ASC LPI
3. Dubarenco, V.V., Artemenko, Y.N, Kuchmin, Y.: Dynamic management of objects on the basis of hexapods. In: X1 Proceedings of the All-Russian Congress of the Fundamental Problems of Theoretical and Applied Mechanics, Kazan on 20–24 August 2015
4. Hagan, Y.T., Karpenko, A.P.: Mathematical modeling of kinematics and dynamics of the robotic arm of the "trunk". Mathematical models of the manipulator section as "hexapod" parallel kinematics type. In: Science and Education: Electronic Scientific and Technical Publication, www.technomag.edu.ru, November 2009. http://technomag.bmstu.ru/doc/133731.html
5. Volkomorov, S.V., Karpenko, A., Leletko, A.M.: Optimization of angular and linear dimensions of one- and two-section manipulators based on parallel kinematics. Science and Education: Electronic Scientific and Technical Publication. www.technomag.edu.ru, August 2010. http://technomag.edu.ru/doc/154452.html

Smart Electromechanical Systems in Electric Power Engineering: Concept, Technical Realization, Prospects

Aleksandr N. Shilin, Aleksey A. Shilin and Sergey S. Dementiev

Abstract *Problem statement*: Reducing the accident rate of overhead power lines (OHL) is one of the priority areas for the development of the electrical power industry. The reliability of overhead power lines depends on a large number of various factors, among which, it is necessary to highlight extreme climatic loads, leading to frequent failures of OHL due to wire breakage. Prompt recognition of emergency line mode, fault localization, and isolation of the damaged area network are the basis for the prompt restoration of power supply. To successfully solve this problem, it is proposed to use a "smart" electromechanical system (SEMS), namely, it is a switching device mounted on the OHL tower and combined with a neuro-computer information processing unit. The neurocomputer, based on processing of information from sensors of current, voltage and meteorological parameters, and in the event of an accident, controls the switching of power lines. The difference between SEMS and a conventional vacuum switch with an electric drive with a constant setpoint is the automatic correction of the setpoint value, which depends on external factors. The purpose of the research: The development of an algorithm for functioning of a circuit breaker with a neurocomputer, ensuring a reliable operation of the circuit breaker to isolate the damaged section of the OHL. *Results*: The concept of SEMS is proposed in the form of an information-measuring and control system combined with a vacuum switching device and assuming the use of an information processing unit based on an artificial neural network. *Practical significance*: Modern switching devices implemented with a rigid logic and a constant setpoint value don't possess the ability of adaptation with the environmental conditions. The proposed smart electromechanical system lacks this disadvantage due to the use of a neurocomputer, which allows taking into account the

A. N. Shilin (✉) · A. A. Shilin · S. S. Dementiev
Volgograd State Technical University, Volgograd, Russia
e-mail: eltech@vstu.ru

A. A. Shilin
e-mail: shilin.jr@gmail.com

S. S. Dementiev
e-mail: c165tc34@yandex.ru

© Springer Nature Switzerland AG 2019
A. E. Gorodetskiy and I. L. Tarasova (eds.), *Smart Electromechanical Systems*,
Studies in Systems, Decision and Control 174,
https://doi.org/10.1007/978-3-319-99759-9_20

location of the switching device, the time of year and external climatic factors. The introduction of SEMS as an element in the smart grid will allow to increase the reliability of power supply systems by reducing the time of recovery of accidents.

Keywords Reliability of power supply · Accidents on OHL · Fault location Smart electromechanical systems (SEMS) · Artificial neural networks Smart grids

1 Introduction

The main indicator of the country's energy system is its reliability, and the most unreliable elements of the power system are overhead power lines (OHL). This problem is especially topical for Russia, since Russia's overhead lines are very long and are located in different geographic and climatic conditions, which are in most cases very complex. Therefore, the promptness of determining the type of emergency mode of the OHL and switching off the line or switching consumers to the reserve line allows to reduce the downtime and thereby increase the reliability index—the availability factor or the probability of failure-free operation.

At present, in many developed countries of the world, work is underway to create smart electric networks. Smart grids represent a complex of technical means working in an automatic mode and detecting the weakest and most dangerous parts of the networks. If necessary, these networks change their characteristics and scheme in order to prevent an accident and reduce losses. Smart grids are based on the idea of the parallelism of the transmission of electricity and information flows [1] and the use of modern information-measuring and control systems. One of the main elements of smart networks are relay protection and automation devices, however, existing devices do not allow reliable disconnection of the damaged sections of the OHL.

2 Criteria of Smartness of Relay Protection

The concept of smart relay protection has been known since the late 1990s, and the meaning that is attached to this definition has changed many times over the past few years. As noted in [2], to date, in most cases, this term is used to characterize produced microprocessor protection devices. In a number of publications, one can come across the proposition that microprocessor devices are smart systems, because by upgrading the software and using more promising algorithms have the potential for further improvement and expansion of functionality [2].

However, it is impossible to improve the efficiency of the device only by updating its software and without communication with the external environment. To improve the intelligence of the relay protection, a smart tower has been

Fig. 1 The scheme of a
"smart" tower: 1—current
sensors, 2—voltage sensors,
3—switching unit,
4—information processing
unit, 5—block of
meteorological sensors
(automatic weather station),
6—soil conductivity sensor

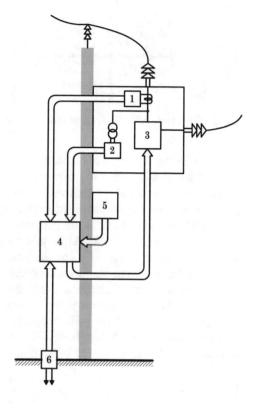

developed, which is a smart electromechanical system (SEMS) for emergency
control automation (Fig. 1).

The tower of the overhead power line is equipped with a switching device
controlled by a neurocomputer unit processing the information obtained from
sensors of current, voltage and meteorological parameters.

"Adaptability" of the device to environmental conditions. When designing of
relay protection (RP) of a classical configuration, regardless of the type of relay
used (microprocessor or mechanical), the averaged data on the conductivity of the
soil, the physical properties of the conductors and atmospheric conditions are used,
while the values of these indicators are assumed unchanged. Such an assumption is
the reason for the incorrect functioning of the protection, namely a false line
shutdown or, in contrast, failure due to a mismatch between the protection settings
of the current state of the electrical circuit of a short circuit. As an example, we'll
provide the organization of an overcurrent protection (OCP) that is not adaptive to
external operating conditions.

In a radial grid with one-sided power supply (Fig. 2) the overcurrent protection
is set on each line. Protection of the line farthest from the power supply has the
lowest operating current and the smallest time delay. The protection of each subsequent line has a time delay greater than that of the previous protection. For

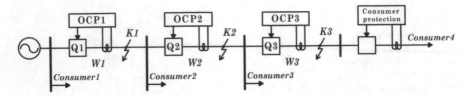

Fig. 2 The scheme of an overcurrent protection of a grid with a single-sided power supply

example, in the case of a short circuit (SC) at the point of SC, the fault current passes through all sections of the network located between the source and the fault location. As a result, all protections located on this site come into operation (1, 2, 3). However, by the condition of selectivity, only protection 3 installed on the damaged line should work. When performing an overcurrent protection with a time delay increasing from the consumer to the source, protection 3 will start before the others, by cutting the line W3, and the protections 1 and 2 will return to the initial position without having time to initiate disconnection.

The key parameter that characterizes the effectiveness of relay protection is its sensitivity. The sensitivity test is carried out at the minimum short-circuit current if damaged at the end of the protected area. The coverage area of the overcurrent protection must cover the protected line and provide redundancy protection for the next section. The sensitivity of protection is estimated by the coefficient

$$K_{sens} = \frac{I_{k.min}}{I_{trip.prot.}}, \tag{1}$$

where $I_{k.min}$—for networks with an isolated neutral—two-phase short-circuit current, for grids with a deadly grounded neutral—one-phase short-circuit current, A; $I_{trip.prot.}$—protection tripping current, A.

According to the current regulations, protection is effective when the sensitivity coefficient on the redundant section is at least 1.2 [3].

The first stage of relay protection design is the calculation of resistances of sections of overhead lines. For the low-voltage electric networks of alternating current with a frequency of 50 Hz and voltage of 6 kV shown of Fig. 2 it can be asserted with sufficient accuracy that the resistance of the OHL is determined by the Carson formula [4]:

$$\underline{Z} = \left(R_0 + 0.05 + j0.145 \lg \frac{20.85}{r_{eq}\sqrt{50 \cdot \gamma \cdot 10^{-9}}} \right) l, \tag{2}$$

where R_0 is specific resistance, Ohm/km; r_{eq}—equivalent wire radius with the account of the surface effect, mm; γ—the specific electric conductivity of the soil, S/m (siemens per metre); l—total length of wire in OHL spans, km.

As noted in [5], the inductive resistance of the OHL doesn't depend on the height of the wire suspension.

The active resistance R_0 is a function of temperature and is characterized by the relation [4]:

$$R_{0t} = R_{020}[1 + \alpha_R(t_{wire} - 20)], \tag{3}$$

where R_{020} is the tabulated value of conductor resistance at wire temperature 20 °C, Ohm/km; t_{wire}—current wire heating, °C; α_R—temperature coefficient of electrical resistance, 1/deg.

The length of the sagging linear wire is also unstable due to a change in its temperature with fluctuations in climatic conditions and current load [4]:

$$l_t = l_{20}[1 + \alpha_l(t_{wire} - 20)], \tag{4}$$

where l_{20} is the length of the OHL wire at a temperature of 20 °C, km; α_l—temperature coefficient of linear elongation, 1/deg.

The operating current of the overcurrent protection is found from the condition [3]

$$I_{trip.prot.}^{OCP} \geq \frac{K_{det}K_{self-l.}}{K_{reset}} I_{max.work}, \tag{5}$$

where K_{det} is the coefficient of detuning, $K_{self-l.}$—coefficient of self-loading of the motor load; K_{reset}—reset coefficient; $I_{max.work}$—the maximum operating line current, the value of which is calculated on the base of the relation

$$I_{max.work} = \frac{\sum S_{cons.}}{\sqrt{3} \cdot U_{nom.}}, \tag{6}$$

where $\sum S_{cons}$ is the total load of consumers connected to the line, kVA; U_{nom}—nominal voltage of the electric network, kV.

In this example, overcurrent protection of the head section of the line is performed on the basis of the classic current relay with $K_{det} = 1.2$ and $K_{reset} = 0.85$, among the consumers the small motor load predominates ($K_{self-l.} = 1.7$), and the active and reactive resistances of the power system are 2 and 5 Ω, respectively. The remaining parameters of the grid shown in Fig. 2 are given in Table 1.

By using the expressions (1)–(6), can determine the change in the sensitivity of the protection of the main section of the networks with the nonadaptive set point in the redundant sections with variations in the wire temperature from −5 to 95 °C and the soil state from dry ($\gamma = 0.01 \times 10^{-2}$ S/m) to wet (3.5×10^{-2} S/m) taking into account the uneven heating of the wire and the moistening of the earth in zones W1–W3.

Considering that the parameters of wire temperature and soil conductivity are random variables distributed according to the normal law, n = 90 examples of

Table 1 Electrical grid parameters

Parameters/grid sites	W1	W2	W3
Power of transformers, kVA	100	100	63
Wire type	Steel-aluminum	Aluminium	Steel-aluminum
Wire diameter, mm	8.4	5.1	5.6
R_{020}, Ohm/km	0.7774	1.8007	1.7818
l_{20}, km	16.5	8.1	3.2
$\alpha_R \times 10^{-3}$, 1/grad	4.03	4.03	4.03
$\alpha_l \times 10^{-6}$, 1/grad	19.2	24.0	19.2

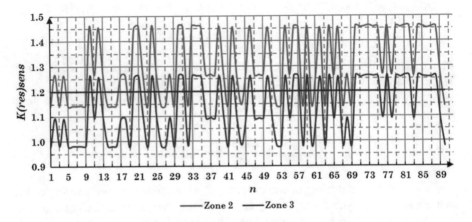

Fig. 3 Shewhart control card of the network protection relay operation

possible combinations of conductor temperatures and soil conductivity on three sections of the network were generated using a computer. The obtained data samples are used to analyze the quality of the relay protection operation, which graphically can be represented as a Shewhart control card (Fig. 3).

It is obvious from Fig. 3, the sensitivity of the classical, non-adaptive relay protection of the main part of the network in the reserved areas is not constant, and in 29% of cases in the W2 area and 59% of the cases in the W3 area the sensitivity of relay protection is insufficient, which can lead to failure of this protection in case of failure of the main protection zones W2, W3. Thus, the influence of external factors leads to an error in the operation of the relay protection, which is the reason for the decrease in its sensitivity and the violation of the selectivity of the over-current protection units.

From the foregoing it follows that the necessary condition for the implementation of "smart" relay protection is the value of the setpoint, calculated taking into account the temperature of the wire and the conductivity of the soil.

The use of an overhead module for measuring wire temperature in the SEMS emergency control system requires the establishment of a wireless communication link between the information processing unit and the sensor, as well as its power supply.

The temperature of the wire depends on the current flowing through it and the atmospheric conditions: air temperature, wind speed and direction, solar radiation. The equation connecting these parameters and heating the wire can be used to indirectly characterize its temperature regime [5]:

$$t_{wire} = t_{air} + \frac{5.018\left[(I/S)^2 N_a^2 d_a^4 (R_{020}/d)\right](1+0.0052 t_{air}) + 8.926 \cdot \varepsilon \cdot K_H \cdot W_r \sin \delta}{(t_{air}+273)^{3.3}(1-d_a/d) \times 10^{-6} + 1977 K_\theta \cdot v^{0.6}/d^{0.4}}$$

(7)

where t_{air} is the air temperature, °C; I—current load, A; S—cross-section of the wire, mm^2; N_a—number of strands of wire of current-carrying part of the wire; d_a— diameter of aluminum wire, mm; d—diameter of the wire, mm; ε—the degree of blackness of the body (for oxidized aluminum, 0.6); K_H—a coefficient that takes into account the effect of the elevation of the OHL route above sea level (for moderate climate zone of Russia 1.0); W_r—intensity of solar radiation, W/m^2; δ— active angle of sunlight, deg.; v—wind speed, m/s; K_θ—correction factor for the angle of attack.

The active angle of inclination of the sun's rays is calculated by the formula [5]

$$\delta = \arccos[\cos h_{sun} \cdot \cos(180 - \psi)],$$

(8)

where h_{sun} is the angular height of the Sun above the horizon, deg.; ψ—geographical angle of the OHL (orientation in relation to the meridian in the range 0°– 180°).

Also, the correction factor for the angle of attack K_θ entering into Eq. (7) is found as follows [5]:

$$K_\theta = 0.5443 - 5.2164 \times 10^{-4}\theta + 2.3128 \times 10^{-4}\theta^2 - 1.939 \times 10^{-6}\theta^3,$$

(9)

where θ is the angle between the wind and the wire surface, deg.

Due to the fact that the heating of the wire is a function of several variables $t_{wire} = f(I, t_{air}, W_r, v, \theta)$, the absolute error of the indirect method of monitoring the wire temperature will be determined by the errors of the respective sensors:

$$\Delta t_{wire} = \frac{\partial f}{\partial I}\Delta I + \frac{\partial f}{\partial t_{air}}\Delta t_{air} + \frac{\partial f}{\partial W_r}\Delta W_r + \frac{\partial f}{\partial v}\Delta v + \frac{\partial f}{\partial \theta}\Delta \theta,$$

namely, taking into account (7)–(9)

$$\Delta t_{wire} = \frac{2A_4}{I(A_1 - B_1)} \Delta I + \left(\frac{3,3A_1(t_{air} + 273)(A_3 + A_4)}{(A_1 - B_1)^2} - \frac{0,026I^2N_a^2R_{020}d_a^4}{S^2d(A_1 - B_1)} + 1 \right) \Delta t_{air}$$

$$+ \frac{A_3}{(A_1 - B_1)W_r} \Delta W_r - \frac{0,6(A_3 + A_4)A_2}{(d \cdot v)^{0,4}(A_1 - B_1)} \Delta v - \frac{v^{0,6}(A_3 + A_4)A_5}{d^{0,4}(A_1 - B_1)} \Delta \theta,$$

where

$$A_1 = (t_{air} + 273)^{3.3}(1 - d_a/d) \cdot 10^{-6}$$
$$A_2 = -3.833 \cdot 10^{-3}\theta^3 + 0.457\theta^2 - 1.031\theta + 1076.081$$
$$A_3 = 8.926 \cdot \varepsilon \cdot K_H \cdot W_r \sin \delta$$
$$A_4 = 5.018I^2N_a^2R_{020}d_a^4(1 + 0.0052t_{air})/(S^2d)$$
$$A_5 = -0.0120\theta^2 + 0.9140\theta - 1.031$$
$$B_1 = v^{0.6}A_2/d^{0.4}$$

Changing conductor parameters of an OHL requires updating its simulation model and corresponding "re-training" of the information processing unit (IPU) at certain intervals with a given discreteness. In this regard, the implementation of IPU in the form of a neurocomputer, trained to recognize the type of emergency modes and the distance to the fault, looks organic [6].

A neurocomputer is a computing device implemented on a software (virtual neurocomputer) or hardware (on the basis of a computer neuroprotection), reproducing the principles of organization and algorithms of information processing inherent in biological nervous systems [7]. Simulating the low-level structure of the brain, an artificial neural network (ANN) also has the ability to generalize the data obtained as a result of "learning".

Thus, Fig. 1 SEMS anti-damage automatics differs from a classical-style recloser by the presence of meteorological parameters sensors, as well as the use of a neurocomputer information processing unit.

This SEMS functions as follows:

Step 1. Interrogation of meteorological parameters sensors and exchange of meteorological information between SEMS at the line sections to collect data on the current state of the network.

Step 2. Compilation of the OHL calculation scheme (calculation of the current line parameters of each section of the line), taking into account the obtained data.

Step 3. Modeling of possible emergency modes (single-phase short-circuit as a result of wire breakage, double-phase short-circuit as a result of wire clipping, double-phase-to-ground short circuit, three-phase short-circuit) at different distances from the power supply. The goal of this stage is the formation of a data for the neural network. Obviously, as the distance to the accident site x increases, the current signals I and the voltage U recorded by the corresponding sensors will also change.

Step 4. Training of the neurocomputer.

In this case, an artificial neural network can have a perceptron architecture with n-th number of hidden layers (Fig. 4). The number of layers and the type of the activation function of neurons are determined exclusively empirically at the stage of designing the neuro-unit, by repeatedly comparing networks of different configurations with the goal of optimizing the perceptron by parameters such as speed and error of training [7].

As can see from Fig. 4, the analysis of phase voltage and current signals on the line requires the presence of six neurons of the input layer, whereas the classification of the accident and the determination of the distance to damage requires four output neurons (three neurons code the type of accident (000—no damages, 001—single-phase short-circuit, 010—double-phase short-circuit, 011—double-phase-to-ground short circuit, 100—three-phase short-circuit), and fourth neuron indicates distance x to damage).

The number of neurons in the hidden layer of the perceptron can be found by the rule of the pyramid [8].

Then, in the case of a perceptron with one hidden layer (Fig. 5a), this number of neurons is calculated by the formula [8]:

$$N_{hid} = \sqrt{N_{input} \cdot N_{output}},$$

where N_{input}, N_{output} are the number of neurons in the input and output layers, respectively.

The number of neurons in the two hidden layers of the perceptron (Fig. 5b) is determined as follows [8]:

$$N_{hid1} = N_{output} \cdot s^2,$$
$$N_{hid2} = N_{output} \cdot s,$$

where

$$s = \sqrt[3]{\frac{N_{input}}{N_{output}}}.$$

Fig. 4 ANN scheme for classification of accidents and location of damage

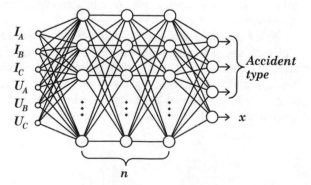

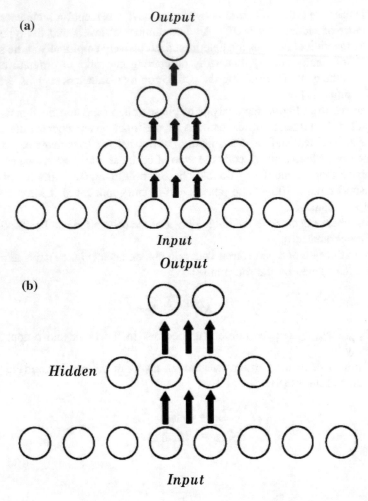

Fig. 5 ANN Structure of the perceptron: **a** three-layer perceptron, **b** four-layer perceptron

Thus, in accordance with the rule of the pyramid, the use of a three-layer perceptron requires five neurons of a hidden layer, and a four-layered one requires five and four.

At the training stage, the synaptic weights of the neural network are calculated. For each set of input parameters (current and voltage signals) from the training sample ("data"), the required value of the output of the neural network (three-digit damage code and distance to the accident site) is known. This process can be considered as a solution to the optimization problem. Its goal is to minimize the error function on the training set by selecting the values of the synaptic weights [9].

Step 5. Feeding to the input of the neural network of real (at the time of the accident) current and voltage signals on the line, the classification of the accident and the location of the damage, taking into account the terrain and the sagging of the wires. In the event that the distance to the accident site corresponds to the main protected area, a trip command with a minimum time delay is given. Accidents on the reserved areas are also accompanied by a trip command, but with a longer time delay to observe the principle of selectivity.

Self-diagnostics and the possibility of interaction of several devices. Automatic diagnostics of the SEMS state means detection of faults in communication lines in the electromagnetic drive control system of the circuit breaker, as well as checking its power supply system.

In a present time, a large number of methods for self-diagnosis of the electric drive control system are known, which can be divided into two groups: implying the detection of hidden defects, for example, breaking the communication line, using contactless and contact fault sensors, and also by sequential feeding and analysis of test signals [10].

Currently, capacitive energy storage devices are used to supply high-speed drives of electrical apparatuses, which make it possible to obtain large-amplitude current pulses (10 kA and higher), and there is an objective need for the presence of charge sensors, the integrity of the supply lines, and the state of the battery. The integrity of the winding of the electric drive is checked by briefly applying voltage to it and monitoring its current flow.

The need for the interaction of several emergency control devices is dictated not only by the need for more accurate simulation of the electric grid, taking into account the changing and uneven distribution of climatic effects on the OHL, but also the ability to quickly reconfigure the grid protection relay structure in case of a component failure in any section—namely, inclusion of the reserve zone into the main zone, protected by the SEMS network, which is closer to the power source.

3 Conclusions

In modern switching devices, the block for information processing and control of the switch drive implemented on networks logic doesn't have the ability to adapt to environmental conditions, which is the reason for the unreliability of the operation of the protection device. The proposed "smart" electromechanical system (SEMS) lacks this disadvantage due to the use of a neurocomputer, changes the value of the relay protection setting on the base of the measured information of the parameters of the influencing factors. "Smart tower", containing a group of smart electromechanical systems (SEMS) of emergency control automatics, allows to ensure normal group interaction of the blocks of overcurrent protection and, accordingly, to increase the accuracy of registration of emergency conditions. The introduction of SEMS as an element of the smart networks will increase the reliability of power supply systems by reducing the recovery time of accidents.

The paper presents the results of studies performed under the program Erasmus +#573879-EPP-1-2016-1-FREPPKA2-CBHE-JP "Internationalisation of Master Programs In Russia and China in Electrical Engineering".

References

1. Fardiev, A., et al.: Status and prospects for the development of energy. On the innovative project "smart grid. Energetics Tatarstan **3**, 5–12 (2010)
2. Lachugin, V., et al.: Principles of constructing smart relay protection of electrical grids. Proc. Acad. Sci. Energetics **4**, 28–37 (2015)
3. Bulychev, A., Navolochny, A.: Relay protection in distribution electrical grids; a manual for practical calculations. ENAS, Moscow (2011)
4. Margolin, N.: Resistance of Overhead Power Lines. United Scientific and Technical Publishing House, Moscow-Leningrad. The main edition of energy literature (1937)
5. Zarudsky, G.: Calculation of the temperature of wires of EHV transmission lines on the basis of the method of criterial experiment planning. Vestnik MEI **1**, 85–90 (1997)
6. Dementiev, S.: Diagnosis of the state of overhead power lines with the use of a neurocomputer. Energy Resour Saving Ind. Transp. **2**(19), 21–26 (2017)
7. Shilin, A., Shilin, A., Dementiev, S.: The use of artificial neural networks for forecasting accidents of overhead power lines during the ice-covered period. Electro. Electr. Eng. Electr. Power Ind. Electrotech. Ind. **2**, 15–21 (2016)
8. Masters, T.: Practical neural network recipes in C++. Academic Press Inc, Boston (1993)
9. Osovsky, S.: Neural networks for information processing. Finance and Statistics, Moscow (2002)
10. Osipov, O.: Technical diagnostics of automated electric drives. Energoatomizdat, Moscow (1991)

Three-Axis Electromechanical Drive of the Robotic Complex for Monitoring Shells and Their Assembly

Aleksandr N. Shilin and Dmitriy G. Snitsaruk

Abstract *Problem statement*: The big rotational shells are the basic parts in aerospace engineering, nuclear industry, oil and gas engineering, as well as in other areas of technology. The rotational shells can be large (up to 10 m in diameter) and are made of metal sheet on roll bending machines. Then, from these parts, the assembly of hull products is carried out. Main problem of technological process is the operative and accurate measurement of their geometric parameters in the process of their manufacture. The second problem is the complexity and difficulty of assembly and installation operations in the manufacture of products. These problems have a negative influence on the quality, technical and operational characteristics and the cost price of the products. To solve these problems it is convenient to use robotic complexes with machine vision system which should determine the position of the controlled robot relative to the detail, measure the geometrical parameters of the workpiece and to arrange the relative positions of parts for the assembly of the product. The main blocks of the robotic complex are the robot's machine vision system, which controls a three-axis electromechanical drive, which is a smart electromechanical system (SEMS). SEMS should move the video camera in three coordinates to ensure the measurement of the geometric parameters of the part with minimal error. After processing the measurement results, the SEMS must perform the movement of the parts to the assembly position based on the optimal arrangement of the parts. *Purpose of work*: Development algorithms and technical implementation of SEMS of a robotic complex for controlling geometric parameters of large rotational shells and assembly of hull products with improved technical and operational characteristics and with lower cost. *Results*: The SEMS parameters were reviewed and analyzed. The design and algorithm of operation of the SEMS of the robotic complex are developed. The possibility of obtaining the required accuracy and speed of the developed robotic complex is analyzed. The software for control of SEMS was developed and a laboratory sample of the robot was manufactured, the

A. N. Shilin (✉) · D. G. Snitsaruk
Volgograd State Technical University, Volgograd, Russia
e-mail: eltech@vstu.ru

D. G. Snitsaruk
e-mail: norzes@mail.ru

© Springer Nature Switzerland AG 2019
A. E. Gorodetskiy and I. L. Tarasova (eds.), *Smart Electromechanical Systems*,
Studies in Systems, Decision and Control 174,
https://doi.org/10.1007/978-3-319-99759-9_21

tests of which were confirmed by the results of theoretical studies. *Practical meaning*: The developed robotic complex with SEMS is intended for automation of technological processes in various industries that produce the large-sized cylindrical products, as well as for joining cylindrical elements of aerospace systems.

Keywords Control of geometric parameters · Shell of rotation
Robotic measuring complex · Three-axis drive · SEMS · Machine vision
system

1 Introduction

The estimation of the technological accuracy of large-sized shells based on the following characteristics: mathematical expectation and dispersion, which follows from the stochastic nature of the technological process [1]. Deviation of the mathematical expectation from the nominal size is a systematic error by which the control action is performed, and which is eliminated by the influence on the technological process. The standard deviation of parts is comparable to the permissible category of permits for compliance.

2 Main Equations of the Rotational Shell Shape

To calculate the control effect on the technological process in the production of cylindrical hollow parts, it is necessary to perform measurements at individual points of a certain number of parts (Fig. 1) [2]. The mathematical expectation and variance of the current radius of a part measured relative to a certain reference axis are calculated using the expressions [2]:

$$\bar{r} = \frac{1}{N} \sum_{l=1}^{q} \sum_{j=1}^{n} \sum_{i=1}^{m} r_{lji}, \tag{1}$$

$$\sigma_r^2 = \frac{1}{N} \sum_{l=1}^{q} \sum_{j=1}^{n} \sum_{i=1}^{m} \left(r_{lji} - \bar{r} \right)^2 \tag{2}$$

where

q	number of parts in a batch;
m	number of sections along the length of the part;
n	number of measurements in the section along the perimeter of the workpiece;
$N = qmn$	total number of measurements in a batch;

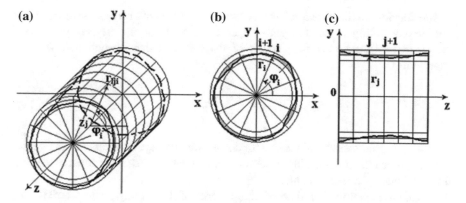

Fig. 1 Schemes for the location of the measured spatial coordinates of the shell of revolution

r is the unit measurement of the size of the current radius on the l-th part in the j-th section along the length and at the i-th point along the perimeter of this section.

Expressions (1) and (2) allow to determine the general characteristic of the accuracy of parts and do not allow to investigate the influence of individual factors on the accuracy of manufacturing parts.

All the factors that affect the accuracy of manufacturing parts can be divided into three groups:

1. the factors causing the deviation of the current radius from its mean value in each cross-section;
2. the factors causing the deviation of the current radius from its average value in each longitudinal section;
3. factors causing a spread of the values of the current radius of individual parts taken.

Errors in the cross section of shells are functionally related to the quality of the sheet material, the initial shape of the workpiece, the design of the bending machine, and the algorithm for controlling the of shaping process. The error in the longitudinal section is determined by the position of the shell relative to the rollers, by the adjustment of the counter pressure mechanism, which should exclude the deformation of the bending machine rolls during the straightening.

From the analysis of errors in the transverse and longitudinal sections, it follows that there is no correlation between them, and therefore these errors can be measured separately by means of different measuring devices. Information on the error in the longitudinal section can be used to control the backpressure mechanism. The basic information for controlling the editing process is the error in the cross section since the shape of the section is necessary for calculating the control algorithm for the editing process.

Information on the individual parts errors in the batch after manufacturing can be used in assembling shells of assembly operations, for example, the average value of the current radius in the j-th section of the l-th part can be obtained with the expression:

$$\bar{r} = \frac{1}{n} \sum_{i=1}^{n} r_{lji} \tag{3}$$

With the arrangement of shells with decreasing average diameters in the vessel, the laboriousness of assembly work is reduced when installing internal devices from the side of a larger diameter.

Thus, the isolation of individual components of the accuracy characteristic will allow using the information obtained in the process control.

When analyzing the errors in the cross section of the shells of revolution, it is convenient to use the spectral method based on the expansion of the detail profile function in polar coordinates in the Fourier series [1]:

$$r(\varphi) = R + \sum_{k=1}^{\infty} A_k \cos(k \cdot \varphi + \varphi_k), \tag{4}$$

where

φ is the alternating angle formed by the current dimension with the polar axis;
R is the radius of the mean circle, defined as the average value of the function $r(\varphi)$:

$$R = \frac{1}{2\pi} \int_{-\pi}^{\pi} r(\varphi) d\varphi; \tag{5}$$

A_k and φ_k is the amplitude and initial phase of the k-harmonic characterizing the shape error.

The number of the harmonic, from which the microgeometry starts, limits the higher harmonic component k. With the hardware implementation of the spectral method, the radius determining problem of the mean circle and the center coordinates is relatively easy to solve, since the amplitude and phase of the first harmonic determine the displacement of the center relative to the measurement base.

When representing the profilogram as a sum of harmonic components, using the Nyquist theorem, the question of choosing the discretization step is relatively easy to solve. In addition, with hardware implementation, the problem of filtering harmonic components that cannot be corrected in roller bending machines can be solved.

When processing the results of measurements in digital information and measurement systems, numerical integration methods are used.

The coefficient k determines the degree of faceting of the profile. Based on the spectral analysis of the profilograms of the batch of shells, the distribution of the average amplitudes of the harmonic components (Fig. 2) was obtained, from which

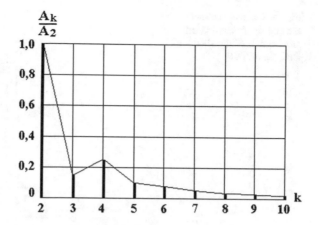

Fig. 2 Distribution of the mean values of the amplitudes of the harmonic components of the profilograms of the batch of shells

it follows that the second harmonic component ellipse predominates in the production of shells.

The predominance of ellipse is explained by the features of the technological process [3]. The ellipse shape of the profile has a property, namely the centers of the circle and the rectangle into which the ellipse fits, which can be used in the design of the center of the cross section of the workpiece.

3 Analysis of Methods for Measuring the Profilograms of Shells

At present, for two-coordinate measurements, technical of vision systems (TVS) are widely used that allow the intelligent processing of information. The main element of the vision system is the optoelectronic matrix converter. To measure the profile of the cross section of the shell—profilogram, it is advisable to use the TVS, since for the measurement process to be performed it is necessary to first determine the measurement base—the center of the cross section. The disadvantage of this projection method is the possibility of measuring only in the end plane of the shell. In the optoelectronic system for profilograms measuring, the input image $f(x, y, t)$ must be converted in the end face of the shell x, y to the output image $g(x', y', t)$ in the image analysis plane of the system x', y'. For simplicity of analysis, the coordinate systems x, y and x', y' are normalized and in this case both images are described by functions of the same arguments $f(x, y, t)$ and $g(x, y, t)$ [3].

Consider the projection method for the profilograms measuring of shells.

When using a matrix converter as a photodetector (Fig. 3), spatial sampling of the image is performed. When a fixed image of the profilogram is transmitted in the time interval $[0, T_k]$, the entire image $f(x, y)$ is projected onto the converter input. The spatial discretization of the image $g(x, y)$ can be represented as a set of light

Fig. 3 Optical-electronic method for measuring the profilogram of the shell with the help of TVS

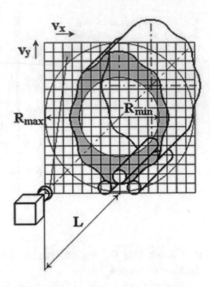

"spots" whose centers are at points with coordinates 0, 0; 0, Δy; ...; 0, $M\Delta y$; Δx; 0; Δx, $M\Delta y$; ...; $P\Delta x$, 0; ...; $P\,\Delta x$, $M\Delta y$ [3], i.e.

$$g(x,y) = \sum_{i=0}^{P} \sum_{j=0}^{M} g_{ij}(x,y) \tag{6}$$

where $g_{ij}\,(x,\,y)$ is the image of the spot with the number ij.

When the image is transferred in parallel, the matrix converter is connected to the information processing device by means of $P \times M$ channels. After the transmission and processing of information in the time interval $[T_k, \infty]$, it is stored (registered). The intensity at the center of each of the spots is related to the intensity of the corresponding point of the input image and the exposure interval T_k [3].

$$g_{ij}(i\Delta x, j\Delta y) \cong \int_{0}^{T_k} f(i\Delta x, j\Delta y) rect(t, T_k) dt = f(i\Delta x, j\Delta y) T_k \tag{7}$$

where $rect(\alpha, A) = \begin{cases} 1 \; at \; \alpha \in [0,A]; \\ 1 \; at \; \alpha \notin [0,A]; \end{cases}$—special function.

In the transmission of a moving image, the statement of the problem changes as follows. In the time interval $[0, NT_k]$, the input of the system is fed with the image $f\,(x,\,y,\,t)\,rect\,(t,\,NT_k)$. In this interval, both the transmission and the perception of the transmitted image $g\,(x,\,y,\,t)\,rect\,(t,\,NT_k)$ at the receiving end of the system must be realized.

The transition from the transmission of still images in a system with $P \times M$ channels to the transmission of moving images is associated with the exclusion of the integration operation at the receiving end [3]:

$$g(x, y, t) = \sum_{i=0}^{P} \sum_{j=0}^{M} g_{ij}(x, y, t) \qquad (8)$$

where $g_{ij}(i\Delta x, j\Delta y, t) \approx f_{ij}(i\Delta x, j\Delta y, t)$.

When the field is angularly scanned, the distance to the part being scanned changes and the maximum change is determined by the expression

$$\Delta_{\max} = \sqrt{L^2 + R_{\max}^2} - L \qquad (9)$$

For example, with the ratio of the measurement scheme parameters $R_{max} = L$, the maximum change value is $\Delta_{max} \approx 0.41L$, which is the source of the measurement error of the profilograms. It should be noted that this error could be automatically compensated.

4 The Control Scheme of a Robotic Complex

The control scheme is as shown in (Fig. 4). On a solid foundation, the parts prepared for inspection are fixed. The robot moves from the part to the part along the rails. The main measuring transducer is a digital camera, with which you can determine the coordinates of the center by the method of a rectangle, and then remove the profilogram from the image taken from the center of the part. The rectangle method consists in the fact that the detail representing an ellipse fits into a rectangle formed by tangents to the profile of the part (Fig. 5), after which the coordinates of the center of the resulting rectangle coincide with the center of the cross section of the part.

When a part enters the camera's scope, it is recognized, and the analysis of the last frame is performed, along which the positions of the tangents to the profile of the part are determined [4].

The industrial robot is marked in the positive direction of the abscissa axis until the image of the part appears in the camera's scope of view. Then the robot's computer recognizes the part image and searches for the tangents to the found part profile.

After finding the tangents, the coordinates of the center are calculated:

$$x_c = \frac{x_{max} - x_{min}}{2}; y_c = \frac{y_{max} - y_{min}}{2}.$$

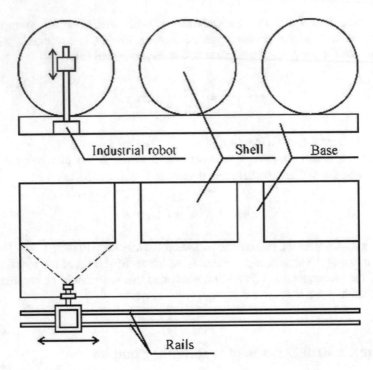

Fig. 4 Scheme of control

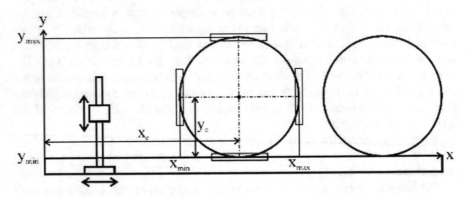

Fig. 5 Scheme of center find

After the coincidence of the optical axis of the camera coincides with the center of the image of the cross-section of the part, the profilogram is recorded along the image. The information about the profilogram is transferred to the server for further analysis. The method of the described rectangle allows reducing the search time of the center, since the resulting image is not analyzed in a pixel-by-pixel manner, as, for example, in the case of the geometric center of gravity search method [4].

Fig. 6 Algorithm of the profilogram registration process

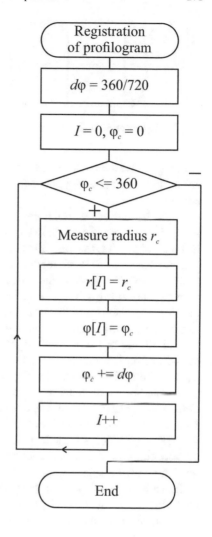

In addition, the movement of the camera along one coordinate is not associated with the movement along another coordinate, i.e. there are no cross-links, which allows improving the dynamic characteristics of robot drives. The advantage of this method is that all processes (moving the camera in two coordinates) occur in parallel, carrying out a group non-synchronous interaction of electromechanical systems within one industrial robot. The parallelism of the processes makes it possible to substantially reduce the search time of the center and, therefore, to reduce the monitoring time.

Then the profilogram is controlled (Fig. 6), and it is precisely the dependence of the radius vector r_i on the angle of rotation φ with respect to a certain center of

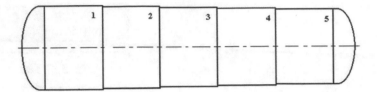

Fig. 7 The sequence of the cylindrical apparatus assembly

the image. The advantage of the method of registering an image profilogram is the ability to control accuracy, in contrast to the electromechanical scanning of the part contour.

After the end of the monitoring process, the removed profilogram is sent to the server for data processing and visualization, and the robot moves to the next detail. If the robot reaches the end of the rails, the process stops, and the robot is set to its original position. Several robots can be connected to one server (depending on production volumes), working independently of each other.

After the control of the series of parts is completed, their parameters (part name, profilogram, diameter of the average circle, date, etc.) are entered in the database for further processing with the purpose of optimizing assembly operations using a special algorithm [5]. In the proposed algorithm for optimizing the assembly of cylindrical bodies, the current radius vectors and the radius of the mean circle were used as the initial information (Fig. 7).

The algorithm contains two steps: the first assumes the arrangement of shells in order of decreasing the average radius along the length of the housing, which facilitates the installation of devices inside the housing from the large diameter side; the second provides a mutual arrangement of the joined shells with a minimum sum of deviations of the radius vectors in all angular coordinates, which ensures a minimum edge displacement.

In addition, the developed program considers several requirements imposed by OST 26.291-87 on welded joints of shells. Thus, the withdrawal of f edges in butt-welded joints should not exceed $f \leq 0.1S + 3$ mm.

5   Conclusions

Thus, the developed robotic complex will automatically control the geometric parameters of the shells, covering all the details manufactured in the enterprise. Now, a metrological analysis of the optical measurement scheme has been carried out [6] and patents for utility models for several individual robot nodes have been obtained [7, 8].

In the same way, the control algorithm was tested using a laboratory prototype and the main electromechanical components of the complex were developed.

The article presents the research results for the Erasmus + program No. 573879-EPP-1-2016-1-FREPPKA2-CBHE-JP "InternationaliSation of master Programs in Russia and China in Electrical Engineering".

References

1. Tochnost' proizvodstva v mashinostroenii i priborostroenii [The accuracy of production in engineering and instrument-making]/ Pod red. A.N. Gavrilova. – Moskow: Mashinostroenie [Mech. Eng.], 567s (1973)
2. Stoletnij, M.F., Klempert, E.D.: Tochnost' trub [The accuracy of the pipes]. – Moscow: Metallurgiya [Metallurgy] 240s (1975)
3. Smirnov, A.Y.A., Men'shikov G.G.: Skaniruyushchie pribory [Scanning devices]. – Leningrad: Mashinostroenie [Mech. Eng.] 145s (1986)
4. Shilin, A.N.: Adaptivnaya optiko-ehlektronnaya sistema kontrolya profilogramm obolochek vrashcheniya [Adaptive optoelectronic system for monitoring profilograms of shells of revolution]/ A.N. Shilin, D.G. Snitsaruk// Pribory i sistemy. Upravlenie, kontrol', diagnostika [Dev. systems. Manag. Control Diagn.]. **2**, C. 40–45 (2018)
5. Shilin, A.N.: Avtomatizaciya opredeleniya optimal'nyh uslovij sborki korpusov neftegazovogo oborudovaniya [Automation of the determination of optimal conditions for assembling the hulls of oil and gas equipment]/ A.N. Shilin, S.A. Petrov, V.P. Zayarnyj// Sborka v mashinostroenii, priborostroenii [Assembl. Machine Build. Instrum. Mak.]. **6**, C. 10–14 (2010)
6. Shilin, A.N.: Metrologicheskij analiz ustrojstva pozicionirovaniya optiko-ehlektronnogo pribora kontrolya krupnogabaritnyh obolochek vrashcheniya [Metrological analysis of the positioning device of the optoelectronic device for monitoring large-sized shells of revolution]/ A.N. Shilin, D.G. Snitsaruk// Pribory i sistemy. Upravlenie, kontrol', diagnostika [Dev. Syst. Manag. Control Diagn.]. **5**, C. 39–45 (2017)
7. P. m. 171730 Russian Federation, IPC G01B11/08 Opticheskoe ustrojstvo dlya izmereniya diametrov krupnogabaritnyh detalej [Optical device for measuring the diameters of large-sized parts]/ D.G. Snitsaruk, A.N. Shilin; VolgGTU [VSTU]. (2017)
8. P. m. 159150 RF, IPC G01B11/00. Optiko-ehlektronnoe ustrojstvo dlya izmereniya razmerov obechaek [Optical-electronic device for measuring the dimensions of shells]/ D.G. Snitsaruk, A.N. Shilin; VolgGTU [VSTU] (2016)

Diagnosis in SEMS Based on Cognitive Models

Vladimir V. Korobkin and Anna E. Kolodenkova

Abstract *Problem statement*: SEMS is a complex dynamic object in the operational phase of which it is likely that abnormal situations may occur in which the state of the equipment goes beyond the normal functioning, which can subsequently lead to an accident. Despite the importance of the need and importance of effectively solving the problems of diagnostics of SEMS, at the present time there is no single approach to solving similar problems taking into account the variety of emergent contingencies. Therefore, the actual task is the development of methods, algorithms and special diagnostic tools that allow predicting the development of defects, diagnose processes and recognize violations of normal operation at an early stage of their development to ensure the efficiency, reliability and safety of the operation of SEMS in real time. *Results*: cognitive to diagnose SEMS in conditions of interval uncertainty and fuzzy initial data, cognitive and fuzzy cognitive modeling is used to reflect the problems of SEMS in a simplified form (in the model), to investigate possible scenarios for the emergence of risk situations at an early stage of their development, and to find ways to resolve them in the model of the situation. As an example, a fuzzy cognitive model of SEMS diagnostics is proposed. Pessimistic and optimistic scenarios of possible development of risk situations, developed with the help of impulse simulation, are given and their brief analysis is given. The system indices of the fuzzy cognitive model are calculated, allowing to identify which of the factors have the greatest impact on SEMS and vice versa; To search for the best values of factors reflecting the normal operation of SEMS.

V. V. Korobkin (✉)
Acad. Kalyaev Scientific Research Institute of Multiprocessor Computer Systems,
Taganrog, Russia
e-mail: vvk@niimvs.ru

A. E. Kolodenkova
Department Chair «Information Technologies», Samara State Technical University,
Samara, Russia
e-mail: anna82_42@mail.ru

© Springer Nature Switzerland AG 2019
A. E. Gorodetskiy and I. L. Tarasova (eds.), *Smart Electromechanical Systems*,
Studies in Systems, Decision and Control 174,
https://doi.org/10.1007/978-3-319-99759-9_22

Practical significance: the ability to systematically take into account the long-term consequences of possible abnormal situations and identify side effects that allow to take into account the multifactory of the process of diagnosing SEMS in the process of operation. The task of identifying possible risks in general, and at the operational stage in particular, should be an important part of the diagnostics of SEMS equipment.

Keywords SEMS · Information-control systems · Cognitive and fuzzy cognitive modeling

1 Introduction

SEMS is a cyberphysical system that is a complex dynamic object, in the operational phase of which it is likely that abnormal situations may arise as a result of deviations from the normal functioning of mechanical, electronic equipment and software. Such a deviation can lead to the failure of the SEMS itself, and in general the device where it is used. What is characteristic for such systems is to identify abnormal situations, just as it is impossible to determine the state and response of the system to such situations at the design stage of SEMS.

Therefore, an urgent task is the development of methods, algorithms and special diagnostic tools that allow predicting the development of defects, diagnose processes and recognize violations of normal operation at an early stage of their development [1, 2].

Despite the importance of the need and importance of effectively solving the problems of diagnosing SEMS, at the present time there is no single approach to the solution of such problems, taking into account the variety of emergent supernumerary situations as a result of the action of external factors. The existing methods of diagnostics are not able to generate rational typical solutions because they are based on heterogeneous "heavyweight" mathematical tools, adapted to a variety of private paradigms [3], which adversely affects the speed and energy efficiency of the information management system in SEMS, even when data acquisition and processing are parallelized.

Given the urgency of the problem, new diagnostic technologies are of great interest, with the identification and analysis of risk situations, by building a set of models that fill the "cognitive distance" between the automated subject area and the quality of the diagnosis of SEMS.

In this regard, it is proposed to use cognitive and fuzzy cognitive modeling for the multi-factoring of the SEMS diagnosis process in the process of exploitation for diagnosing, identifying risk situations and recognizing the disruption of normal operation at an early stage of their development [4].

2   Cognitive and Fuzzy Cognitive Modeling of the Diagnostic Process

Given the importance, complexity and multidimensionality of the process of diagnosing SEMS in conditions of interval uncertainty and fuzzy initial data, it is advisable to apply cognitive and fuzzy cognitive modeling.

In this work, cognitive modeling is a tool for modeling the SEMS diagnosis process, including methods, techniques, algorithms designed to solve interrelated systemic problems (searching for cycles of the cognitive model, searching for eigenvalues, topological analysis, setting control actions, establishing current values of indicators and initial impulses), which allows to analyze the diagnosis process, as well as clarify clear cognitive models (PCM).

Fuzzy cognitive modeling is a tool for modeling the SEMS diagnosis process, including methods, methods, algorithms designed to solve interrelated system tasks (topological analysis, factor interference, fuzzy cognitive model (NCM) training, setting of control actions, establishing current values of indicators and initial impulses), which allows to analyze the process of diagnosability, as well as to specify the NAC.

The generalized scheme of the methodology for cognitive and fuzzy cognitive modeling of the SEMS diagnosis process consists of seven steps [4].

Step 1 Identification of factors (target, control), the relationships between them, as well as setting their values.

At this stage, there is an analysis of the problem, the definition of the purpose and objectives of the study, as well as the cognitive structure of the knowledge of the leader and executors of the project.

Step 2 Setting the values of factors and the relationships between them.

Step 3 Processing factor values and relationships between them (processing of undefined raw data).

Step 4 Building a clear and/or fuzzy cognitive model.

A clear cognitive model is a cognitive map (a sign-oriented oriented graph), which is obtained by structuring the knowledge of data based on theoretical representations, initial data, the application of various expert methods [5, 6]:

$$G = \langle V, E \rangle \tag{1}$$

where

$V = \{vi\}$ is the vertex set, $i = 1, \ldots, h$, h is the number of vertices;
E is the binary relation on V (the arcs (links) between the vertices vi and vj). The elements eij, $eij \in E$ ($i, j = 1, \ldots, h$) characterize the direction and strength of the influence between the vertices vi and vj, eij = e (vi, vj).

A fuzzy cognitive model is a fuzzy cognitive map in which vertices represent factors, and edges are fuzzy cause-effect relationships between factors (2.2) [6, 7]:

$$G_{неч} = <V, W> \tag{2}$$

Where

$V = \{vi\}$ is the vertex set, $vi \in V$, $i = 1, ..., h$, h is the number of vertices;
$X = \{xvi\}$ is the set of vertex parameters, $i = 1, ..., h$. Each vertex is assigned one parameter;
W, fuzzy cause-effect relationships between vertices. The elements wij, $wij \in W$ characterize the direction, the force of influence between the vertices vi and vj and possess the properties described in [8].

Note that at this stage, not one clear and/or fuzzy cognitive model can be constructed, but a set of clear G and/or fuzzy cognitive models Gnech.

Note that in the general case, while the work is not carried out with the CNM and/or NCC as a mathematical model, we operate with the term "factors". As soon as the work with the ChMK and/or NCC began, we use the term "top".

Step 5 Computational experiment of a clear and/or fuzzy cognitive model, including analysis of model structures, impulse simulation and scenario analysis.

In the case of impulse simulation on a clear and/or unclear cognitive model, scenarios can be used to construct various scenario scenarios for the development of situations related to the diagnosis of SEMS, in order to reduce negative trends and/or enhance positive trends.

To carry out the impulse simulation on the CMM and/or NAC, it is necessary to investigate the dependence of the change in the parameters of the vertices on the time xvi (t), $t = 1, 2, 3, ...$

The process of propagation of the perturbation along the graph G is determined by the expression described in [9, 10]. The scenarios generated by the perturbations give an answer to the question: "What will happen at time t $(n + 1)$, if …?".

Carrying out a computational experiment by impulse simulation requires its preliminary planning. Planning refers to the choice of vertices, in which disturbing influences are to be introduced.

To solve the problem of choosing the best scenario, you need to obtain a sufficient number of implementations of each scenario, and then conduct several series of impulse simulation. Further it is necessary to apply expert estimates or mathematical methods of comparison [11–13].

Step 6 Analysis of the results.

After the analysis of the results, it is proposed to make a choice, namely, to make a decision on adjusting/not adjusting the initial CKM G and/or NAC G$_{nec}$ or about the development of a new CKM G$_{(new)}$ and/or NAC$_{Gnich (nov)}$.

During the analysis, you can add or remove factors; establish new links between factors; change the values of factors.

Step 7   Conclusion on the diagnosis of ICS SEMS, taking into account the risk situations and the recognition of a violation of normal operation at an early stage of their development.

3   Construction and Computational Experiment of the Diagnostic Model of the Information Control System (ICS) in SEMS

As an example for the diagnosis of ICS SEMS, NCM was developed (Fig. 1).

Here v1—number of tasks; v2—system performance; v3—number of ICS SEMS diagnosis results estimates (estimates can be obtained using different approaches and methods); v4—completion of diagnostics (failure to obtain necessary data, sudden interruption of system operation); v5—economy (work with the least cost); v6—reliability of ICS SEMS (the system is in a working state for a certain length of time), v7—security and protection of ICS SEMS (the property of the system to function properly without manifesting various negative consequences for people and the environment); v8—external factors affecting SEMS (seismic impacts, climatic effects, flood, fire, loss of electricity supply); v9—the number of

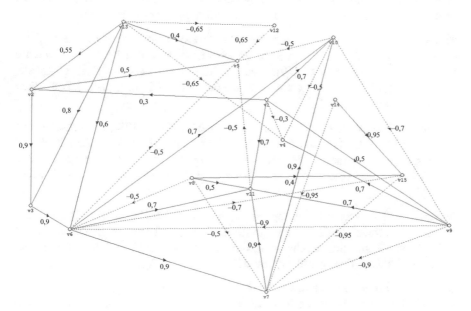

Fig. 1 Fuzzy cognitive model of diagnostics of distributed information-control systems

errors in the calculations; v10—time of diagnosis of SEMS; v11—the cost of diagnosing SEMS (financial resources spent on diagnosis); v12—number of iterations; v13—number of parameters; v14—violation of normal operation; v15—emergency situation.

Interpret the connections shown in Fig. 1, can be as follows: the conversion of percentage changes in causes into percent changes in the effects. For example, the relation v6 → v7 with a weight of 0.9 means that if the value of the vertex parameter v6 increases (decreases) by 10%, then the value of the vertex parameter v7 will increase (decrease) (9%) (10% * 0, 9 = 9%). For example, the relation v6 → v15 with a weight of −0.7 means that if the value of the vertex parameter v6 decreases by 10%, then the value of the vertex parameter v15 increases by 7% (10% * 0.7 = 7%).

Note that technical indicators are hidden at the tops of v6 and v7, and also the most common (key) factors characterizing the development of defects and risk situations arising at the early stage of their development are considered in the NCM. In subsequent stages of ICS SEMS diagnostics, when constructing NCM, factors that characterize the individuality of objects can be used.

In order to develop valid ICS SEMS diagnostics results, the structural stability analysis of the model in the form of NCM is carried out. In this paper, it is proposed to understand the degree of survivability of a fuzzy graph under structural stability [14]. The analysis of the structural stability of the NCM showed that the degree of structural stability of the model under study is at the level of 0.48, which classifies the model as a medium-stable one.

Carrying out a computational experiment by impulse simulation requires its preliminary planning. Below are the fragments of the results of the pulse simulation of the scenarios of the NCM development and their analysis is given.

Scenario 1 This scenario corresponds to a "pessimistic" scenario. The impulse arrives at three vertices. Let's ask the question: "What will happen if we increase the number of tasks by q1 = 10%, reduce the number of ICS SEMS diagnostic score estimates by q3 = −10% and decrease the ICS SEMS diagnostic process time by q10 = −10%?" (Fig. 2).

Recommendations an increase in the number of tasks by 10%, a decrease in the number of ICS SEMS diagnostics results (−10%) and a decrease in the ICS SEMS diagnostic process time (−10%) leads to a sharp decrease in the recognition of a violation of normal operation at an early stage of their development. ICS SEMS to (−19%).

Scenario 2 This scenario corresponds to an "optimistic" scenario. The impulse enters two vertices. Let's ask the question: "What will happen if the number of ICS SEMS diagnostics results is increased by q3 = 10% and the ICS SEMS diagnostic process time is reduced by q11 = −10%?" (Fig. 3).

Recommendations an increase in the number of ICS SEMS diagnostic score scores by 10% and a decrease in the ICS SEMS diagnostic process time (−10%) leads to

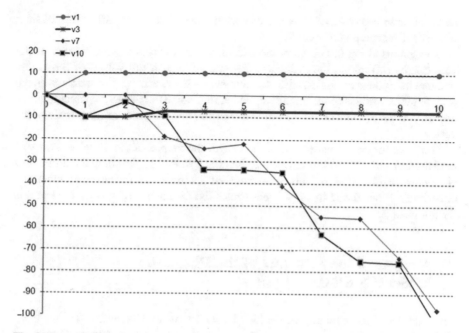

Fig. 2 Scenario No. 1

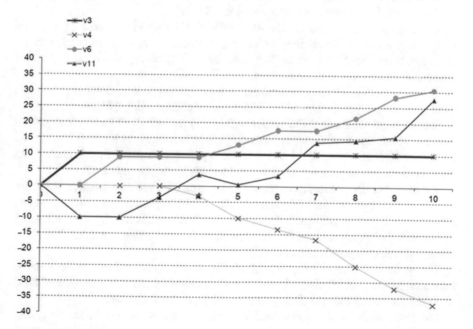

Fig. 3 Scenario No. 2

an increase in detection of a failure of normal operation at the early stage of their ICS SEMS development by 9%.

The graphs along the abscissa indicate simulation cycles n (specific time intervals can be set), along the ordinate axis—the change in the values of the vertex parameters (you can understand the numbers characterizing the rate of signal growth at the vertices of the cognitive model, or change the vertex in percent, which allows you not to think about the units of measure and the order of magnitude of the vertices).

Note that for better perception of the image, the graphs of the impulse processes contain four vertices. At the same time, we note that the graphs show the number of simulation cycles that reflects the trends of changes. A further computational experiment showed that in subsequent cycles the tendencies of growth or decrease do not change.

4 Analysis of the Structure of the Diagnostic Model ICS SEMS

The analysis of the structure of the NCM consists of three stages.

The first stage is topological analysis, which allows one to determine the mutual influences of concepts against each other, as well as to receive information about implicit mutual influences between concepts [15, 16].

Topological analysis showed that the vertices v5, v6, v11 can be chosen as the target factors. In this case, the vertices v1, v2 and v8 are recommended to be excluded from the NCM, since they can be a "stumbling block" for effective interaction of vertices with each other.

The second stage is the evaluation of mutual influence of the vertices on each other (the system parameters of the NCM are calculated) [17, 18].

Studies have shown that the vertices v3 (0.81) and v2 (0.73) exert the greatest positive influence on the target vertex v7. The vertexes v14 (−0.95) and v15 (−0.95), which are at a high level, have a negative influence, this indicates that these vertices can negatively affect the values of the target vertex v7.

We note that the peaks v2 (0.19), v3 (0.10) exert the greatest positive influence on the nanocomposite, and they practically do not experience an inverse strong influence. A little less positive influence on the NAC is rendered by the vertices v3 (0,08), v13 (0,08). By influencing the above mentioned vertices, you can "turn" the whole system in a positive direction.

The peaks v11 (0,2), v3 (0,13), v10 (0,13), v6 (0,08) are subject to the greatest influence from the NAC side. There is a high probability that the influence of NFC on these peaks can extinguish any negative impact from outside.

The third stage is the guaranteed detection of factor values in the intervals, recognition of a violation of normal operation at an early stage of their development of ICS SEMS with limited resources available, using the NQM training procedure described in [19, 20].

Studies have shown that after four iterations a steady state appeared, determined by the vector of variables: Xout = (0.6572, 0.7945, 0.8945, 0.4548, 0.5728, 0.7488, 0.2113, 0.6588; 0.5719, 0.8851, 0.9448, 0.5196, 0.6597, 0.6589, 0.7549). However, the values of the vertices v7 (0.2113), v10 (0.8851), and v11 (0.9448) did not fall within the given interval, therefore, the NCM training procedure was used.

Then, calculations were carried out for the same initial values of the parameters of the vertices X (0) = (0.0003, 0.64, 0.08, 0.95, 1, 0.77, 0, 0.55, 0.55, 0, 1; 0,99; 0,003; 0,93; 0,6; 0,4) with application of updating of values of links at each iteration.

As a result, after five iterations, the steady state of the NCM appeared, determined by the vector of variables: Xout = (0.9777, 0.9901, 0.9950, 0.9280, 0.9472, 0.9824, 0.6229, 0.9778, 0, 9691, 0.9946, 0.9976, 0.9489, 0.9778, 0.9778, 0.9880) whose values are v6 (0.9824), v7 (0.6229), v10 (0.9946), v11 (0.9976) satisfy the initial constraints.

5 Conclusion

The main advantage of the proposed model of the diagnostic process is the ability to systematically take into account the long-term consequences of possible contingencies and identify side effects that allow to take into account the multifactority of the process of diagnosing SEMS in the process of operation. The task of identifying possible risks in general, and at the operational stage in particular, should be an important part of the diagnostics of SEMS equipment.

Analysis of the structure of the NCM made it possible to identify the factors that exert the greatest influence on the entire system and vice versa; factors that are most affected by the system, as well as the best values of the parameters of factors reflecting the recognition of a violation of normal operation at an early stage of their development, taking into account available resources. Recommendations are provided for justifying the choice of target and control vertices of NCM in conditions of uncertain initial data and information for management actions aimed at improving the NCM.

Acknowledgements This work was financially supported by Russian Foundation for Basic Research, Grant No. 17-08-00402.

References

1. Yurkov, N.K.: A systematic approach to the organization of the life cycle of complex technical systems. Reliability and quality of complex systems. J. Res. Pract. **1**, 27–35 (2013)
2. Bushueva, M.E., Belyakov, V.V.: Diagnostics of complex technical systems. In: Novgorod, N. (eds.) Proceedings of the First Workshop on the NATO Project SfP-973799 Semiconductors "Development of Radio-Resistant Semiconductor Devices for

Communication Systems and Precision Measurements Using Noise Analysis", pp. 63–99 (2001)

3. Semenov, S.S.: The main provisions of the system analysis in assessing the technical level of complex systems using the expert method. Reliab. Qual. Complex Syst. **4**, 45–53 (2013)

4. Krioni, N.K., Kolodenkova, A.E., Korobkin, V.V., Gubanov, N.G.: Intelligent decision-making support system using cognitive modeling for project feasibility assessment on creating complex technical systems. Int. J. Appl. Bus. Econ. Res. **14**(10), 7289–7300 (2016)

5. Roberts, F.S.: Discrete Mathematical Models Applied to Social, Biological and Ecological Problems. Nauka, Moscow (1986)

6. Dickerson, J., Kosko, B.: Virtual worlds as fuzzy cognitive maps. Virtual Reality Annual International Symposium, pp. 471-477 (1993)

7. Kosko, B.: Fuzzy cognitive maps. Int. J. Man-Mach. Stud. **1**, 65–75 (1986)

8. Silov, V.B.: Strategic Decision Making in a Fuzzy Environment. INPRO-RES, Moscow (1995)

9. Kulba, V.V., Kononov, D.A., Kovalevsky, S.S.: Scenario analysis of dynamics of behavior of socio-economic systems. IPP RAS, Moscow (2002)

10. Casty, J.: Large Systems: Connectivity, Complexity and Disasters. Mir, Moscow (1982)

11. Novichikhin, A.V., Ulankin, A.N.: Methodical features of project programming for the development of enterprises in the resource region (on the example of the coal industry). Min. Inf. Anal. Bull. **3**, 332–337 (2011)

12. Sadovnikova, N.P.: Application of the cognitive modeling for analysis of the ecological and economical efficiency of the urban planning project. Internet-Vestnik VolgGASU **5**, 1–4 (2011)

13. Kazanin, IYu.: Research of social and economic security of Rostov Region, cognitive modeling of development strategy. News SFU Tech. Sci. **3**, 12–16 (2009)

14. Bozhenyuk, A.V., Ginis, L.A.: Application of fuzzy models to the analysis of complex systems. Control Syst. Inf. Technol. **51**(1.1), 122–126 (2013)

15. Atkin, R.H., Casti, J.: Polyhedral Dynamics and the Geometry of Systems, RR-77-6. International Institute for Applied Systems Analysis, Laxenburg (1977)

16. Gorelova, G.V., Zakharova, E.N., Radchenko, S.A.: Investigation of semi-Structured Problems of Socio-economic Systems: A Cognitive Approach. RSU, Rostov-on-Don (2006)

17. Korostelev, D.A., Lagerev, D.G., Podvesovsky, A.G.: Application of fuzzy cognitive models to formation of a set of alternatives in decision-making problems. Vestnik Bryansk State Techn. Univ. **4**, 77–85 (2009)

18. Borisov, V.V., Kruglov, V.V., Fedulov, A.S.: Fuzzy Models and Networks. Goryachaya Linia - Telecom, Moscow (2007)

19. Kolodenkova, A.E.: Evaluation of the feasibility of the project on the creation of information control systems using the fuzzy cognitive model training procedure. Vestnik USATU, **20**(2 (72)), 123–133 (2016)

20. Papageorgiou, E.I.: Unsupervised hebbian algorithm for fuzzy cognitive map training. In: Papageorgiou, E.I., Stylios, C.D., Groumpos, P.P. (eds.) Proceedings of the 5-th International Workshop on Computer Science and Information Technologies. – Ufa, vol. 1., pp. 209–216 (2003)

Part V
Information-Measuring and Control Soft and Hardware

Robotic Assembly and Servicing Space Module Peculiarities of Dynamic Study of Given System

P. P. Belonozhko

Abstract *Problem statement*: prospects of space robotics are significantly determined by the need in automation of assembly and servicing operations. One of possible solution to this problem is group usage of emerging SEM-systems, that are robotic assembly and servicing space modules (RASSM) equipped with a manipulator (or manipulators) that are capable of moving themselves in space and adapted for contact with objects being assembled (maintained) [1–8]. One of key features of such modules is presence of dynamic modes in which not only cargo can be moved with a help of a manipulator relative to the bottom, but also the bottom itself can be moved [1, 3, 6–9]. At that there is a possibility of its own inertia movements by internal degrees of freedom of the RASSM emerging both in absence of external forces and moments and control actions in joints of the manipulator. Studying these movements are of interest from the point of view of combination of controlling a separate RASSM as well as a group of interacting RASSM. At that studying the given system which dynamic equations in certain conditions can be deduced in the form of Routh equations turns to be effective. It is convenient to interpret the full mechanical energy of the given system as sum of kinetic and potential components. Formulation of the source system kinetic energy as a function of joints position is an important stage in this process. For a generic case of a three-dimensional system of two solid bodies connected by a certain massless mechanism with an intentionally alterable structure there is kinetic energy derivation as a function of position and cargo speeds relative to the bottom as well as vector projections of the constant moment of momentum in the inertia frame onto axes of the reference frame connected with a movable bottom. *Practical significance*: obtained results are of interest from the point of view of quality examination of peculiarities of the three-dimensional variant of proper inertia movements of the system by manipulator degrees of freedom at a system non-zero moment of momentum, as well as from the point of view of usage of advanced computerized tools for scientific research (particularly, symbolic mathematics system) while deducing equations of the given system dynamics for various kinematic schemes of the manipulation mechanism.

P. P. Belonozhko (✉)
Bauman Moscow State Technical University, Moscow, Russia
e-mail: byelonozhko@mail.ru

© Springer Nature Switzerland AG 2019
A. E. Gorodetskiy and I. L. Tarasova (eds.), *Smart Electromechanical Systems*,
Studies in Systems, Decision and Control 174,
https://doi.org/10.1007/978-3-319-99759-9_23

Keywords Robotic assembly and servicing space modules · Manipulator
Kinetic energy · Equations of the given system dynamics

1 Introduction

For systems of the class under consideration at modeling of dynamics the possibility to neglect mass of the manipulation mechanism in comparison with masses of the movable bottom-space module (SM) and payload (PL) is quite important [1, 3, 7–9]. In this case the system momentum is defined by movement of SM and PL relative to a certain inertia frame. This movement can be viewed as one of two components: SM movement relative to the inertia frame system and PL movement relative to SM. In its turn, PL movement relative to SM is defined by an intentional change of massless manipulation mechanism structure connecting SM and PL, i.e. law of joint position changes. The article shows that in certain conditions that are satisfied for quite a wide class of movement modes of practical interest the system SM-PL kinetic energy is a quadric form of relative speeds with ratios that are functions of relative coordinates, i.e. it is fully defined by PL movement relative to SM. The obtained result is of interest from the point of view of studying PL movement dynamics relative to SM by means of studying a certain given system which independent dynamics equations in certain conditions can be obtained in the form of Routh equations.

2 Problem Statement

Let us consider the SM and PL system free in inertial space. As indicated above, we are going to consider SM and PL connected with a help of a certain massless controlled mechanism (manipulator). Let us also assume the possibility to ignore all external forces and moments affecting the SM-manipulator-PL system. Then the system mass centre is moving relative to a certain inertial frame in straight lines and evenly. Let us introduce into consideration a non-rotating reference frame XYZ with the origin in the mass centre of the system that will be inertial as well. The values referred to the given reference frame will be accompanied by the 'inr' ('inertial') index. Let us connect the SM mass centre to the reference frame $X_bY_bZ_b$ with the origin in the SM mass centre and axes directed along the SM inertia main central axes. The values referred to the given reference frame will be accompanied by the 'b' ('base') index. Let us connect the PL mass centre to the reference frame $X_pY_pZ_p$ with the origin in the PL mass centre and axes directed along the PL inertia main central axes. The values referred to the given reference frame will be accompanied by the 'p' ('payload') index.

The expression for the kinetic energy of the motion of the system under consideration with respect to the inertial coordinate system XYZ has the form

$$T = \frac{1}{2}\left(m_b \bar{v}_b^{inr} \cdot \bar{v}_b^{inr} + \bar{\omega}_b^{inr} \cdot \bar{\theta}_b \cdot \bar{\omega}_b^{inr} + m_p \bar{v}_p^{inr} \cdot \bar{v}_p^{inr} + \bar{\omega}_p^{inr} \cdot \bar{\theta}_p \cdot \bar{\omega}_p^{inr}\right), \quad (1)$$

where

m_b SM mass;
m_p PL mass;
$\bar{\theta}_b$ SM inertia tensor relative to its mass centre;
$\bar{\theta}_p$ PL inertia tensor relative to its mass centre;
$\bar{v}_b^{inr}$ linear speed of SM mass centre relative to XYZ;
$\bar{v}_p^{inr}$ linear speed of PL mass centre relative to XYZ;
$\bar{\omega}_b^{inr}$ SM angular speed relative to XYZ;
$\bar{\omega}_p^{inr}$ PL angular speed relative to XYZ;

On the basis of general dynamics laws we have

$$m_b \bar{v}_b^{inr} + m_p \bar{v}_p^{inr} = 0, \quad (2)$$

$$m_b \bar{r}_b^{inr} \times \bar{v}_b^{inr} + \bar{\theta}_b \cdot \bar{\omega}_b^{inr} + m_p \bar{r}_p^{inr} \times \bar{v}_p^{inr} + \bar{\theta}_p \cdot \bar{\omega}_p^{inr} = \bar{L}^{inr}, \quad (3)$$

where

$\bar{r}_b^{inr}$ radius vector of SM mass centre (origin of $X_b Y_b Z_b$) relative to system mass centre (origin of XYZ);
$\bar{r}_p^{inr}$ radius vector of PL mass centre (origin of $X_p Y_p Z_p$) relative to system mass centre (origin of XYZ);
$\bar{L}^{inr}$ kinetic moment of system relative to origin XYZ.

It is plain that the relative to the origin XYZ kinetic moment $\bar{L}^{inr}$ is a constant vector defined by the initial conditions of motion.

Let us also write apparent formulas

$$\bar{v}_p^{inr} = \bar{v}_b^{inr} + \bar{\omega}_b^{inr} \times \bar{r}_p^b + \bar{v}_p^b, \quad (4)$$

$$\bar{\omega}_p^{inr} = \bar{\omega}_b^{inr} + \bar{\omega}_p^b, \quad (5)$$

$$\bar{r}_p^{inr} = \bar{r}_b^{inr} + \bar{r}_p^b, \quad (6)$$

where

$\bar{r}_p^b$ radius vector of PL mass centre (origin of $X_p Y_p Z_p$) relative to SM mass centre (origin of $X_b Y_b Z_b$);

$\bar{v}_p^b$ linear speed of PL mass centre (origin of $X_p Y_p Z_p$) relative to $X_b Y_b Z_b$;

$\bar{\omega}_p^b$ PL angular speed relative to $X_b Y_b Z_b$;

Correlations (1)–(6) are written in an invariant form. The «×» symbol in formulas (3) and (4) stands for vector product of vectors. The «·» symbol in formulas (1) and (3) stands for scalar product of two vectors (in formula (1) scalars $\bar{v}_b^{inr} \cdot \bar{v}_b^{inr}$ and $\bar{v}_p^{inr} \cdot \bar{v}_p^{inr}$), or scalar product of a vector and a tensor on the left or right (in formula (3) vectors $\bar{\theta}_b \cdot \bar{\omega}_b^{inr}$ and $\bar{\theta}_p \cdot \bar{\omega}_p^{inr}$). Components $\bar{\omega}_b^{inr} \cdot \bar{\theta}_b \cdot \bar{\omega}_b^{inr}$ and $\bar{\omega}_p^{inr} \cdot \bar{\theta}_p \cdot \bar{\omega}_p^{inr}$ in formula (1) are scalars that present the result of scalar multiplication of vectors $\bar{\omega}_b^{inr}$ and $\bar{\omega}_p^{inr}$ by vectors $\bar{\theta}_b \cdot \bar{\omega}_b^{inr}$ and $\bar{\theta}_p \cdot \bar{\omega}_p^{inr}$ (or vectors $\bar{\omega}_b^{inr} \cdot \bar{\theta}_b$ and $\bar{\omega}_p^{inr} \cdot \bar{\theta}_p$ equal to them) correspondingly.

Let us switch to the matrix form of formulas (1)–(6). Let us present the vectors and tensors that are included in these formulas as coordinate matrices in $X_b Y_b Z_b$. Hereinafter the «·» symbol stands for matrix product.

$$T = \frac{1}{2} m_b V_b^{inr\,b^T} \cdot V_b^{inr\,b} + \frac{1}{2} \Omega_b^{inr\,b^T} \cdot \theta_b^b \cdot \Omega_b^{inr\,b} + \frac{1}{2} m_p V_p^{inr\,b^T} \cdot V_p^{inr\,b} + \frac{1}{2} \Omega_p^{inr\,b^T} \cdot \theta_p^b \cdot \Omega_p^{inr\,b},$$

$$\tag{7}$$

$$m_b V_b^{inr\,b} + m_p V_p^{inr\,b} = 0, \tag{8}$$

$$m_b \hat{R}_b^{inr\,b} \cdot V_b^{inr\,b} + \theta_b^b \cdot \Omega_b^{inr\,b} + m_p \hat{R}_p^{inr\,b} \cdot V_p^{inr\,b} + \theta_p^b \cdot \Omega_p^{inr\,b} = L^{inr\,b}, \tag{9}$$

$$V_p^{inr\,b} = V_b^{inr\,b} - \hat{R}_p^{bb} \cdot \Omega_b^{inr\,b} + V_p^{bb}, \tag{10}$$

$$\Omega_p^{inr\,b} = \Omega_b^{inr\,b} + \Omega_p^{bb}, \tag{11}$$

$$R_p^{inr\,b} = R_b^{inr\,b} + R_p^{bb} \tag{12}$$

$$\hat{R}_p^{inr\,b} = \hat{R}_b^{inr\,b} + \hat{R}_p^{bb}. \tag{13}$$

Here

$V_b^{inr\,b}$ column matrix of vector coordinates $\bar{v}_b^{inr}$ in $X_b Y_b Z_b$;

$V_p^{inr\,b}$ column matrix of vector coordinates $\bar{v}_p^{inr}$ in $X_b Y_b Z_b$;

V_p^{bb} column matrix of vector coordinates $\bar{v}_p^b$ in $X_b Y_b Z_b$;

$\Omega_b^{inr\,b}$ column matrix of vector coordinates $\bar{\omega}_b^{inr}$ in $X_b Y_b Z_b$;

$\Omega_p^{inr\,b}$ column matrix of vector coordinates $\bar{\omega}_p^{inr}$ in $X_b Y_b Z_b$;

Ω_p^{bb} column matrix of vector coordinates $\bar{\omega}_p^b$ in $X_b Y_b Z_b$;

$\hat{R}_b^{inr\,b}$ skew-symmetric matrix put in correspondence to matrix column $R_b^{inr\,b}$ of vector coordinates $\bar{r}_b^{inr}$ in $X_b Y_b Z_b$;

$\hat{R}_p^{inr\,b}$ skew-symmetric matrix put in correspondence to matrix column $R_p^{inr\,b}$ of vector coordinates $\bar{r}_p^{inr}$ in $X_b Y_b Z_b$;

$\hat{R}_p^{bb}$ skew-symmetric matrix put in correspondence to matrix column R_p^{bb} of vector coordinates $\bar{r}_p^{b}$ in $X_b Y_b Z_b$;

θ_b^{b} square matrix of tensor component $\bar{\theta}_b$ in $X_b Y_b Z_b$, constant diagonal matrix as axes $X_b Y_b Z_b$ coincide with SM inertia main general axes;

θ_p^{b} square matrix of tensor component $\bar{\theta}_p$ in $X_b Y_b Z_b$, elements of this symmetric matrix depend on angular position $X_p Y_p Z_p$ relative to $X_b Y_b Z_b$;

$L^{inr\,b}$ column matrix of vector coordinates $\bar{L}^{inr}$ in $X_b Y_b Z_b$.

Skew-symmetric matrices $\hat{R}_b^{inr\,b}$, $\hat{R}_p^{inr\,b}$ and $\hat{R}_p^{bb}$ are introduced to present vector multiplication of vectors in the matrix form the following way. Let us have

$$\bar{c} = \bar{a} \times \bar{b}. \tag{14}$$

Coordinate column matrices in a certain coordinate system of vectors included in (14)

$$A = [a_1, a_2, a_3]^T, \quad B = [b_1, b_2, b_3]^T, \quad C = [c_1, c_2, c_3]^T. \tag{15}$$

At that, according to the definition of vector product we have

$$c_1 = a_2 b_3 - a_3 b_2, \quad c_2 = a_3 b_1 - a_1 b_3, \quad c_3 = a_1 b_2 - a_2 b_1. \tag{16}$$

Introducing a skew-symmetric matrix $\hat{A}$ put in correspondence to the vector A

$$\hat{A} = \begin{bmatrix} 0 & -a_3 & a_2 \\ a_3 & 0 & -a_1 \\ -a_2 & a_1 & 0 \end{bmatrix}, \tag{17}$$

we can write vector product (14) in the matrix form

$$C = \hat{A} \cdot B. \tag{18}$$

Matrix presentation (18) of vector product (14) is used at transition from formula (3) to formula (9) and from formula (4) to formula (10).

For vectors, tensors and matrices included in formulas (1)–(13) the following indications are given. The lower index defines the body under consideration—SM or PL—and may correspondingly take values 'b' ('base') и 'p' ('payload'). The first upper index in formulas (1)–(13) defines in relation to which reference frame position, speed or angular speed of the corresponding body are considered and may take values 'inr' (XYZ), 'b' ($X_b Y_b Z_b$) and 'p' ($X_p Y_p Z_p$). The second upper index in formulas (7)–(13) defines in which coordinate system vector coordinates and tensor components are present. According to the type of the current research the given index takes only 'b' value. i.e. all vectors and tensors are presented by projections in $X_b Y_b Z_b$.

The objective is to transform formula (7) with usage of formulas (8)–(13) so that it includes only those matrices that characterize PL position relative to SM ($\hat{R}_p^{bb}$ and θ_p^b), linear speed of the PL mass centre relative to SM (V_p^{bb}) and PL angular speed relative to SM (Ω_p^{bb}) as well as the column matrix $L^{inr\,b}$.

3 Eliminating Absolute Linear Speeds

Let us consider the first and third components (7). With (8) taken into account we have

$$\frac{1}{2} m_b V_b^{inr\,b^T} \cdot V_b^{inr\,b} + \frac{1}{2} m_p V_p^{inr\,b^T} \cdot V_p^{inr\,b} = \frac{1}{2} m_b (V_b^{inr\,b^T} - V_p^{inr\,b^T}) \cdot V_b^{inr\,b}. \qquad (19)$$

From (8) and (10) we obtain

$$V_b^{inr\,b^T} - V_p^{inr\,b^T} = \Omega_b^{inr\,b^T} \cdot \hat{R}_p^{bb^T} - V_p^{bb^T}, \qquad (20)$$

$$V_b^{inr\,b} = \frac{m_p}{m_p + m_b} (\hat{R}_p^{bb} \cdot \Omega_b^{inr\,b} - V_p^{bb}),$$
$$V_p^{inr\,b} = -\frac{m_b}{m_p + m_b} (\hat{R}_p^{bb} \cdot \Omega_b^{inr\,b} - V_p^{bb}). \qquad (21)$$

Let us substitute (20) and the first equation (21) in (19). Taking into account $V_p^{bb^T} \cdot \hat{R}_p^{bb} \cdot \Omega_b^{inr\,b} = \Omega_b^{inr\,b^T} \cdot \hat{R}_p^{bb^T} \cdot V_p^{bb}$ and $\hat{R}_p^{bb} = -\hat{R}_p^{bb^T}$, obtain

$$\frac{1}{2} m_b V_b^{inr\,b^T} \cdot V_b^{inr\,b} + \frac{1}{2} m_p V_p^{inr\,b^T} \cdot V_p^{inr\,b} = (\frac{1}{2} \tilde{m}(V_p^{bb^T} \cdot V_p^{bb} - 2V_p^{bb^T} \cdot \hat{R}_p^{bb} \cdot \Omega_b^{inr\,b}$$
$$- \Omega_b^{inr\,b^T} \cdot \hat{R}_p^{bb} \cdot \hat{R}_p^{bb} \cdot \Omega_b^{inr\,b}), \qquad (22)$$

where $\tilde{m} = \frac{m_p m_b}{m_p + m_b}$.

Substituting (22) and (7), we obtain

$$T = \frac{1}{2} \tilde{m} V_p^{bb^T} \cdot V_p^{bb} - \tilde{m} V_p^{bb^T} \cdot \hat{R}_p^{bb} \cdot \Omega_b^{inr\,b}$$
$$+ \frac{1}{2} \Omega_b^{inr\,b^T} \cdot (\theta_b^b - \tilde{m}\hat{R}_p^{bb} \cdot \hat{R}_p^{bb}) \cdot \Omega_b^{inr\,b} + \frac{1}{2} \Omega_p^{inr\,b^T} \cdot \theta_p^b \cdot \Omega_p^{inr\,b}. \qquad (23)$$

4 Eliminating Absolute Angular Speeds

Then it is necessary to eliminate from (23) $\Omega_b^{inr\,b}$ and $\Omega_p^{inr\,b}$, having expressed them through V_p^{bb}, Ω_p^{bb} and $L^{inr\,b}$. Taking into account equations (12), (13) and (21), let us write (9) as

$$\tilde{m}\hat{R}_p^{bb}(V_p^{bb} - \hat{R}_p^{bb} \cdot \Omega_b^{inr\,b}) + \theta_b^b\Omega_b^{inr\,b} + \theta_p^b(\Omega_b^{inr\,b} + \Omega_p^{bb}) = L^{inr\,b}. \qquad (24)$$

From (24) can be used to express SM absolute angular speed

$$\Omega_b^{inr\,b} = A_0^{-1} \cdot (L^{inr\,b} - \theta_p^b \cdot \Omega_p^{bb} - \tilde{m}\hat{R}_p^{bb} \cdot V_p^{bb}), \qquad (25)$$

where

$$A_0 = \theta_b^b + \theta_p^b - \tilde{m}\hat{R}_p^{bb} \cdot \hat{R}_p^{bb}. \qquad (26)$$

As matrices θ_b^b and θ_p^b are symmetric and matrix $\hat{R}_p^{bb}$ is skew-symmetric, we can write

$$A_0^T = \theta_b^{b^T} + \theta_p^{b^T} - \tilde{m}\hat{R}_p^{bb^T} \cdot \hat{R}_p^{bb^T} = A_0, \qquad (27)$$

i.e. matrix A_0 is also symmetric, and, consequently, $A_0^{-1^T} - A_0^{-1}$. So we have

$$\Omega_b^{inr\,b^T} = (L^{inr\,b^T} - \Omega_p^{bb^T} \cdot \theta_p^b + \tilde{m}V_p^{bb^T} \cdot \hat{R}_p^{bb}) \cdot A_0^{-1}. \qquad (28)$$

Having substituted (12) in (23), we obtain

$$T = \frac{1}{2}\tilde{m}V_p^{bb^T} \cdot V_p^{bb} + \frac{1}{2}\Omega_p^{bb^T} \cdot \theta_p^b \cdot \Omega_p^{bb}$$
$$+ (\Omega_p^{bb^T} \cdot \theta_p^b - \tilde{m}V_p^{bb^T} \cdot \hat{R}_p^{bb}) \cdot \Omega_b^{inr\,b} + \frac{1}{2}\Omega_b^{inr\,b^T} \cdot (\theta_b^b + \theta_p^b - \tilde{m}\hat{R}_p^{bb} \cdot \hat{R}_p^{bb}) \cdot \Omega_b^{inr\,b}. \qquad (29)$$

Substituting (25) and (28) in (29) after simple manipulations we obtain

$$T = \frac{1}{2}\tilde{m}V_p^{bb^T} \cdot V_p^{bb} + \frac{1}{2}\Omega_p^{bb^T} \cdot \theta_p^b \cdot \Omega_p^{bb}$$
$$- \frac{1}{2}B_0^T \cdot A_0^{-1} \cdot B_0 + \frac{1}{2}L^{inr\,b^T} \cdot A_0^{-1} \cdot L^{inr\,b} + \frac{1}{2}B_0^T \cdot A_0^{-1} \cdot L^{inr\,b} - \frac{1}{2}L^{inr\,b^T} \cdot A_0^{-1} \cdot B_0, \qquad (30)$$

where

$$B_0 = \theta_p^b \cdot \Omega_p^{bb} + \tilde{m}\hat{R}_p^{bb} \cdot V_p^{bb},$$
$$B_0^T = \Omega_p^{bb^T} \cdot \theta_p^b + \tilde{m}V_p^{bb^T} \cdot \hat{R}_p^{bb^T} = \Omega_p^{bb^T} \cdot \theta_p^b - \tilde{m}V_p^{bb^T} \cdot \hat{R}_p^{bb}. \tag{31}$$

As $B_0^T \cdot A_0^{-1} \cdot L^{inr\,b} = L^{inr\,b^T} \cdot A_0^{-1} \cdot B_0$, formula (30) is written as

$$T = \frac{1}{2}\tilde{m}V_p^{bb^T} \cdot V_p^{bb} + \frac{1}{2}\Omega_p^{bb^T} \cdot \theta_p^b \cdot \Omega_p^{bb} - \frac{1}{2}B_0^T \cdot A_0^{-1} \cdot B_0 + \frac{1}{2}L^{inr\,b^T} \cdot A_0^{-1} \cdot L^{inr\,b}. \tag{32}$$

Substituting (31) in (32), we finally obtain

$$T = \frac{1}{2}V_p^{bb^T} \cdot A_1 \cdot V_p^{bb} + V_p^{bb^T} \cdot A_2 \cdot \Omega_p^{bb} + \frac{1}{2}\Omega_p^{bb^T} \cdot A_3 \cdot \Omega_p^{bb} + L^{inr\,b^T} \cdot A_0^{-1} \cdot L^{inr\,b}, \tag{33}$$

where

$$A_1 = \tilde{m}(E + \tilde{m}\hat{R}_p^{bb} \cdot A_0^{-1} \cdot \hat{R}_p^{bb}), \tag{34}$$

$$A_2 = \tilde{m}\hat{R}_p^{bb} \cdot A_0^{-1} \cdot \theta_p^b, \tag{35}$$

$$A_3 = \theta_p^b(E - A_0^{-1} \cdot \theta_p^b), \tag{36}$$

where E—a single 3 × 3 matrix.

From the obtained formula (33) with indications (34)–(36) taken into account one can easily see that while studying proper ballistic movements of systems of the 'SM-manipulator-PL' under consideration by internal degrees of freedom of the manipulator mechanism it is important to differentiate cases

$$\bar{L}^{inr} = 0, \tag{37}$$

and

$$\bar{L}^{inr} \neq 0. \tag{38}$$

In the case of (37) formula (33) is written as

$$T = \frac{1}{2}V_p^{bb^T} \cdot A_1 \cdot V_p^{bb} + V_p^{bb^T} \cdot A_2 \cdot \Omega_p^{bb} + \frac{1}{2}\Omega_p^{bb^T} \cdot A_3 \cdot \Omega_p^{bb}, \tag{39}$$

i.e. system *kinetic energy* is a quadric form of relative speeds with ratios, that are functions of relative coordinates, and is fully defined by PL movement relative to

SM. In [3, 7, 8] by the example of flat movement of a joint torque link it is shown that in this case momentum (39) of the source system may be interpreted as the given system momentum and an independent dynamics equation of relative movement in the form of the Routh equation may be obtained as the Lagrange equation of the 2nd kind for the given system.

In case (38) elements of the column matrix $L^{inr\,b}$ in its general case will be, as obvious, functions of generalized coordinates characterizing angular position of SM in XYZ, and thus this case assumes additional operations. As shown in [3, 7, 8], in a particular case of flat movement at correspondingly chosen reference frames elements of the column matrix $L^{inr\,b}$ will be constant and the kinetic energy (33) of the source system may also be interpreted as the given system kinetic energy having both kinetic and potential components.

5 Conclusion

It is supposed that for the 'SM-manipulator-PL' system there are no external forces and there is a possibility to ignore the mass of the manipulator. Thus, there is kinetic energy derivation as a function of position and cargo speeds relative to the bottom as well as vector projections of the constant kinetic moment in the inertial frame onto axes of the reference frame connected with a movable base. The article shows the main stages of derivation, and, taking into account the awkwardness of the obtained formula, it is the necessary condition of its practical usage. For instance, direct usage of a formula similar to the one given in [9], requires subsidiary formulas obtained in the given study.

The result is of interest from the point of view of analysis of quality peculiarities of movements of the system under consideration, particularly correlation of cases with and without the system kinetic moment. It is also important to note that analysis of peculiarities of manipulation system of a specific type is also possible. In this case coordinates and speeds of PL relative to SM are familiar functions of joint coordinate and speeds that may be substituted directly into the obtained formula.

References

1. Artemenko, Y.N., Karpenko, A.P., Belonozhko, P.P.: Features of manipulator dynamics modeling into account a movable platform. In: Gorodetskiy, A.E. (ed.) Smart Electromechanical Systems. Studies in Systems, Decision and Control, vol. 49, pp. 177–190. Springer International Publishing Switzerland (2016). https://doi.org/10.1007/978-3-319-27547-5_17
2. Artemenko, Y.N., Karpenko, A.P., Belonozhko, P.P.: Synthesis of control of hinged bodies relative motion ensuring move of orientable body to necessary absolute position. In: Gorodetskiy, A.E., Kurbanov, V.G. (eds.) Smart Electromechanical Systems: The Central

Nervous System. Studies in Systems, Decision and Control, vol. 95, pp. 231–239. Springer International Publishing Switzerland (2017). https://doi.org/10.1007/978-3-319-53327-8_16

3. Belonozhko, P.P.: Methodical features of acquisition of independent dynamic equation of relative movement of one-degree of freedom manipulator on movable foundation as control object. In: Gorodetskiy, A.E., Kurbanov, V.G. (eds.) Smart Electromechanical Systems: The Central Nervous System. Studies in Systems, Decision and Control, vol. 95, pp. 261–270. Springer International Publishing Switzerland (2017). https://doi.org/10.1007/978-3-319-53327-8_19

4. Karpenko, A.P., Sayapin, S.A., Dang, X.H.: Dodekapod as universal intelligent structure for adaptive parallel spatial self-moving modular robots. In: Nature-inspired Mobile Robotics, Proceedings of the 16th International Conference on Climbing and Walking Robots and the Support Technologies for Mobile Machines, pp. 163–167. University of Technology, Sydney, Australia, 14–17 July 2013

5. Karpenko, A.P., Sayapin, S.A., Dang, X.H.: Universal adaptive spatial parallel robots of module type based on the Platonic solids. Lecture Notes in Engineering and Computer Science, vol. 2, pp. 1365–1370 (2014)

6. Belonozhko, P.P.: Emerging robotic assembly and servicing space modules. Robot. Tech. Cybern. 2(7), 18–23 (2015)

7. Belonozhko, P.P.: Study of flat inertia movements of space manipulator on movable base as non-linear oscillatory system. Robot. Tech. Cybern. 4(13), 52–58 (2016)

8. Belonozhko, P.P.: Analysis of given non-linear oscillatory system corresponding to one degree of freedom manipulators on movable and gimble-mounted base. Robot. Tech. Cybern. 3(16), 44–52 (2017)

9. Alpatov, A.P., Belonozhko, P.A., Gorbuntsov, V.V., Ivlev, O.G., Chernyavskaya, S.S., Shichanin V.N.: Dynamics of spatially evolved mechanical systems of alterable structure, 256 p. Naukova Dumka, Kyiv (1990)

The Position Monitoring Robotic Platforms of the Radiotelescope Elements on Base of Autocollimation Sensors

Anton A. Nogin and Igor A. Konyakhin

Abstract *Problem statement*: Smart electromechanical systems (SEMS) can be used to control the elements of the radio telescope surface. For this purpose, we can use robotic platforms with a surveillance system. Elements of the surface have mechanisms for adjusting the position in space. The platform, in accordance with the reflected marks, calculates the relative position of the surfaces and commands to the adjustment mechanisms to correct the position in space. When the displacements are extremely small in the observation plane, the images of the mark practically merge and as a result, this error can't be corrected. *Purpose of research*: development of an algorithm capable of measuring the displacement error for small displacements. *Results*: algorithms of image selection in the analysis plane are considered. A mathematical model of the problem of overlapping marks developed. An algorithm for measuring the coordinates of overlapping marks was developed and tested. The algorithm is tested using an autocollimation scheme as a platform surveillance system. *Practical significance*: The developed algorithm allows reducing the zone of inoperability of the device, to increase the accuracy of the adjustment of surfaces. The algorithm opens up new opportunities for SEMS and another measurement system.

Keywords Robot · SEMS · Image processing · Radiotelescopes
Surface control · Measuring systems

A. A. Nogin (✉) · I. A. Konyakhin
Department of Optical-Electronic Devices and Systems, ITMO University,
Saint-Petersburg, Russia
e-mail: anogin@corp.ifmo.ru

I. A. Konyakhin
e-mail: igor@grv.ifmo.ru

© Springer Nature Switzerland AG 2019
A. E. Gorodetskiy and I. L. Tarasova (eds.), *Smart Electromechanical Systems*,
Studies in Systems, Decision and Control 174,
https://doi.org/10.1007/978-3-319-99759-9_24

1 Introduction

When monitoring the position of large objects, it becomes necessary to process information from all sensors located both on a specific element and on the entire object. In this connection, there is a need to develop smart systems and complexes capable of simultaneously receiving information from several sensors, controlling various elements (drives), controlling the change in parameters. All these tasks can be solved by the smart electromechanical systems (SEMS) [1].

To solve the problem of monitoring the position of the elements of the primary mirror telescope was decided to use a robotic platform with a tracking system. The platform monitors the radiation reflected from a special control element mounted on the telescope elements. Elements of the radio telescope are equipped with mechanical drives that allow to correct the slope angles of the element. SEMS manages these drives constantly submitting information to the drives based on data from the tracking system.

The tracking system is realized with the help of an optical-electronic autocollimator capable of measuring the angular position of the elements in space with high accuracy and high speed [2]. The coordinates are measured by recording the movement of the image in the analysis plane caused by the rotation of a special control element mounted on the monitored object [3, 4]. Such systems are installed on oil and gas pipelines, long objects, bridge supports, hangar structures. Using a special control element other than a flat mirror makes it possible to control several parameters at once by creating several images in the analysis plane. This feature creates a situation where, at very small offsets, marks are repositioned, which causes the situation of the system's inoperability.

In this paper we present an algorithm capable of partially solving this situation.

2 Optical-Electronic Autocollimator as an Instrument of Angular Measurements

Optical-electronic autocollimator consists of two parts: a control element (reflector) that is placed on the monitoring object and an autocollimation sensor that is placed on the base object (Fig. 1). A passive reflector with a small optical reduction coefficient is used as a control element. This system allows for the measurement of non-contact and fast, and there is no need for power at the point of control. The sensor consists of emitting optoelectronic channel and receiver with a microprocessor. The emitting channel generates an optical beam that incident on the reflector. An optical-electronic receiver, for example. a CCD image sensor, records the radiation reflected from the control element. The microprocessor processes the resulting frame, and the rotation angles of the object are calculated.

Traditionally, a flat mirror is used as a control element. However, its characteristics (the coefficient of optical reduction, insensitivity to turns around

Fig. 1 Coordinate axis of the measuring system

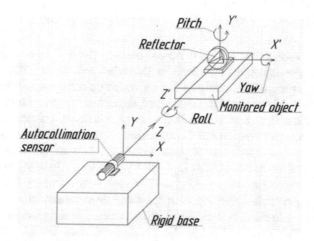

the sighting) do not allow to fully realize the potential of this system. To reveal all the possibilities, special control elements are made in the form of pyramids, tetrahedron, and so on. with a small coefficient of optical reduction and an error in three-edged angles [5, 6]. Such control elements allow to track the rotation around all three axes but introduce the problem of overlapping marks. The importance of this problem is discussed by the first step in calculating the rotation angle of the object. by calculating the displacement of the mark image in the analysis plane [7].

3 Description of the Algorithm

Calculation of mark movement in the analysis plane, when the control element is turned, is the main stage of the operation of the angle measuring autocollimator. It is from these data SEMS formed commands for the drives in order to correct the angular position of the monitored object. When transposing, it is not possible to calculate the center of the marks in order to estimate their displacement. To do this, it is necessary to recognize the mark, that is, to distinguish it from the background and other objects in the image, and also to measure the coordinates of its center.

To solve the problem, a priori known information about the geometric form of the mark is used. In this system, the mark has a circular shape, which makes it stable to noise and invariant to rotation, scale, etc. A circle can be described by an equation with three parameters (x, y, r), where r is the radius of the circle. Thus, the primary task is to detect geometric primitives corresponding to a given combination of parameters.

It was decided to use the Hough transform to detect circles and measure the coordinates of the center [7]. Hough Transform is perfect for this since it is able to work with geometric primitives, noise immunity, and the equation of a circle

$$(x^2 + y^2) = r^2 \qquad (1)$$

allows determining the coordinates of the center i.e. x and y after detection. Using the equation of a circle in Cartesian coordinates will create a three-dimensional space of the accumulation, in which calculations will be performed.

The work of the developed algorithm follows: the program makes a frame from the device and, if it is necessary, makes it grayscale. This is necessary for the correct operation of the algorithm. When conversion was done, the algorithm needs to know the range of possible values for the radius of the mark, as well as the value of the threshold filter for further processing. These values can be pre-recorded in the algorithm when configured for each new situation. To study the algorithm, these parameters were selected once and used consistently in all experiments.

In the next step, the algorithm calculates the gradient value in each direction and then calculates the gradient magnitude from formula

$$\nabla img = \sqrt{(\nabla img_x)^2 + (\nabla img_y)^2} \qquad (2)$$

where ∇img—gradient magnitude, ∇img_x и ∇img_y—gradient horizontally and vertically, respectively. The points that will be used for voting, for the subsequent determination of the center of the circle are detected in the received array. For this purpose, the value is compared with the threshold value (specified together with the range of radii). The pixel background is excluded from the voice due to this.

At the next stage, the accumulation array is formed. The dimension of this array always coincides with the number of variables that define the geometric primitive. Already at this stage, it is possible to select the boundaries of marks. However, the coordinates of the mark centers have fundamental importance for measuring the angle of rotation of the monitoring object, and for their calculation, it is necessary to calculate the local maxima in the space of the accumulation array.

For this purpose, the array is divided into parts. The minimum part size can be one-quarter of the short side of the matrix or 1.5 the maximum possible radius of the mark. Between these values, the minimum is chosen and is assigned as the size of the "area of interest" (Fig. 2).

After the zones of interest were selected, each of them is analyzed for the presence of identical near-standing pixels (not less than 8) with a value above the specified level. Each such group of pixels is assigned its own index. For each such group, there is a local maximum. Among the candidates that have been formed, group by contiguity was formed, and the center of gravity for each group is calculated. This center will be the center of the circle i.e. marks (Fig. 3).

These coordinates are superimposed on the original image and display the detected circles and their centers. Coordinates of the centers are displayed in the command line and can be used for further calculation of the angles.

The next stage of the work was performing functional testing, evaluating the behavior of the algorithm in real conditions, correcting the algorithm in accordance with the revealed differences between the mathematical model and the real image.

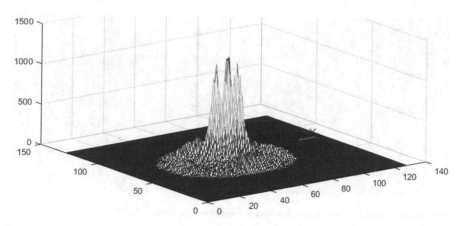

Fig. 2 3D view of accumulation array in area of interest

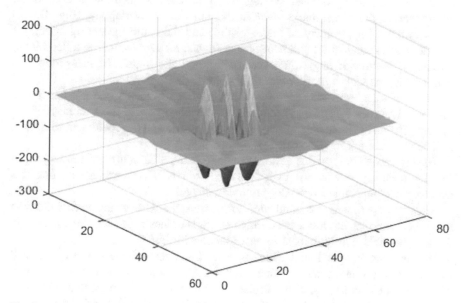

Fig. 3 Accumulation array after local maximum filtering

4 Algorithm Testing Results

After the development of the algorithm and its verification on mathematical models with the help of MatLAB technology, functional testing was carried out using the model of the measuring system. As negative factors on real images, the following phenomena were distinguished: glare from other faces of the control element, non-uniformity of the background, defects at the edges of the photodetector, vignetting.

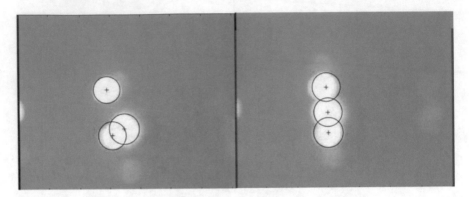

Fig. 4 The results of the testing algorithm for intersection marks

Despite the listed problems, the algorithm correctly defined and measured the coordinates of the image marks, even in the situation of overlapping (Fig. 4).

To evaluate the accuracy of the algorithm and the entire measuring system, it was decided to conduct the following experiments: measurement with a flat mirror, measuring collimation angles, and measuring the twist angle. The measuring system used a lens with a focal length of 100 mm.

In the classical operating mode of the measuring system, when the flat mirror acts as a control element, it is possible to reliably measure the accuracy of the system. The main parameters of the measuring system and the coefficient of optical reduction of the mirror (equal to 2) are known.

Having carried out a series of measurements using a flat mirror, it can be judged that the accuracy of the system is 2.2 arc seconds. It is worth noting that this accuracy is comparable to serially produced devices.

Then an experiment was conducted to measure the twist angle. In this mode, a special control element has a wide range of measurements, in fact 180°. Labels in the analysis field will be moved in accordance with Fig. 5.

In this case, images 1 and 2 move around the center around the circle, and image 3 has no information value, but can create a situation of overriding. Measuring the coordinates of the displacement of marks, we can estimate the angle of twisting. As a result of a series of experiments it was proved that the accuracy of measurements in this mode is 4.45 angular minutes.

The last experiment was to study the mode of measuring the collimation angles with this control element. One of the features of the control element used is the ability to measure collimation angles in a wide range. In this case, the labels move along one axis to the meeting or from each other (Fig. 6).

The accuracy of the measurement in this mode was not estimated, but the possibility of measuring collimation angles in this way was practically proved.

Fig. 5 The movement of marks when measuring angles. 1, 2, 3—the initial position, 1′, 2′, 3′—after the rotation of the control element

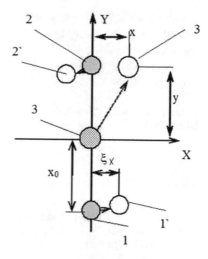

Fig. 6 Extended measuring range. 1, 2—the initial position, 1′, 2′—after the rotation of the control element

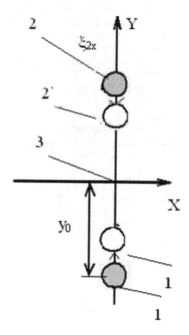

5 Conclusion

The developed algorithm allows to solve the problem of label overlapping when using special control elements. As a result of the experiment, its performance was proved.

The results of the experiment demonstrate the unique capabilities of the developed system in comparison with analogues. This system is single-channel and compact, while its accuracy characteristics are comparable with serially produced products, and the ability to measure the twist angle does not require an additional measuring channel.

The use of such measuring systems in SEMS allows to expand their capabilities and applications.

Acknowledgements This work was financially supported by Government of Russian Federation, Grant 074-U01.

References

1. Gorodetskiy, A.E. (ed.): Smart electromechanical systems. In: Studies in Systems, Decision and Control, vol. 49, 277 p. Springer International Publishing, Switzerland (2016). https://doi.org/10.1007/978-3-319-27547-5_4
2. Mikheev, S.V., Konyakhin, I.A., Barsukov, O.A.: Optical-electronic system for real-time structural health monitoring of roofs. In: Proceedings SPIE 9896, 98961C (2016)
3. Nekrylov, I.S., Korotaev, V.V., Denisov, V.M., Kleshchenok, M.A.: Modern approaches for a design and development of optoelectronic measuring systems. In: Proceedings SPIE 9889, 988920 (2016)
4. Korotaev, V.V.: Deflection measuring system for floating dry docks. Ocean Eng. **117**, 39–44 (2016)
5. Hoang, P., Konyakhin, I.A.: Autocollimation sensor for measuring the angular deformations with the pyramidal prismatic reflector. In: Proceedings SPIE **10231**, 102311I (2017)
6. Konyakhin, I.A., Moiseeva, A.A., Moiseev, E.A.: Configurations of the reflector for optical-electronic autocollimator. In: Proceedings SPIE 9889, 98891S (2016)
7. Nogin, A.A., Konyakhin, I.A.: Problem analysis of image processing in two-axis autocollimator. J. Phys. Conf. Ser. **735**(1), 012045 (2016)

Optic-Electronic System for Measuring Angular Position of SEMS Units on Base of Autoreflection Sensors

Igor A. Konyakhin and Aiganym M. Sakhariyanova

Abstract *Problem statement*: The problem of the relative orientation of SEMS units with the ability to control the angular positions of surfaces and technical objects is becoming actuality nowadays. *Purpose of research*: developing of the structure and metrological parameters of the autoreflection sensor to generate SEMS interaction signals at the robot system for high-precision operations or 3D object angular alignments. Analyzing the method to increase the distance of the pitch and yaw angular measuring for the autoreflection sensor. *Results*: autoreflection sensor using a sight target compose of reflectors designed as plane mirrors or glass tetrahedrons with angles of 90° between reflecting edges allows for generation of SEMS control and interaction signals for steering the platform robots angular positions and its operating arm in all motion modes: movement between target objects, approaching to the stop point, steering during performance of operations. *Practical significance*: The method of increase the work distance and the allowed angular deviation of the autoreflection sensor is described. Universal autoreflection sensor for generation the SEMS interaction signals is researched.

Keywords Robotic platform · SEMS · Image processing · Radiotelescopes Surface control · Measuring systems

1 Introduction

Nowadays the many countries design new radio astronomy instruments, for instant, China (FAST with 500 m aperture), the USA (NRAO Green Bank with 100 m main dish), Italy (SRT with 64 m diameter main dish), Mexico (LMT with 50 m main dish).

I. A. Konyakhin (✉) · A. M. Sakhariyanova
Department of Optical-Electronic Devices and Systems, ITMO University,
Saint Petersburg, Russia
e-mail: igor@grv.ifmo.ru

© Springer Nature Switzerland AG 2019
A. E. Gorodetskiy and I. L. Tarasova (eds.), *Smart Electromechanical Systems*,
Studies in Systems, Decision and Control 174,
https://doi.org/10.1007/978-3-319-99759-9_25

Russian Academy of Science implements a project of the radiotelescope RT70 Suffa with 70 m diameter parabolic main dish and the 3 m elliptical counter-reflector. The distance from top of main dish to counter-reflector is closely 20 m [1].

For control the angular shifts of counter-reflector it is suggested to use smart electromechanical systems (SEMS), like High-Speed Parallel-kinematic Micropositioner with Controller or Hexapods [2]. According to the design the counter-reflector is positioned on a mobile platform of the hexapod. The base platform of hexapod is fixed with stationary rods to the base of radiotelescope main dish (Fig. 1).

For working at the millimeter wave range it is necessary the high quality of the main dish parabolic surface. The root mean square of the point deviation on a surface from theoretical parabola is not more than 0.1 mm.

However, the construction weight and the temperature influence are the reasons of the radiotelescope component deformations. For example, the linear deformations of the main dish surface have got value close to 30 mm.

The required parameters of reflecting components are implemented with help of a main dish surface adaptation system.

The monitoring the position of the points of the main dish is suggested to use The special robotic platform like hexapod SEMS-unit with an optic-electronic tracking system are used for monitoring the position of the points of the main dish (Fig. 1) [3]. Optic-electronic tracking system is set on the mobile part of the robotic platform. This system monitors the radiation marks on the control points of the main dish. Elements of the main dish are equipped with mechanical drives that

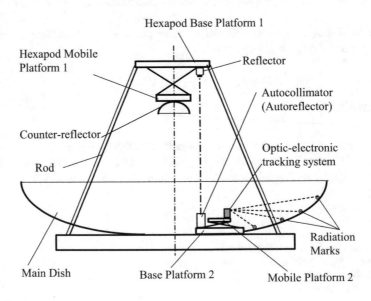

Fig. 1 The measuring systems of radiotelescope

allow correcting its shift. The information to the drives bases on data from the optic-electronic tracking system and the control signals from its robotic platform.

Another important condition for the researching in the millimeter wave range is the few (no more than 2 arc. seconds) deviation the direction of the counter-reflector axis relatively the axis of main dish. Since it is necessary to measure the angular deviation of the base platform of the counter-reflector hexapod relatively base platform of the counter reflector hexapod (Base Platforms 1 and 2 on Fig. 1).

To solve this problem it is suggested to use an optic-electronic autocollimation sensor of the angular displacement [4, 5].

Following the autocollimation technique, a Reflector as a mirror, is set on Hexapod Base Platform 1 and the emitting-receiver Autocollimator is placed on the Base Platform 2 at the top of the main dish (Fig. 1).

2 Structure of the Optic-Electronic Autocollimation Sensor

The optic-electronic autocollimation sensor concludes the autocollimator and reflector [4].

The radiating system of autocollimator includes the laser diode 1 with aperture-mark as the source of radiation, which is placed in the focal plane of objective 2 (Fig. 2). The radiation channel generates the collimation optical beam and directs it on reflector 3. The reflected beam is received by the receiving channel of autocollimator. The receiving channel includes the objective 2, semi-reflecting

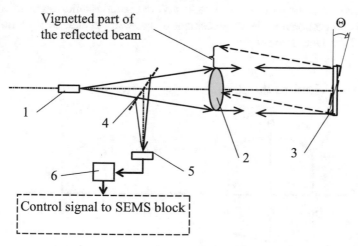

Fig. 2 Structure of the optic-electronic autocollimation sensor: 1—source of optic radiation, 2—objective aperture-mark, 3—reflector, 4—beam splitter r, 5—photo-receiver matrix, 6—microprocessor system

splitter 4 and the photo-receiving matrix 5, which is placed in the focal plane of objective 2. The reflected beam forms on a photo-receiving matrix 5 the image of the aperture-mark 1. The video-frame from photo-receiving matrix 5 is calculated by the digital microprocessor 6. When the reflector 3 rotates on angle Θ, the reflected beam is deviated from the original direction, so the image shifts on the matrix photo-receiver. The value of this shift determined by equation:

$$x = f \cdot tg(K \cdot \Theta) \tag{1}$$

In this equation f is focal length of an objective 2, $K = 2$ for autocollimation sensor.

The microprocessor 6 calculates the video-frames from the matrix photo-receiver and determines the value x. The value Θ of angular rotation is calculated by Eq. (1).

3 Problem of the Angular Measuring on the Long Distances

The work distance L from autocollimator to reflector for the traditional autocollimation sensors isn't more than 10 m, since the shifting of the reflected beam in the plane of the entrance pupil of the receiver objective after the reflector rotation. As result, for some angle of rotation the main part of reflected beam is cut by aperture of the objective. This situation is the "vignetting" of the reflected beam.

The vignetting is the reason of the asymmetrical redistribution the irradiance of image in the matrix photoreceiver as shown in Fig. 3. As a result, the energy axis of the image shifts relative to the geometrical axis. Matrix photoreceiver fixes displacement of the energy center of image, and measured rotation angle of reflecting system is proportional to the displacement of geometry center, so vignetting is the reason of the measurement error [6].

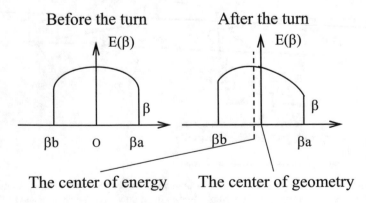

Fig. 3 Scheme of displacement the energy axis relative to the geometrical axis

If the work distance from autocollimator to reflector more than 10 m, vignetting error increase to 50 or 60% of the value Θ. Obviously, autocollimator sensor hasn't to measure the deviation of Hexapod Base Platform 1 on work distance $L = 20$ m. The original autoreflection sensor with long work distance is suggested.

4 Autocollimation and Autoreflection Sensors Schemes

The emitted beam has internal and external areas. The internal area is limited by a conical surface formed by the rays c and b from the rim points of the aperture-mark 1 (Fig. 4). The L_{fr} distance from center of the objective 2 to the point P_{fr} is the distance of "beam formatting". From the outside, the external area is limited by the border rays B, C.

The L_{fr} distance is determined by the equations:

$$L_{fr} = D_2/(2 \cdot tg(\beta_m)); \quad \beta_m = arctg(a/f) \tag{2}$$

In Eq. (2) D_2 is the diameter of objective, β_m—beam divergence angle, a—radius of mark.

The vignetting error for autocollimation sensor is the ignorable value if the L_1 distance to the reflector 3 is smaller than distance L_{fr}: $L_1 < L_{fr}$ and the edges of the reflector 3 are into internal are between rays b, c (Fig. 4, position 3).

In this case working distance of the *autocollimation* sensor is following:

$$L_1 = \frac{D_2 - D_3}{2 \cdot tg(\beta_m)} = \frac{(D_2 - D_3) \cdot f}{2 \cdot a}, \tag{3}$$

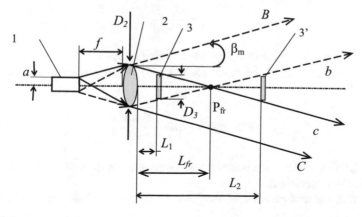

Fig. 4 The structure of the emitted beam: 1—aperture-mark, 2—objective, 3—reflector (mirror) in an autocollimation position, 3'—reflector (mirror) in autoreflection position

In the Eq. (3) D_3 is the diameter of the mirror reflector.

For typical autocollimator $D_2 = 50\,\mathrm{mm}$, $f = 500$ mm, $a = 0.25$ mm, $D_3 = 40\,\mathrm{mm}$ and form Eq. (3) the working distance is only 10 m. It is not enough to control large-scale objects.

Working distance can be increase if mirror reflector 3 is located at distance more than distance L_{fr} $(L_2 > L\phi)$ as shown in Fig. 4, pos. 3′. This case is determined the *autoreflection* sensor with working distance as:

$$L_2 = \frac{D_2 + D_3}{2 \cdot tg(\beta_m)} = \frac{(D_2 + D_3) \cdot f}{2 \cdot a} \qquad (4)$$

The working distance of the autoreflection sensor is more than for autocollimation one, while the coefficient K in Eq. (1) is decreased: $K = 1$. As result, the autoreflection sensor sensitivity is smaller than for autocollimation sensor. Another disadvantage of autoreflection sensor is the large value of the vignetting error. However the vignetting error can be determined by the analytic equation, so the algorithm to decrease this error can be implemented.

5 Analytic Description of Vignetting Error and Synthesis of the Decreasing Algorithm

The distribution of the irradiance in the image on photoreceiver matrix is described by an exact expression [7]:

$$E(r) = \frac{2}{\pi} \cdot \left(\arccos\left(\frac{r}{R_m}\right) - \sqrt{1 - \left(\frac{r}{R_m}\right)^2} \cdot \left(\frac{r}{R_m}\right) \right), \qquad (5)$$

$$R_m = \frac{D_1 \cdot f}{2 \cdot L}, \qquad (6)$$

where $r \leq a$ is the radius of the point on aperture-mark 1.

The approximation of the exact formula (4) is suggested as the exponential function $E(X, Y)$:

$$E(X, Y) = \exp\left(-\frac{X + Y}{k \cdot R_m}\right) \qquad (7)$$

where X, Y are the coordinates of a point on aperture-mark 1, $k = 0.59$—a dimensionless coefficient.

As result, the analytical expression for vignetting error has been synthesized as equation:

Fig. 5 The dependence of vignetting error from the value measuring angle

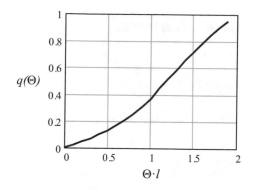

$q(\Theta)$

$\Theta \cdot l$

$$q(\Theta) = \left(\frac{\exp\left(\Theta \cdot l; /_k\right) \cdot (\Theta \cdot l - k - 1) + (k+1)}{\exp\left(\Theta \cdot 1 /_k\right) - 2 \cdot \exp\left(\Theta \cdot 1 /_k\right) + 1} \right) \cdot \frac{1}{l}, \text{ while } \Theta \cdot l \le 1 \qquad (8)$$

$$q(\Theta) = \left(\frac{\exp\left(-\Theta \cdot 1 /_k\right) \cdot \exp\left(2 /_k\right) \cdot (k + \Theta \cdot l - 1) - (k+1)}{\exp\left(-\Theta \cdot 1 /_k\right) \cdot \exp\left(2 /_k\right) - 1} \right) \cdot \frac{1}{l},$$

$$\text{while } 1 < \Theta \cdot l \le 2 \qquad (9)$$

where q—the error of measuring the angle Θ of the reflector 3 rotation, $l = f/R_m$. The graph of function $q(\Theta)$ is shown in Fig. 5.

Based on the analytical description (8), (9) of the measurement error $q(\Theta)$ due to vignetting, the compensation algorithm can be constructed.

Using this compensation algorithm of systematic error due to vignetting reduced to a negligible value, that allows to increase the working distance of autoreflection sensor up to several tens of meters.

6 Conclusion

The advantage of the autoreflection sensor is the longer working distance than autocollimation sensor. Disadvantage of autoreflection sensor is the large value of the systematic measuring error. The reason of this error is the vignetting of the reflected beam. However the analytical expression was found for this systematic error and algorithm to its decreasing was synthesized. The algorithm allows to reduce the measuring error to a negligible value.

The using of the developed autoreflection sensors allows to measure the angle deviation of the base platforms of SEMS-units on the distance to several tens of meters.

Acknowledgements The work is also financially supported by Government of Russian Federation, Grant 074-U01.

References

1. Konyakhin, I.A., Timofeev, A.N., Usik, A.A., Zhukov, D.V.: Optic-electronic systems for measuring the angle deformations and line shifts of the reflecting elements at the rotatable radio-telescope. In: Proceedings SPIE 8082, 80823R (2011). https://doi.org/10.1117/12.89005
2. Gorodetskiy, A.E. (ed.): Smart electromechanical systems. In: Studies in Systems, Decision and Control, vol. 49, 277 p. Springer International Publishing, Switzerland (2016). https://doi.org/10.1007/978-3-319-27547-5_4
3. Konyakhin, I.A., Vasilev, A.S., Petrochenko, A.V.: Electrooptic Converter for Measuring Linear Shifts of the Section Boards at the Main Dish of the Radiotelescope. Studies in Systems, Decision and Control, vol. 49, pp. 269–277 (2016)
4. Gorodetskiy, A.E., Konyakhin, I., Tarasova, I.L., Li, R.: Analysis of Errors at Optic-Electronic Autocollimation Control System with Active Compensation. Studies in Systems, Decision and Control, vol. 49, pp. 251–258 (2016)
5. Konyakhin, I.A., Gorodetskiy, A.E., Hoang, V.P., Konyakhin, A.I.: Optic Electronic Autocollimation Sensor for Measuring Angular Shifts of the Radiotelescope Subdish. Studies in Systems, Decision and Control, vol. 49, pp. 259–267 (2016)
6. Konyakhin, I.A., Smekhov, A.: Survey of illuminance distribution of vignetted image at autocollimation systems by computer simulation. In: Proceedings SPIE 8759, 87593F (2013). https://doi.org/10.1117/12.2014609
7. Konyakhin, I. A., Sakhariyanova, A.M., Li, R.: Design the algorithm compensation of vignetting error at optical-electronic autoreflection system by modelling vignetted image. In: Proceedings SPIE 9889, 988915 (2016)

Intelligent Reflectometer for Diagnostics of Air Transmission Lines

Aleksandr N. Shilin, Aleksey A. Shilin, Nadezhda S. Kuznetsova
and Danila N. Avdeyuk

Abstract *Problem statement*: At a present time, in many countries of the world work is underway to create intelligent electric grids (Smart Grid). Intelligent networks are a complex of technical means that automatically detect the weakest and most dangerous parts of the network, and then change the characteristics and the network design in order to prevent an accident and reduce losses. In addition, the intelligent network should have self-diagnostic and self-repair functions and include sensory, communication and control technologies in order to increase the efficiency of transmission and distribution of energy. Thus, the intelligent network is a self-controlled energy system with minimal human participation. The main technical means are digital control systems that carry out control, management and solution of artificial intelligence tasks. The main elements of electrical networks are switching systems, which must disconnect lines with emergency or pre-emergency modes and connect consumers to backup lines. The main problem is the decision making in switching lines is the reliability of information about the modes. For this purpose, devices are used to determine the type and location of the accident.

Keywords Reflectometer · Electromechanical devices · Introduction
Automatically · SEMS

A. N. Shilin (✉) · A. A. Shilin · N. S. Kuznetsova · D. N. Avdeyuk
Volgograd State Technical University, Volgograd, Russia
e-mail: eltech@vstu.ru

A. A. Shilin
e-mail: shilin.jr@gmail.com

N. S. Kuznetsova
e-mail: artex23@yandex.ru

D. N. Avdeyuk
e-mail: avdprod@yandex.ru

© Springer Nature Switzerland AG 2019
A. E. Gorodetskiy and I. L. Tarasova (eds.), *Smart Electromechanical Systems*,
Studies in Systems, Decision and Control 174,
https://doi.org/10.1007/978-3-319-99759-9_26

1 Introduction

One of the main causes of faults in electromechanical devices and wired communication systems and power transmission systems are conductors of windings of
devices and transmission lines. Therefore, for the detection of faults in windings
and lines, modern diagnostic tools must be used. This problem is especially relevant
for smart electromechanical systems (SEMS), which must have a self-diagnostic
function. At present, there are a lot of different means in the electric power industry
for detecting faults in the winding of devices and transmission lines. The most
widely used devices are those whose operation is based on the location of the
probing pulse on the monitored wire and on the measurement of the delay time of
the reflected pulse from the fault location t. The probe pulse propagates at a speed.
The distance to the accident site is determined by the formula $l = 0.5 \cdot v \cdot t$.

2 Instruments that Implement this Method
Are Called Pulse Reflectometers

As advantages of the reflectometer, it is necessary to note the following: relatively
simple control operation. As a disadvantage of such devices it is necessary to note
the following: the complexity of recording a reflected signal, which can be reflected
not only from damage, but also from other inhomogeneities. Reflectometer
implements an indirect method of measurement, therefore, the accuracy of the
measurement is adversely affected by various external factors. In addition, the
reflected signal is always weakened and distorted, which is the source of the error of
control. In distributed electrical networks, several reflectometer can be used to
detect emergency conditions. With insufficient accuracy of the reflectometer, it may
be possible to obtain SEMS conflicting information from several reflectometer for
one accident. Therefore, increasing the accuracy and reliability of accident registration will ensure the normal operation of the SEMS group.

   The AC wiring line at the high current frequency is a line with distributed
parameters and therefore the speed of the pulse in the line is determined by its
parameters. As a mathematical model of the line, in practical calculations, electric
substitution schemes for an infinitesimal portion of the line are used [1, 2]. It should
be noted that substitution schemes are an approximate model and for this reason,
several variants of substitution schemes have been developed, each of which is
recommended for a particular facility [1, 2]. When modeling overhead power lines,
a replacement circuit is used, shown in Fig. 1.

   The primary parameters of a unit of length of a two-wire line are:

C_0 is the transverse capacitance between the forward and reverse conductors,
L_0 is the inductance of the loop formed by the direct return wires,
R_0 is the longitudinal resistance of the forward and return conductors,

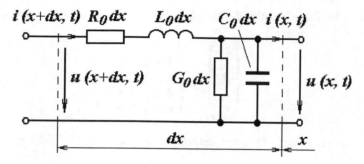

Fig. 1 The scheme of substitution of the elementary section of the line; i—is the instantaneous value of the current, u—is the instantaneous value of the voltage

G_0 is the transverse active conductivity of the insulation leakage between forward and reverse wires.

The characteristic or secondary parameters of a long line are its attenuation coefficients α, phase β and wave impedance Z_B.

Characteristic parameters of the long line through the primary parameters are determined from the equation.

$$\gamma^2 = (R_0 + j\omega L_0) \times (G_0 + j\omega C_0) = (\alpha + j\beta)^2, \tag{1}$$

where γ—is the propagation coefficient of the electromagnetic wave; ω—is the angular frequency.

The phase velocity $v_\phi = \omega/\beta$ of the pulses is determined by the phase coefficient.

$$\beta = \frac{1}{\sqrt{2}} \sqrt{\omega^2 L_0 C_0 - R_0 G_0 + \sqrt{\left(R_0^2 + \omega^2 L_0^2\right)\left(G_0^2 + \omega^2 C_0^2\right)}}. \tag{2}$$

As an example of calculating errors, we set the values of the primary parameters and their deviations (5% of the values of the parameters R_0, L_0, C_0 and 10% of the value of parameter G_0). We will analyze the phase coefficient at 5 frequencies, determine the absolute and relative errors of the phase coefficient [3].

$$R_0 = 14\,\Omega \qquad\qquad \Delta R_0 = 0.7\,\Omega$$
$$G_0 = 5 \times 10^{-6}\,S \qquad \Delta G_0 = 0.5 \times 10^{-6}\,S$$
$$L_0 = 17.8 \times 10^{-8}\,H \qquad \Delta L_0 = 0.89 \times 10^{-8}\,H$$
$$C_0 = 6.35 \times 10^{-9}\,F \qquad \Delta C_0 = 0.317 \times 10^{-9}\,F$$

Let's make the calculation for the phase coefficient at different frequencies:

$$f = \begin{pmatrix} f_1 \\ f_2 \\ f_3 \\ f_4 \\ f_5 \end{pmatrix} = \begin{pmatrix} 50 \\ 1 \times 10^3 \\ 1 \times 10^6 \\ 4 \times 10^8 \\ 8 \times 10^8 \end{pmatrix} \text{Hz.}$$

The phase coefficient for these frequencies is determined by the formula:

$$\beta(\omega) = \frac{1}{\sqrt{2}} \sqrt{\omega^2 L_0 C_0 - R_0 G_0 + \sqrt{\left(R_0^2 + \omega^2 L_0^2\right)\left(G_0^2 + \omega^2 C_0^2\right)}}, \qquad (3)$$

where $\omega = 2\pi f$.

Then we express the deviation of the phase coefficient through the deviations of the parameters R_0, L_0, C_0, G_0.

$$\Delta\beta \approx \frac{\partial f}{\partial R_0}\Delta R_0 + \frac{\partial f}{\partial L_0}\Delta L_0 + \frac{\partial f}{\partial C_0}\Delta C_0 + \frac{\partial f}{\partial G_0}\Delta G_0;$$

$$\Delta\beta = \Delta\beta_R + \Delta\beta_L + \Delta\beta_C + \Delta\beta_G, \qquad (4)$$

where $\Delta\beta$ is the absolute error of the phase coefficient.

For different frequencies, we calculate all the terms of the sum of expression (4).

The total maximum absolute error of the phase coefficient is determined by the formula:

$$\Delta\beta_\Sigma = |\Delta\beta_{R1}| + |\Delta\beta_{G1}| + |\Delta\beta_{C1}| + |\Delta\beta_{L1}|. \qquad (5)$$

We determine the values of the absolute and relative errors of the phase velocity for the same frequencies by the formulas:

$$\Delta v = \frac{\partial v}{\partial\beta}\Delta\beta = -\frac{\omega}{\beta^2}\Delta\beta; \quad \delta_v = \frac{\Delta v}{v}. \qquad (6)$$

The results of calculating the relative errors are shown in the graph (Fig. 2).

Since the result of the control is the distance to the accident, we will calculate the errors to determine the distance of the three distance values S_i = 120, 210, 300 km. By the formulas

$$\Delta S = t \cdot \Delta v; \quad \delta_S = \frac{\Delta S}{S}$$

The results of calculating the values of the relative error can be represented in the form of a graph (Fig. 3), but the absolute values of the errors in determining the distances for given coordinates in the form of a graph (Fig. 4).

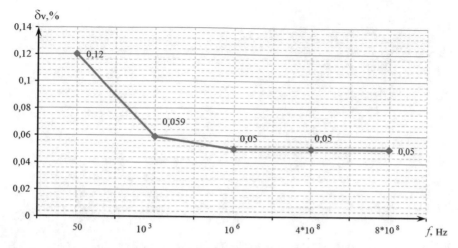

Fig. 2 Dependence of the relative error caused by the deviation of the phase velocity

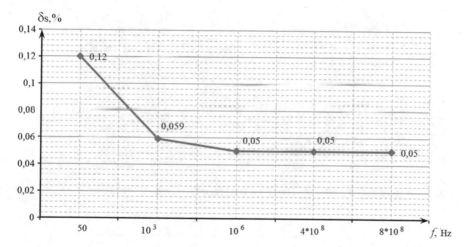

Fig. 3 Dependence of the relative error caused by the deviation of the distance to the fault site

To reduce the measurement error, the instruments must operate at frequencies above 1 MHz. Based on the performed calculations, it can be concluded that, using reflectometers at a low frequency, to reduce the error, it is necessary to introduce a correction for the change in parameters.

The accuracy of the fault location is also affected by the attenuation of the signal, since the amplitude of the pulse decreases and the signal-to-noise ratio decreases. The deviation of the signal attenuation coefficient can be expressed through the deviations of the parameters R_0, L_0, C_0, G_0, respectively.

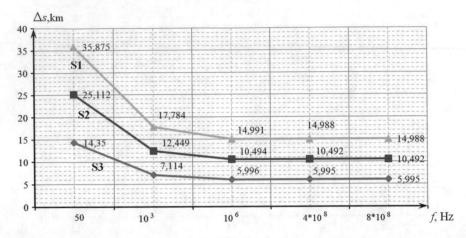

Fig. 4 Dependence of the absolute error on the distance to the accident site

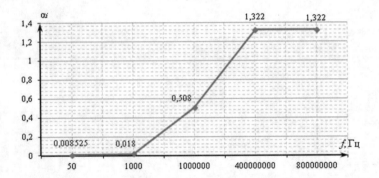

Fig. 5 Graph of the dependence of the attenuation coefficient on the frequency

$$\alpha(\omega) = \sqrt{\frac{1}{2}\left[G_0 R_0 \omega^2 C_0 L_0 + \sqrt{(R_0^2 + \omega^2 L_0^2)(G_0^2 + \omega^2 C_0^2)}\right]}. \tag{7}$$

Let's calculate the attenuation coefficient by the formula (7) at 5 frequencies and construct the dependence graph (Fig. 5).

Then we express the deviation of the damping coefficient through the deviations of the parameters R_0, L_0, C_0, G_0, respectively.

$$\Delta\alpha \approx \frac{\partial f}{\partial R_0}\Delta R_0 + \frac{\partial f}{\partial L_0}\Delta L_0 + \frac{\partial f}{\partial C_0}\Delta C_0 + \frac{\partial f}{\partial G_0}\Delta G_0 = \Delta\alpha_R + \Delta\alpha_L + \Delta\alpha_C + \Delta\alpha_G. \tag{8}$$

The total maximum absolute and relative errors of the damping coefficient are determined by the formulas:

$$\Delta\alpha_{\Sigma i} = |\Delta\alpha_{R_i}| + |\Delta\alpha_{G_i}| + |\Delta\alpha_{C_i}| + |\Delta\alpha_{L_i}|;$$

$$\delta_{\alpha_i} = \frac{\Delta\alpha_i}{\alpha} \cdot 100\%. \tag{9}$$

The above calculations are shown in the graph (Fig. 6).

From the analysis of the calculations results of the errors in the phase coefficients and damping, it follows that the error values depend on the frequency of the harmonic signal. Since the reflectometers use rectangular pulses as a signal, representing the sum of the harmonic components, it is obvious that when the pulse passes through the line, the waveform will be distorted (Fig. 7). The instability of the shape of the reflected signal, which enters the comparator, is the source of the

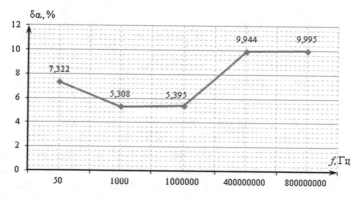

Fig. 6 Dependence of the relative error of the damping factor on the frequency

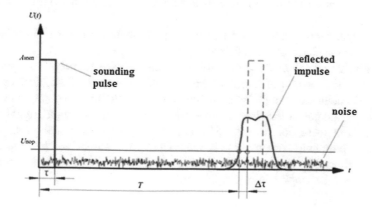

Fig. 7 The scheme of recording the reflected pulse

error $\Delta\tau$. Moreover, the error is due to both the dependence of the phase velocity of the harmonic component on the frequency, and the dependence of the damping coefficient of the harmonic component on the frequency [4].

To quantify the degree of distortion of the pulse shape, we define the pulse shape by the spectral method. Any periodic function $f(t)$ with period T can be represented as a sum of sines and cosines from the argument $n\omega t$ with the help of the Fourier series [2].

$$f(t) = a_0 + \sum_{n=1}^{\infty} (a_n \cos nwt + b_n \sin nwt);$$

$$A_0 + \sum_{n=1}^{\infty} (A_n \sin(nwt + \alpha_n)) \tag{10}$$

where n—is a positive integer; t—is the time; $\omega = 2p/T$ is the angular frequency;

$$a_0 \frac{1}{T} \int_{-T/2}^{T/2} f(\tau)d\tau; \ a_n = \frac{2}{T} \int_{-T/2}^{T/2} f(\tau)\cos n\omega t d\tau; \ b_n = \frac{2}{T} \int_{-T/2}^{T/2} f(\tau)\sin n\omega t d\tau;$$

$$A_n = \sqrt{a_n^2 + b_n^2}; \ \alpha_n = arctg\frac{b_n}{a_n}.$$

Figure 7 shows the probing pulse signal of the reflectometer. Signal strength is determined by the ratio $S = T/\tau$. In Fig. 8 shows the amplitude frequency characteristics of a rectangular pulse for different values of the duty cycle. From these dependences it follows that as the duty cycle increases, the amplitudes in the low-frequency part of the spectrum decrease. It should be noted that the measuring range is determined by the value of the signal period, but to increase the accuracy of the measurement, it is necessary to reduce the pulse width, and increase the pulse duty ratio accordingly.

In Fig. 9 shows the dependence of the amplitude of the first harmonic on the pulse width.

This dependence allows you to choose the frequency range when analytically determining the shape of the reflected signal.

To solve this problem, the superposition method is used, namely, the input signal is represented as a sum of harmonic components. Using the dependencies of the phase and attenuation coefficients on frequency, the harmonic components of the signal at the output of the line are determined, and then the output signal is synthesized from these components. Since a rectangular pulse contains an infinite number of harmonic components, in calculations the number of harmonic components is limited by the duty cycle and the minimum specified amplitude level as a percentage of the amplitude of the first harmonic.

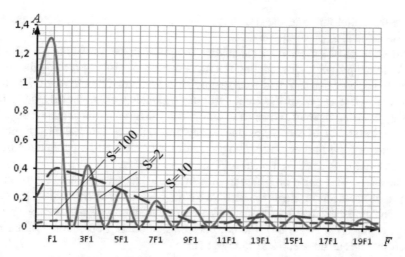

Fig. 8 Amplitude frequency characteristics of a rectangular pulse for different values of the duty cycle

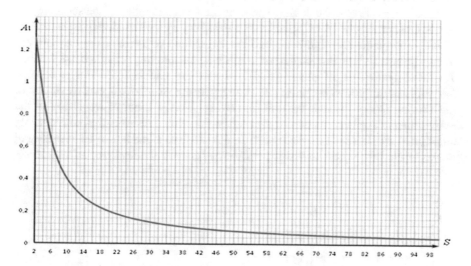

Fig. 9 Dependence of the amplitude of the first harmonic on the pulse width

Next, the coefficients of the Fourier series (10) for a square pulse signal with amplitude U_m, frequency f, and duty cycle S, duration are determined τ. The equation of the impulse signal in the line has the following form:

$$u(x,t) = A_0 e^{-\alpha(0)\cdot x} + \sum_{n=1}^{\infty} A_n e^{-\alpha(n\omega)\cdot x} \cos(n\omega t - \beta(n\omega)x + \psi_n), \qquad (11)$$

where the initial phase of the n-th harmonic in view of the cosine decomposition will be defined as $\psi_n = \alpha_n - \pi/2$.

An example of a method for calculating the pulse shape was implemented using the *Mathcad* program for modeling the transmission of probing rectangular pulses with an amplitude of 2 V and a frequency of 50 Hz via 110 kV wires. For fixed values of u, the graphs of the input and output signals were obtained (Fig. 10).

The proposed method allows us to analytically determine the shape of the pulse and the instability of the shape due to the influence of external factors on the parameters of the line and, accordingly, reasonably choose the parameters of the reflectometer with improved metrological characteristics.

The speed of a pulse propagation in a line depends on external factors and in existing instruments is taken into account by the coefficient of shortening, the value of which is established approximately. Devices of this type are widely used in power engineering because of the not complicated control operation and are produced by various domestic and foreign firms.

Obviously, in order to improve the accuracy of measure the distance of the fault location in power lines, it is necessary to choose the value of the shortening factor more reliably, which depends on external factors. In the developed reflectometer, automatic correction of the passing speed of the local and reflected pulses along the power line carried out. The Correction is carried out by the functional dependencies of long-line parameters that enter into the expression of speed, from external climatic factors. Values that characterize climatic factors are measured by sensors. Technical realization of phase velocity correction can be carried out through the information processing unit or with the purpose of minimal modernization of the existing device through the unit for setting the coefficient of truncation of the reflectometer. The instrument also uses a new functional receiver that continuously amplifies the pulses in accordance with the law inverse to the law of the damping of pulses in the line, which makes it possible to improve the accuracy of registering

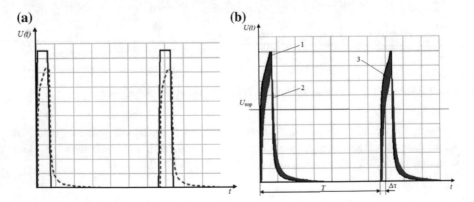

Fig. 10 a Graphics of the input and output signals; **b** 1—output signal at nominal line parameters, 2—output signal with deviation of line parameters from nominal, 3—signal blur due to instability of line parameters

the reflected pulses by increasing the signal-to-noise ratio. All calculation and control operations are performed using a digital device.

The functional block diagram of the device is shown in Fig. 11. The reflectometer contains four main functional blocks: a pulse generator unit (PGU), a sensing and recording unit (SRU), an adaptation unit (AU), an information processing unit (IPU).

The PGU controls the synchronous processes of measuring, adapting and converting the time of the pulse signal into a digital signal. SRU controls the process of probing and recording pulses, automatically correcting the attenuation of the amplitude of the reflected pulses. The AU performs the adaptation of the threshold for the operation of the reflected pulse when it is registered in order to minimize the distance measurement error. SPU carries out the time transformation of the pulse signal into digital, calculation of the pulse speed in the line from the measured values of climatic factors and functional dependencies, as well as the signal attenuation coefficient for the functional receiver.

The connections between the devices and the sides are shown in Fig. 11.

The clock generator (CG) is connected to a probe pulse generator (PPG), whose output through the switch (C) is connected to a functional receiver (FR), which corrects the gain of the reflected pulse as a function of the time of its return, and, correspondingly, the distance to the accident site. The output of the functional receiver is connected via a commutator (C) and a rectifier (R) to the first input of the comparison element (CE). The output of the comparison element is connected to the originator of the counting and zeroing pulses (OCZP), whose output is connected to the input of the pulse generator for recording and zeroing (PGRZ) and to the first input of the threshold voltage generator (TVG). The PGRZ output is connected to the second input of the TVG, and the probe pulse generator (PPG) is connected to the third input. The output of the TVG is connected to the second

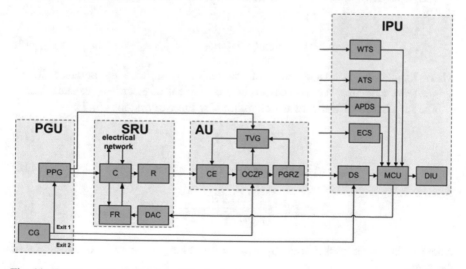

Fig. 11 Functional block diagram of the device

input of the EC. A distance sensor is connected to the PGRZ output, which is connected to a microcontroller unit (MCU), to whose analog inputs the sensors are connected: the wire temperature sensor (WTS), the air temperature sensor (ATS), the air permeability dielectric sensor (APDS), and the earth conductivity sensor (ECS). The output of the microcontroller unit (MCU) is connected through a digital-to-analog converter (DAC) to the input of the functional receiver of the FR and to the digital indication unit (DIU). The low-frequency output 1 of the CG is connected to the OCZP, and the high-frequency output 2 is connected to the DS.

The reflectometer operates as follows. The microcontroller unit MCU can set two modes of operation: the mode of determining the type and place of the accident, the mode of setting the phase velocity or the coefficient of electromagnetic wave shortening in the transmission line. In the mode of setting the shortening factor, the blocks of the road accident, WTS, ATS, APDS, ECS and MCU work. In the mode of determining the type and location of the damage, all blocks of the structural diagram work, except the sensors.

The phase velocity is determined, respectively, using the expression [1, 2]

$$v_\phi = \omega / \beta \ ,$$ (12)

where β the phase coefficient of the long line is determined by the formula [1, 2, 5]:

$$\beta(\omega) = \frac{1}{\sqrt{2}} \sqrt{\omega^2 L_0 C_0 - R_0 G_0 + \sqrt{\left(R_0^2 + \omega^2 L_0^2\right)\left(G_0^2 + \omega^2 C_0^2\right)}}.$$ (13)

The deviations of the parameters of the line R_0, L_0, C_0, G_0, caused by the influence of climatic factors, cause a deviation of the phase coefficient and, respectively, the phase velocity.

The dependence of the active resistance on the wire temperature is determined [6, 7]:

$$R_{0t} = R_{020}\left(1 + \alpha\left(t_{pr} - 20°\right)\right),$$ (14)

where R_{020} is the tabulated value of the resistivity at wire temperature 20 °C; t_{pr}—wire temperature, °C; α—temperature coefficient of electrical resistance, $\Omega/°$.

The impedance of the wire is determined by the expressions [6, 7]:

$$Z_{np} = R_{np} + jX_L;$$ (15)

$$R_{np} = \left(R_{020}(1 + \alpha(t - 20°)) + \pi^2 f \times 10^{-4}\right);$$ (16)

$$X_L = 29f \lg \frac{0.178}{r_{np}\sqrt{f\gamma \times 10^{-9}}} \times 10^{-4},$$ (17)

where f—is the network frequency; r_{pr}—wire radius; γ—specific conductivity of the earth.

The relative permittivity of gases depends on the temperature at constant pressure. For dry air, the TK in the temperature range from −60 to +60 °C can be considered constant and approximately equal to 2×10^{-6} °C^{-1}. However, rain and snow have a significant effect on the specific capacity of the line.

The transverse conductivity G_0 is due to the loss of active power ΔP due to imperfect insulation (leakage on the insulator surface and in the insulator material) and ionization of air around the conductor due to corona discharge [8].

Conductivity is determined by the general formula for the shunt:

$$G_0 = \Delta P_{\text{кор}}/U^2_{\text{ном}} \tag{18}$$

where $\Delta P_{\text{кор}}$—losses of active power per crown, kW;

$U^2_{\text{ном}}$—is the line voltage, kV.

In air lines of all voltages, losses through insulators are small even in areas with heavily polluted air, so they are not taken into account.

The phenomenon of coronation in the overhead line occurs only when the electric field strength at the wire surface is exceeded, kV$_{\text{max}}$/cm:

$$E = (0.354 \cdot U)/(r \cdot \lg(d_{\text{cp}}/r)) \tag{19}$$

The critical value is about 17–19 kV/cm.

The conditions for coronation occur in a 110 kV overhead line and a higher voltage.

The attenuation coefficient is determined using formula (7).

According to the parameters measured by the sensors: the temperature of the wire of the accident WTS, the air temperature of the ATS, the permittivity of the air APDS, and the specific conductivity of the ground ECS and the specified functional dependences, the MCU calculates the propagation velocity of the pulse in the line or the coefficient of shortening. The MCU also calculates the resistance value for the line attenuation coefficient, which is sent through the DAC to the input of the functional transformer. Then, based on the calculated values, the distance to the accident site is determined.

3 Conclusions

The conducted studies of the error sources in the reflectometer and the functions of the external factors influence on its error made it possible to create an intelligent reflectometer that automatically reduces the external factors influence on the accuracy of the measurement. Using the reflectometer in relay protection of power lines will allow creating smart electromechanical systems (SEMS) for switching lines and diagnosing devices. Increasing the reliability of determining the location and type of accident using an intelligent reflectometer in an electrical network will reduce the probability of inconsistent information from several reflectometers for one accident.

The paper presents the results of studies performed under the program Erasmus +#573879-EPP-1-2016-1-FREPPKA2-CBHE-JP "Internationalisation of Master Programs In Russia and China in Electrical Engineering".

References

1. Kaganov, Z.G.: Electric Circuits with Distributed Parameters and Chain Diagrams, 248 p. Energoatomizdat, Moscow (1990)
2. Bessonov, A.A.: Theoretical Foundations of Electrical Engineering: Electric Chains, 528 p. High School, Moscow (1978)
3. Shilin, A.N., Shilin, A.A., Artyushenko, N.S.: Calculation of errors of OTDRs for monitoring transmission lines. Control Diagn. (9), 52–59 (2015)
4. Shilin, A.N., Shilin, A.A., Artyushenko, N.S.: Analysis of signal waveform distortion during local monitoring of power lines. Control Diagn. (7), 44–49 (2017)
5. Characteristic parameters of a long line [Electronic resource]. Radio. Access mode: http://www.radioforall.ru/2010-01-19-08-09-28/597-2010-01-20-09-05-28. Date of circulation: 04/06/2014
6. Margolin, N.F.: Resistance of Overhead Transmission Lines. Mosoblpoligraf, Moscow (1937)
7. Pospelov, G.E., Ershevich, V.V.: Influence of the temperature of wires on the losses of electric power in active resistances of wires of overhead transmission lines. Electricity (10), 81–83 (1973)
8. Gerasimenko, A.A., Kinev, E.S., Chupak, T.M.: Electric Power Systems and Networks: A Summary of Lectures, 279 p. IPK SFU, Krasnoyarsk (2008)

Intelligent Low-Frequency Electromechanical Vibration Transducers

Aleksandr N. Shilin, Aleksey A. Shilin, Mikhail N. Sedov
and Mohanad N. Mustafa

Abstract *Problem Statement*: To improve the reliability of equipment in the power engineering and industry using the diagnosing of machines conditions and predicting their performance. Diagnostics of rotating machines use vibration control methods, which allow to determine the degree of wear and consequently the technical condition of machines. When diagnosing equipment with rotational frequencies of a one Hz, inertial electromechanical transducers are used. In some designs of such transducers the electromagnetic suspension with parametric feedback allowing to control rigidity of the oscillating system is used. Parametric feedback in such devices is realized on analogue elements. The instability of the parameters of these elements and the presence of dry friction in the guides is the source of the error of the electromechanical transducer. Leakage currents in the capacitor of the feedback unit limit the low-frequency boundary of the frequency range of the transducer. Parametric feedback on analog elements does not allow the automatic adaptation of the characteristics of the transducers. Therefore, to expand the functionality and improve the accuracy of control, it is advisable to use smart electromechanical systems (SEMS). The purpose of the study: to develop a smart electromechanical system (SEMS) with electromagnetic guides and a digital unit for adaptive control of parametric feedback, which makes it possible to extend the frequency range to the low-frequency region. Using an electromagnetic suspension in the transducer will reduce the dry friction in the guides and, consequently, increase the sensitivity of the transducer. Results: the use of digital controllable parametric feedback SEMS allows to control the characteristics of the transducer

A. N. Shilin (✉) · A. A. Shilin · M. N. Sedov · M. N. Mustafa
Volgograd State Technical University, Volgograd, Russia
e-mail: eltech@vstu.ru

A. A. Shilin
e-mail: shilin.jr@gmail.com

M. N. Sedov
e-mail: sobborn2001@mail.ru

M. N. Mustafa
e-mail: moohanadnm72@yahoo.com

© Springer Nature Switzerland AG 2019
A. E. Gorodetskiy and I. L. Tarasova (eds.), *Smart Electromechanical Systems*,
Studies in Systems, Decision and Control 174,
https://doi.org/10.1007/978-3-319-99759-9_27

depending on the measured parameters and characteristics of information signals. The use of digital devices will increase the frequency range of the transducer to a low-frequency region, since the length of storage of information in digital devices is not limited. The use of SEMS in the information-measuring system ensures the most optimal interaction between the transducers. Practical significance: the proposed intelligent electromechanical converter has improved metrological characteristics and therefore can monitor and diagnose equipment in various fields of human activity. This problem is particularly relevant in hydropower.

Keywords Vibration transducers · Seismic sensors · Electromagnetic suspension Frequency characteristics of vibration transducers

1 Introduction

At present, one of the urgent problems in the power engineering and industry is the increase in the reliability of the operation of various equipment, since the emergency regimes of some technological processes can lead to technogenic catastrophes. An effective method to improve the reliability of equipment is to diagnose the condition of machines and to predict their performance. For the diagnosis of rotor spinning machines are widely used vibration control methods. The nature of the vibration process of the machine depends on the wear and tear of the machine parts and therefore controlled parameters of the vibration process can determine the degree of wear and the corresponding technical condition of the machines. Depending on the frequency range of the vibration process, various parameters are used as sources of information. For example, in the low-frequency range, the vibration displacement parameter is measured, in the mid-frequency range—vibration velocity, high-frequency—vibration acceleration [1]. devices are widely used in the measurement technology, using piezoelectric sensors, which measure vibration acceleration, and vibration velocity and vibration displacement are defined by means of integrators. Such devices operate in the frequency range 10–10,000 Hz [2], however, in practice, arises a need to diagnose slow-moving equipment with rotational frequencies of a unit of Hz, for example, hydroelectric power stations HPS. Thus, vibration diagnostics of equipment at low frequencies is currently an urgent problem. A possible solution to this problem is the use of an inert mass on a stiffly-controlled electromagnetic suspension using parametric feedback (Fig. 1).

Fig. 1 System with electromagnetic suspension: *N, S*—movable magnet, Sr—position sensor, Td—transducer, *R*—a resistor connected to a coil in the field of a magnet NS

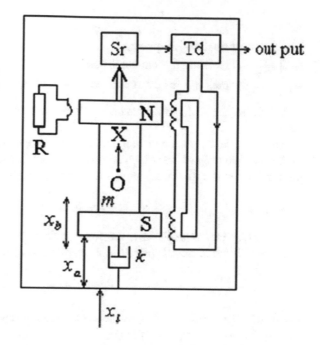

2 Vibration Transducers

The relationship between the measured input variable *xl*, the displacement of the magnet *xb* and the output parameter *xa* is determined by the difference:

$$x_a = x_b - x_l \tag{1}$$

Taking into account expression (1), the differential equation connecting the input quantity *xl* and the output quantity *xa* will take the form [3, 4]:

$$m\ddot{x}_a + k\dot{x}_a + cx_a = -m\ddot{x}_l \tag{2}$$

where *k* is the coefficient of viscous friction; *c* is the stiffness coefficient of the suspension of the magnetic mass.

From the Eq. (2) the expression for the amplitude-frequency characteristic (AFC) of the system is obtained:

$$|G(j\lambda)| = \frac{\lambda^2}{\sqrt{(1-\lambda^2)^2 + 4D^2\lambda^2}} \tag{3}$$

where:

$\lambda = \frac{\omega}{\omega_0}$ normalized frequency;

$D = \frac{k}{2m\omega_0}$ attenuation system coefficient;

$\omega_0 = \sqrt{\frac{c}{m}}$ expression for the natural oscillation frequency of the system.

We define the stiffness c of such an electromagnetic suspension of the moving mass.

To obtain the output signal and the feedback signal, a position sensor is attached to the movable magnet NS, which produces a signal in the form of a voltage directly proportional to the displacement of the magnet, which is converted into a current:

$$I = S_r x_a \tag{4}$$

According to Ampere's law, the force acting on the length of the conductor element with the current I, placed in a magnetic field is

$$dF = I[\vec{dl}\vec{B}] \tag{5}$$

where $\vec{dl}$—the conductor length element vector, drawn in the direction of the current; $\vec{B}$—vector of magnetic field.

In our case, the lines of the magnetic field are perpendicular to the conductor:

$$dF = IBdl \tag{6}$$

Integrating along the entire length of the conductor, the force will be equal to

$$F = 2IB\pi rN \tag{7}$$

where r—radius of the coil, S_r—sensor coefficient, N—the number of its turns.

So, the force acting on the moving sensor system is equal to

$$F = 2B\pi rNS_r x_a \tag{8}$$

Thus, the electromagnetic suspension stiffness coefficient defined by the expression:

$$c = 2B\pi rNS_r \tag{9}$$

The use of a mechanical damper at low frequencies is difficult, since the forces arising in it are not linear, so a coil is introduced for damping the system (Fig. 1), moving along with the moving mass in the magnetic field and closed to the resistor R [5–7]. We define the expression for the damping coefficient D for such a damping system.

The flux of magnetic induction vector through the surface described by the movement of the magnet relative to the coil is equal to

$$\Phi = \int_S BdS = 2B\pi r_1 x_a \tag{10}$$

where r_1—coil radius, n—number of coil turns.

Then the Electromotive force (EMF) of induction at the ends of the coil is defined by the Faraday law:

$$E = -\frac{d\Phi}{dt} = -2B\pi r_1 n\dot{x}_a \tag{11}$$

The current in the circuit is equal to

$$I_1 = \frac{E}{\sqrt{X_L^2 + R^2}} \tag{12}$$

where X_L—impedance of the coil. The force acting on the coil is

$$F = -4\frac{B^2\pi^2 r_1 r\dot{x}_a n}{\sqrt{X_L^2 + R^2}} \tag{13}$$

So, the coefficient k for electromagnetic damping is

$$k = -4\frac{B^2\pi^2 r_1 rn}{\sqrt{X_L^2 + R^2}} \tag{14}$$

Substituting now the values of physical quantities $B = 0.5$ T, $r_1 = 0.01$ m, $r = 0.012$ m, $N = 100$, $R = 0.1$ Ω, $X_L = 0.01$ Ω, $n = 10$ in the expressions for the coefficients k and c, we construct the amplitude-frequency characteristic of the system (Fig. 2).

The damping coil closed with a resistor leads to non-linearity of the forces arising here, depending on the position of the magnet, due to a change in the reactance of the coil. In addition, in this case it is rather difficult to determine the

Fig. 2 Amplitude-frequency characteristic (AFC) of a system with electromagnetic suspension and damping by means of a coil and a resistor

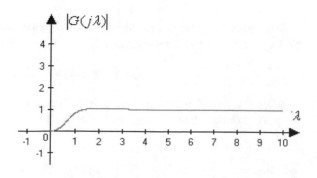

Fig. 3 Sensor with
controlled electromagnetic
suspension: Sr—sensor,
NS—movable magnet,
Pn—proportional element,
Dn—differentiating element,
Gc—current generator

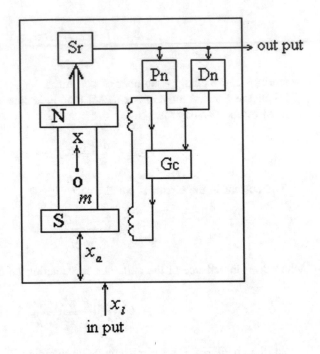

parameters of the damping device. Therefore, we will simulate and study the ideal
differentiating element included in the feedback link in parallel to the proportional
element (Fig. 3).

Then the differential equation of mass motion will be written in the form:

$$m\ddot{x}_a + c(t) \otimes x_a = -m\ddot{x}_l \tag{15}$$

Here the coefficient c depends on time, and therefore the operation of convo-
lution of functions is introduced into the differential equation:

$$c(t) \otimes x_a = \int_0^t c(t)x_a(t - \tau)d\tau \tag{16}$$

Since the transducer is a parallel connected proportional and ideal differentiator,
the expression for the transducer transfer function will look as follows:

$$C(p) = 2B\pi rN(k_0 + pk_1) \tag{17}$$

After the transformations, taking into account expression (17), the transfer
function of the entire electromagnetic system will be determined by the expression:

$$G(p) = \frac{-mp^2}{mp^2 + 2B\pi rNk_1 p + 2B\pi rNk_0} \tag{18}$$

The natural vibration frequency of the mobile mass in the system is

$$\omega_0 = \sqrt{\frac{2B\pi rNk_0}{m}}. \tag{19}$$

The attenuation coefficient is equal to

$$D = \frac{B\pi rNk_1}{m\omega_0}. \tag{20}$$

When tuning the instrument, first the gain k_0 of the proportional link is calculated to obtain the required resonant frequency (19), and then the coefficient k_1 in expression (20) to obtain the value $D = 0.6$, at which uniform frequency response is provided.

The expression for the amplitude-frequency response of a system with an ideal differentiating link is as follows:

$$|G(j\lambda)| = \frac{\lambda^2}{\sqrt{(1-\lambda^2)^2 + 4D^2\lambda^2}} = \frac{\lambda^2}{\sqrt{(1-\lambda^2)^2 + 4\left[\frac{B\pi rNk_1}{m\omega_0}\right]^2 \lambda^2}} \tag{21}$$

In Fig. 4. The amplitude-frequency characteristic of the system.

From the graph it follows that the introduction of parametric feedback allows to ensure uniformity of the frequency characteristic of the sensor in the low-frequency region of the measuring range. The method of choosing the parameters of the system is quite simple.

With the implementation of various methods of damping, the question of accounting for the inertia of the differentiating link may arise, i.e. modeling of the real differentiating link. In this case, the transfer function of the converter will have the following form:

Fig. 4 Amplitude-frequency characteristic (AFC) of a system with proportional and ideal differentiating links

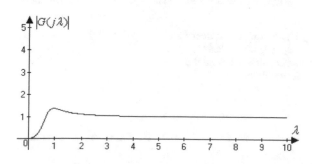

$$C(p) = 2B\pi r N(k_0 + \frac{k_1 pT}{pT + 1}).$$ (22)

The transfer function of the entire system:

$$G(p) = \frac{-mp^2}{mp^2 + \frac{2p^2 T^2 B\pi r Nk_1}{p^2 T^2 - 1} - \frac{2pTB\pi r Nk_1}{p^2 T^2 - 1} + 2B\pi r Nk_0}$$ (23)

The expression for the amplitude-frequency characteristic has the following form:

$$|G(j\omega)| = \frac{m\omega^2}{\sqrt{\left[-m\omega^2 + \frac{2\omega^2 T^2 B\pi r Nk_1}{\omega^2 T^2 + 1} + 2B\pi r Nk_0\right]^2 + \left[\frac{2\omega TB\pi r Nk_1}{\omega^2 T^2 + 1}\right]^2}}$$ (24)

Its graph is shown in Fig. 5.

From the graph it follows that the inertia of the differentiating link worsens the uniformity of the amplitude-frequency characteristic at low frequencies. In addition, the task of selecting the system parameters becomes more complicated. Thus, the realization of the parametric feedback of the electromagnetic system with an ideal differentiating link makes it possible to ensure the uniformity of the amplitude-frequency characteristic at low frequencies.

Obviously, for an electromechanical transducer with feedback, it is advisable to use a digital PID controller that has great functionality and can programmatically select the optimal parameters depending on the conditions and requirements for the measuring device. So, for example, depending on the frequency range of the object's oscillations, the digital PID controller will allow determining the parameters of the transducer with uniform frequency response. To calculate the transducer parameters that provide the necessary metrological characteristics, the digital PID controller must contain a digital computing device. Thus, the electromechanical transducer with feedback with the digital PID controller and the calculating unit is an smart electromechanical system (SEMS). Usually, several transducers are used to control the oscillations of extended objects, which are combined into one information and measuring system (IMS). SEMS group of information-measuring system will ensure the best possible interaction between the transducers.

Fig. 5 Amplitude-frequency response with a real differentiating link in the feedback loop

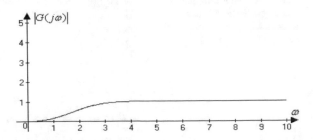

Fig. 6 Block diagram of PID
controller

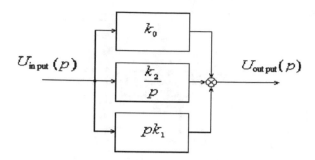

The parameters of the PID controller can be obtained from an analog model. Here is a technique for constructing a digital PID controller (Fig. 6).

The transfer function of the analog PID controller is given by

$$W(p) = \frac{U_{input}(p)}{U_{output}(p)} = k_0 + pk_1 + \frac{k_2}{p} \tag{25}$$

We transform the analog transfer function of the regulator into a discrete form by the z-form method. To do this, we replace the analog variable p with the discrete variable z by the rule

$$p = \frac{z-1}{Tz}. \tag{26}$$

After the conversion, we get the expression:

$$W(z) = \frac{\frac{k_1}{T} + k_0 + Tk_2 + z^{-1}(-k_0 - \frac{2k_1}{T}) + \frac{k_1}{T}z^{-2}}{1 - z^{-1}}. \tag{27}$$

For the convenience of filter synthesis, expression (27) can be suitably reduced to the following form [8]:

$$W(z) = \frac{a_0 + a_1 z^{-1} + a_2 z^{-2} + \cdots + a_m z^{-m}}{1 + b_1 z^{-1} + b_2 z^{-2} + \cdots + b_h z^{-h}}. \tag{28}$$

We define coefficients for powers of z:

$$a_0 = \frac{k_1}{T} + k_0 + Tk_2, a_1 = -k_0 - \frac{2k_1}{T}, a_2 = \frac{k_1}{T}, \tag{29}$$

$$b_0 = 1,$$
$$b_1 = -1.$$

The block diagram of the digital PID controller is shown in Fig. 7 [8].

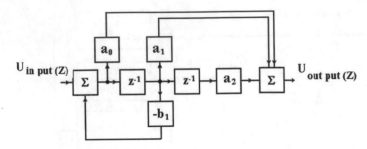

Fig. 7 Block diagram of the digital PID controller

To find the response to the input action at the clock instants of time, the difference equation is used [8]:

$$u_{output}[n] = \sum_{k=0}^{m} a_k u_{input}[n-k] - \sum_{k=1}^{h} b_k u_{output}[n-k] \tag{30}$$

3 Conclusions

Smart Electromechanical transducer allows measurement of low-frequency oscillations in a wide frequency range with the necessary accuracy. The use of digital elements allows to expand the low-frequency boundary of the range, and the use of digital control systems within SEMS ensures the maximum possible measurement accuracy and interaction with other information-measuring system converters.

References

1. Klyuev, V.V., Sosnin, F.R., Filinov, V.N., et al.: Measurements, Monitoring, Testing, and Diagnostics, 464 pp. Mechanical Engineering, Moscow (1996)
2. Tokarev, S.S.: New developments of research and production company "VISCONT". In: Martynova, E.V. (ed.) Energy Resource Efficiency and Energy Saving in the Republic of Tatarstan. VII International Symposium, 152—159 pp, Kazan, 5–7 Dec 2006
3. Kraus, M., Voshi, E.: Measuring Information Systems, 310 pp. Mir, Moscow (1975)
4. Ash, et al.: In: Obukhov, A.S. (ed.) Frants (trans.) Sensors of Measuring Systems: in 2 kn, 424 pp. Mir, Moscow (1992)
5. Shilin, A.N., Sedov, M.N.: Determination of the error of a vibration transducer with an electromagnetic suspension. Control Diagn. (5), 60–63 (2010)

6. Shilin, A.N., Sedov, M.N.: Modeling of a vibration transducer with an electromagnetic suspension. Devices (12), 41–44 (2008)
7. Sedov, M.N., Shilin, A.N.: Vibration measuring device. Patent 95832 of the Russian Federation, MPK G 01 H 11/02. GOU VPO VolgGTU (2010)
8. Shilin, A.N., Krutyakova, O.A.: Digital modeling of electrical and electronic devices: monograph. Russian Academy of Natural Sciences, RAE Publishing House. Publishing House "Academy of Natural History", Moscow

Printed in the United States
By Bookmasters